Monographs in Mathematical Economics

Volume 8

This series, consisting of research monographs and advanced textbooks for graduate students, is designed to bring together mathematicians interested in obtaining challenging new stimuli from economic theories and economists seeking effective mathematical tools for their research.

The scope of the series includes but is not limited to:

- Economic theories in various fields based on rigorous mathematical reasoning
- Mathematical methods (e.g., analysis, algebra, geometry, probability) motivated by economic theories
- Mathematical results of potential relevance to economic theory
- Historical study of mathematical economics

Comparable, existing monographs series are mainly organized from the viewpoint of "users" of mathematics, and are thus of limited interest to mathematicians. This series in contrast aims at genuine interaction between economists and mathematicians.

Most economic phenomena are described by (1) optimizing behaviors of agents (consumers, firms, governments, etc.) and (2) equilibria generated through the interactions of these agents. Consequently, the most basic mathematical subjects for economics fall into two categories:

- Optimization theory, static and dynamic. Basic linear / nonlinear programming, calculus of variations, optimal control, dynamic programming etc. provide key mathematical tools. Modern developments in convex analysis, nonlinear functional analysis, set-valued analysis, and non-smooth analysis form indispensable foundations.
- Equilibrium theory. Existence in finite- / infinite-dimensional settings, characterization, computational algorithms, dynamic adjustment processes, correspondence with economic structures.

Some fixed-point theorems and variational inequalities were created to solve the existence problem of equilibria. The stability theory and the viability theory of differential equations play important roles in the dynamic aspects of the equilibrium theory. Further, differential topology is a basic tool for the geometric analysis of equilibrium manifolds.

Economic phenomena significant in the real world often have a dynamic character. For instance, business cycles and movements of asset prices are governed by dynamic laws expressed in terms of differential or difference equations. Nonlinear analysis, harmonic analysis, and stochastic calculus play significant roles in solving dynamic economic problems. Econometric methods, including time series analysis and forecasting, also require sophisticated mathematical tools. Finally, significant developments in game theory and interaction with mathematical logic should be mentioned.

All of these topics and phenomena come within the scope of this series, aimed at cooperation between economists and mathematicians and the development of their respective disciplines.

Akira Yamazaki

Mathematical Economics of General Equilibrium

Analysis with Measurable and Topological Structures

Springer

Akira Yamazaki
Professor Emeritus
Hitotsubashi University
Tokyo
Japan

ISSN 2364-8279 ISSN 2364-8287 (electronic)
Monographs in Mathematical Economics
ISBN 978-981-96-9496-9 ISBN 978-981-96-9497-6 (eBook)
https://doi.org/10.1007/978-981-96-9497-6

This Springer imprint is published by the registered company Springer Nature Singapore Pte Ltd.
The registered company address is: 152 Beach Road, #21-01/04 Gateway East, Singapore 189721, Singapore

Dedicated to
Yukiko Yamazaki,
my dear companion in life,
to express my gratitude for her unwavering
support and encouragement.

Preface

Modern economic theory, and in particular general equilibrium analysis, is based on an economic model introduced by Walras [122]. It is the seminal contribution to the study of market equilibrium of economies with perfectly competitive markets. Walras established the formal theoretical framework and basis for the simultaneous determination of prices and transactions in competitive markets with interdependent influences. A mathematical model of an economy with perfectly competitive market transactions in this framework is known as a *Walrasian model.* Before Walras, Adam Smith [115] presented a systematic, comprehensive treatise on an economic system consisting of competitive markets, emphasizing the freedom of market transactions, so-called "laissez-faire," as well as the interrelationships among markets and economic variables in determining their values and quantities. It provided an important foundation for the Walrasian economic model.

In reality, all economic activities are directly or indirectly related to transactions of a large number of commodities in markets. Market transactions associated with the valuation of commodities, which are the "prices" quoted in market transactions, are assumed to reach an "equilibrium" level through the interaction of transactions of these commodities in markets. Prices along with transaction quantities at equilibrium are determined simultaneously for all commodities traded in markets. From this point of view the Walrasian model is also called the *general equilibrium* model. The theory of general equilibrium analysis is sometimes called *value theory*, because general equilibrium prices reflect the market valuation of commodities in transactions.

Despite the work of Walras, general equilibrium theory lacked a rigorous mathematical foundation for analysis, as all the leading economic theorists of the time, including Walras himself, were content to count the number of variables and equations to ensure the existence of market equilibrium. The question of the determinacy of equilibrium in the strict sense was not explicitly addressed as a subject of study for more than half a century. A. Wald, who attended K. Menger's seminar in Vienna in the 1930s, presented the first rigorous formulation and proof of the existence of equilibrium in a Cassellian variant of the Walrasian system. After the appearance of von Neumann's work on game theory and Kakutani's fixed point theorem, a number of mathematical theoretical economists were motivated to establish a solid mathematical formulation and foundation for general equilibrium theory using topological and convex structures. Mathematical proofs of the existence of competitive equilibrium were given independently in the 1950s by several economic theorists, including McKenzie, Arrow and Debreu, Nikaido, Gale, and Kuhn (see [88, 27, 124] and NOTES APPENDIX, p. 219, for further accounts on this topic). Their work used topological and convex analysis. Since then, more specific topological structures and measurable structures have been introduced to describe the basis of economic activity.

The purpose of this book is to present a mathematical economic theory of general equilibrium of economies with perfectly competitive markets of commodity transactions in the Walrasian model. The mathematical analysis in this book is based on topological and measurable structures that are essentially incorporated so as to be appropriate for expressing or representing the characteristics of economic agents, commodities in market transactions, and market prices for valuation in markets.

A groundbreaking contribution to the field of modeling an ideal type of perfectly competitive markets was made by Aumann [6, 8], who introduced an atomless measure space of economic agents into the general equilibrium model of perfectly competitive markets. An authoritative account of work involving an atomless measure space of economic agents and topological structures of characteristics of economic agents was given by Hildenbrand [55]. A distinctive aspect of this book is its focus on a non-convex commodity space within a Euclidean space for mathematical analysis of general equilibrium, diverging from the prevailing approach of utilizing a Euclidean space itself, particularly in the context of finite-dimensional commodity spaces. Accordingly, the present book is limited in scope to a finite number of different commodities in market transactions. It is acknowledged that there are numerous studies in general equilibrium analysis that consider an infinite number of commodities. However, the primary objective of our presentation is to formally delineate an analytical framework wherein commodities are not required to be perfectly divisible nor purely indivisible (cf. p. 72). This allows for the commodity space itself to deviate from convexity and a regularized lattice-like structure. Furthermore, the non-convexity may result from the idiosyncrasies of individual consumption patterns. The introduction of a non-convex commodity space as a foundation for general equilibrium analysis in an infinite-dimensional space is beyond the scope of this book.

This book is based on lectures I gave to advanced graduate students in economics at two universities, Hitotsubashi University and Keio University in Tokyo. It is intended for graduate students in economics or applied mathematics who wish to gain a comprehensive understanding of the fundamental mathematical concepts that underlie economic theory, including general equilibrium theory, analytical methods in measure theory, and topology.

Kunitachi, Tokyo, July 2024 *Akira Yamazaki*

Acknowledgements

The content of the mathematical analysis in this book owes great intellectual indebtedness to the pioneering work of Aumann and other subsequent works by Debreu, Hildenbrand, Shmeidler, and others, and especially to the book by Hildenbrand [55]. I am more than grateful to these esteemed scholars for their invaluable contributions. Nevertheless, I must also acknowledge that I have omitted references to a number of relevant contributions related to the contents of this book.

I am grateful to Toru Maruyama for inviting me to publish this book. I appreciate the valuable and encouraging advice of the editors, Shigeo Kusuoka and Toru Maruyama, and the anonymous reviewers for improving the manuscript. I would like to thank the graduate students and young researchers who attended my lectures, providing me with stimulating responses and reactions. Finally, I would like to acknowledge my indebtedness to my teacher, L. W. McKenzie, who introduced me to general equilibrium analysis in economic theory.

Contents

Introduction

This book presents a unified treatment of the general equilibrium analysis of economies with perfectly competitive markets based on the Walrasian model. The mathematical analysis in this book uses topological and measurable structures that are essentially built into the model to provide an appropriate means of expressing the characteristics of economic agents that form a basic part of economies along with commodities in market transactions and market prices for valuation in markets.

The book is divided into Part I and Part II. Part II begins with the mathematical analysis of general equilibrium in perfectly competitive economies. As a mathematical preparation for Part II for most economics students, Part I collects some basic mathematical concepts and facts that are later needed and used in the theoretical economic analysis in this book. Readers who are well prepared in elementary set theory, introductory probability and measure theory, topological analysis, and linear analysis can skip Part I entirely, except for the confirmation of the mathematical symbols used in the book. It is helpful to consult the list of mathematical symbols, p. 7. Others may start with Part I and skip sections that relate to mathematics they already know.

Part I is intended as a mathematical primer for readers in economics. In general, it is written in such a way that no mathematical knowledge beyond introductory college mathematics is required. The mathematical preliminaries include set theory, topological spaces, Euclidean spaces, measurable structures, and linear topological spaces. Note that the content of this part is organized in a way that the mathematical concepts, facts and properties presented largely follow the order in which they are used or required in mathematical economic analysis in Part II. In particular, this means that the order may not seem natural from a mathematical point of view; moreover, some content may seem too trivial to be presented. As for readers who may not feel well prepared for some of the topics in Part I, it is certainly uncomfortable to start reading the texts by going through the mathematical content only. Under these circumstances a suggested way of going through the texts is to familiarize oneself with elementary set theory concepts and symbols in Section 1.1, and then to review the content in Section 1.2 (Linear Spaces) and Section 1.5 (Euclidean Spaces). After that, readers may proceed to Part II, and refer back to Part I if they feel the need to familiarize themselves with related mathematical concepts and results.

Part II is the presentation of the mathematical analysis of general equilibrium for economies with perfectly competitive markets. Some readers may wish to start the book at Part II and refer back to Part I whenever it is necessary to review mathematical concepts and related well-known mathematical results referred to in the text. Part II consists of seven chapters from Chapter 2 to Chapter 8. Chapter 2 introduces the basic components of general equilibrium analysis. In the first section, the space of commodities, or *commodity space*, is specified as a non-empty closed subset of a finite-dimensional Euclidean space. Then *commodity vectors, divisible commodities, indivisible commodities, consumption sets, consumption vectors, preference relations, initial endowments* and *consumption characteristics* are succinctly defined. In the next section, several concepts, including *budget sets* and *demand sets*, related to the behavior of economic agents in markets in terms of their optimal market transactions are detailed. Then, some properties of preference relations are identified, which exclude possibilities of absence of optimal market transactions. A set of all specific preference relations, called the *universal preference set*, is defined and its characterization is given in the final section.

Participants of economic activities are called *economic agents*, and primary economic agents in the analysis of this book are consumers. Economies, with or without production possibilities introduced in later chapters, are defined on the space of consumption characteristics, which consists of preference relations on consumption sets and vectors in the commodity space as endowments of individual agents. Chapter 3 is concerned with the introduction of mathematical structures, appropriate from an economic point of view, in the space of consumption characteristics. In the first section the *topology of closed convergence* on the space of subsets of a metric space is described and its mathematical properties are shown. In the next two sections, topological and measurable structures are defined on the universal preference set. Then, topological properties of subsets of the universal preference set, each of which has one of the specific properties noted in Chapter 2, are clarified.

In Chapter 4 the continuity property of demand relations with respect to prices is studied. It is one of the key factors in the mathematical analysis of general equilibrium of market transactions. The source of their potential "discontinuity" is identified among the points in consumption sets. The set of these source points as a subset of each consumption set is characterized with respect to a price vector and is called the *critical set of the consumption set.* A related concept of *critical sets in the wealth space* is similarly defined. We then clarify the mathematical properties of these critical sets. The continuity properties of individual demand relations are examined in relation to the critical sets of the consumption set and the wealth space. Another issue concerning demand relations is the possibility of emptiness of demand sets. It is closely related to the question of the survival of economic agents without sufficient wealth to participate in market transactions to obtain commodities in their consumption sets. As a means to circumvent the survival issue as well as the potential discontinuity of demand relations, an additional concept of *subdemand* is introduced in the final section.

Chapter 5 begins with a mathematical formulation of economies with trade and consumption of economic agents. After the first section, the aggregate demand relation, which summarizes the total behavior of economic agents in perfectly competitive markets in an economy, is formulated as the *mean demand* of an economy. In the remaining sections of the chapter, *aggregation effects* on aggregate or total demand induced by the summation of individual demands are studied. Two different sources of aggregation effects are considered. One is called *direct effects of aggregation* and the other is called *economic effects of aggregation.* The direct effect is a simple effect resulting from the mere fact of summing individual demands over a "large" population of economic agents. In contrast, the economic effect is the one caused by the distribution of diversified characteristics in the population. Aggregation effects will play an important role in the discussion of economic equilibria in later chapters.

Chapter 6 is about Walrasian equilibria of economies. It focuses on a fundamental question in economic analysis concerning the existence of market equilibria of economies. The economies considered in this chapter are abstracted from their production possibilities, thereby bringing the redistribution of existing and produced commodities to the forefront of the analysis. The production possibilities of economies will be taken into account in the following chapters. In the first section, a formal statement of the definition of a *Walrasian equilibrium* as the equilibrium of economies with perfectly competitive markets is given. An auxiliary equilibrium concept of *subequilibrium* is also introduced to overcome the potential non-survival of economic agents and the lack of continuity property of the demand relation. Then, in the following section, a proof of the existence of subequilibrium is given. The last section proves the existence of a Walrasian equilibrium. The appendix to Chapter 6 refers to some of the literature showing the background of the survival problem in general equilibrium models.

General equilibrium analysis in previous chapters is abstract from the production possibilities open to economic agents. In Chapter 7 the supply side of economies, apart from the initial endowments of commodities, is formalized in terms of production possibilities in competitive markets. For this purpose, the product space, which is different from the commodity space where the consumption characteristics of individual economic agents are defined, is introduced at the beginning and production is defined as technological transformations. Technological transformations are specified by a *production set* in the product space. Then, some analytical properties of production sets are explained. On the basis of these properties, two types of production, namely *Hicksian production* and *Walrasian production*, are discussed, and a *Walrasian production economy* is defined as a Walrasian economy with Walrasian production. The last half of the chapter proceeds with the general equilibrium analysis of perfectly competitive markets with Walrasian production, beginning with the definition of *Walrasian equilibrium for a Walrasian production economy* and *production subequilibrium for a Walrasian production economy.* After a proof of the existence of production subequilibrium for a Walrasian production economy, the existence of Walrasian equilibrium for a production economy is shown on the basis of the existence of production subequilibrium for a Walrasian production economy with tighter restrictions.

Walrasian equilibrium is the classical standard conceptualization of equilibrium in economies where transactions in goods and services take place in perfectly competitive markets. Chapter 8 examines another classical equilibrium concept, originally introduced by Edgeworth, in relation to Walrasian equilibrium. The Edgeworth equilibrium is based on perfect competition in market transactions without the imposition of a market institutional framework that produces uniform prices for identical, indistinguishable goods and services traded in markets. The first section explains how Edgeworth expresses competition in transactions among market participants in terms of "contracts" and "recontracts" on traded items. The allocations for a given economy obtained by Edgeworthian competition are now called the *core allocations* for an economy in general equilibrium analysis. The set of all core allocations is called the *core of an economy*.

In the following section, an auxiliary core concept of *subcore* is also introduced to recover the well-known Aumann's equivalence theorem between Walrasian and core allocations for an atomless economy. Then, the equivalence between the subcore allocations and the subequilibrium allocations for a Walrasian production economy is proved. In view of this "subequivalence" result, the equivalence between Walrasian and core allocations for a Walrasian production economy is given with a stricter requirement.

Notation in Economic Analysis

(References are to the pages on which the notation or the symbols are defined.)

ℓ : The number of distinguishable commodities. 71

$\mathbb{G} \subset \mathbb{R}^\ell$: Commodity space 71

$x = (x_1, \ldots, x_\ell) \in \mathbb{G}$: A commodity vector. 72

$\mathbf{e}_i$: $\mathbf{e}_i = (0, \ldots, 0, \overset{i}{1}, 0, \ldots, 0)$, one unit of i-th commodity, i-th unit vector. .. 72

$\mathcal{D}_{\mathbb{G}} \subset \{1, \ldots, \ell\}$: The set of all perfectly divisible commodities. 72

$X \subset \mathbb{G}$: Consumption set 73

$x \in X$: A consumption vector. 73

$\mathcal{X}$: Universal set for consumption sets 75

$\underline{\beta}, \overline{\beta}$ 75

$(X, \succ), \succ \subset X \times X$: A preference relation or preferences on X. 76

$(X, \precsim), (X, \sim), x \succ y, x \precsim y,$
$x \sim y, x \not\succ y, x \not\precsim y, x \not\sim y$ 76

$\succ(x), \succsim(x)$ 76

$e \in \mathbb{G}$: The initial endowments or endowments of an agent. 77

$((X, \succ), e), ((X, \succ), w)$:
The consumption characteristics or characteristics of an agent. 77

$p \in \mathbb{R}^\ell$: A price vector or prices. .. 78

$B(X, w, p)$: The budget set for X, w, and p. 79

$B \subset \mathcal{X} \times \mathbb{R} \times \mathbb{R}^\ell \times \mathbb{G}$: The budget set relation. 79

$D(X, \succ, w, p)$: The demand set for $(X, \succ)$ at (w, p) 81

$D = \{(X, \succ, w, p, x) \mid x \in D(X, \succ, w, p)\}$:
The demand relation. 81

$\mathcal{P}$: The universal preference space. 83

$\mathcal{P} \times \mathbb{G}, \mathcal{P} \times \mathbb{R}$: The space of economic agents' characteristics or consumption characteristics. 83

$\mathcal{P}_{FD}$ 84

$\mathcal{P}_t$ 84

$\mathcal{P}_{ut}$ 85

$\mathcal{P}_{co}$ 85

$\mathcal{P}_{Ins}$ 87

Some particular mathematical symbols used

Part I
Mathematical Preliminaries

Chapter 1
Some Basic Facts and Properties in Mathematical Analysis

Abstract Some elementary mathematical concepts and facts related to some areas of mathematical analysis — set theory, topological spaces, Euclidean spaces, measurable structures, and linear topological spaces — are given in this preliminary chapter. After introducing various concepts in each of these topics, mathematical properties and facts are stated as "mathematical property" or "mathematical fact." The distinction between them is somewhat arbitrary, and yet the term "facts" tends to be used when they are more or less immediate or straightforward compared to "properties." Some of them are classified as "Mathematical Theorem" or "Mathematical Proposition," in which case proofs are given. In various parts of this book they are used or needed in the development of a theory presented.

1.1 Set Theory

Set theory is said to provide the language of mathematics in the sense that most modern mathematical analysis is based on set theory. In reviewing a minimal number of concepts and notations of set theory, we like to adopt what seems to be standard practice at the level of this book, and follow the framework of naive set theory, e.g. as in Naive Set Theory by Paul Halmos [46], without going into a complicated subject of logic and axiomatic set theory. This does not mean, however, that we do not refer to any of the axioms of set theory, as it will be necessary to recall some of them in our later analysis.

Elementary Notation and Operation of Sets

Sets, Membership, Formation, Subsets

The mathematical concept of a *set* is taken here as a primitive concept. It may be used as the foundation for all known mathematics. The terms "*set*," "*family*," "*collection*,"

A. Yamazaki, *Mathematical Economics of General Equilibrium*, Monographs in Mathematical Economics 8, https://doi.org/10.1007/978-981-96-9497-6_1

and "*class*" may be used synonymously. Another primitive and undefined concept is that of *membership* of sets. The symbol $\in$ denotes membership. Accordingly, $x \in S$ means that x is a *member* or an *element* of the set S. It is sometimes read as "x is in S" or "x belongs to S." The symbol $\notin$ denotes the negation of $\in$, so $x \notin S$ means x is *not an element* of S. Given two elements x, y of S, $x = y$ expresses that they are the same elements, while $x \neq y$ expresses that they are different (or distinct) elements. Instead of writing $x \in S, y \in S, z \in S$, we may write $x, y, z \in S$.

A set can be formed by using braces and making a statement that must be true for any possible member to be in the set. Specifically, $\{x \in S \mid \ldots \text{(statement about } x) \ldots\}$ denotes the set of all elements of S for which the corresponding statement is true. A set T consisting only of elements of S is called a *subset* of S and is written $T \subset S$; in other words, if every element of a set is an element of S, it is a subset of S. In this case, we can say that T is *included in* S or S *includes* T. Two sets T and S are *identical* or *equal*, written $T = S$, if they have the same elements, that is, if $T \subset S$ and $S \subset T$ hold simultaneously. The set without any elements is called the *empty set* and is denoted by $\emptyset$. For example, for any set S, $\{x \in S \mid x \neq x\} = \emptyset$. The empty set is a subset of every set S, i.e. $\emptyset \subset S$ for every S, because the statement "$\emptyset \subset S$" cannot be false, since there are no elements in $\emptyset$. The set consisting of all subsets of S is called the *power set* of S, and is denoted by $\mathbb{P}(S)$. If $x \in S$, then $\{x\} \in \mathbb{P}(S)$.

Algebra of Sets: Union, Intersection, Difference

Let S and T be two sets. The *union of sets* S and T, denoted by $S \cup T$, is the set of all elements belonging to S or, in a non-exclusive sense, to T, i.e. $S \cup T = \{x \mid x \in S \text{ or } x \in T\}$. The *intersection of sets* S and T, denoted by $S \cap T$, is the set of all elements that belong to both S and T, i.e. $S \cap T = \{x \mid x \in S \text{ and } x \in T\}$.Two sets S and T are said to be *disjoint* or, synonymously, *non-intersecting* if $S \cap T = \emptyset$. On the other hand, sets S and T are said to be *intersecting* if they have a common element, so $S \cap T \neq \emptyset$. The *difference* of S and T, denoted by $S - T$, consists of the elements of S that do not belong to T. If T is a subset of S, $S - T$ is usually called the *complement* of T in S. $S - T$ is sometimes written as $S \setminus T$. We have $S - T = S \setminus T = S - (S \cap T)$. The *symmetric difference* of S and T, denoted by $S \Delta T$, is defined by $S \Delta T = (S - T) \cup (T - S)$.

Mathematical Fact 1.1 [Algebra of Sets] *Let S, T, U, and V be sets. Then,*

(1) $S - (S - T) = S \cap T$;
(2) *(Commutative laws)* $S \cap T = T \cap S$ *and* $S \cup T = T \cup S$;
(3) *(Associative laws)* $S \cap (T \cap U) = (S \cap T) \cap U$ *and* $S \cup (T \cup U) = (S \cup T) \cup U$;
(4) *(De Morgan's laws)* $S - (T \cap U) = (S - T) \cup (S - U)$ *and*
$S - (T \cup U) = (S - T) \cap (S - U)$.

Family of Sets, Partition of a Set

Let $\mathcal{F}$ be a family of sets, i.e., a set whose elements are sets; then the *union* of the sets in $\mathcal{F}$, $\cup \mathcal{F} = \cup_{F \in \mathcal{F}} F = \{x \mid x \in F \text{ for some } F \in \mathcal{F}\}$, is the set consisting of the elements belonging to at least one F in $\mathcal{F}$. Its *intersection*, $\cap\mathcal{F} = \cap_{F\in\mathcal{F}}F = \{x \mid x \in F \text{ for every } F \in \mathcal{F}\}$, is the set consisting of the elements belonging to every F in $\mathcal{F}$. If $\mathcal{F}$ is a family of subsets of a set S, and if $\mathcal{F} = \emptyset$, then $\cup\mathcal{F} = \emptyset$ and $\cap\mathcal{F} = S$. A family $\mathcal{F}$ of sets is called a *disjoint family* if no two elements of $\mathcal{F}$ intersect. Given a family $\mathscr{C}$ of non-empty subsets of a set S, it is said to be a *partition* of S if it is a disjoint family and if its union is equal to S so that $S = \cup\mathscr{C}$ and $T \cap T' = \emptyset$ for any $T, T' \in \mathscr{C}$ with $T \neq T'$. Each element of S belongs to only one member of the partition $\mathscr{C}$.

Mathematical Fact 1.2 [Algebra of Sets with Family of Sets] *Let T be a subset of a set S and $\mathcal{F}$ a family of subsets of S. Then,*

(1) *(Distributive laws)* $T \cup (\cap\mathcal{F}) = \cap\{T \cup F \mid F \in \mathcal{F}\}$ *and* $T \cap (\cup\mathcal{F}) = \cap\{T \cup F \mid F \in \mathcal{F}\}$.
(2) *(De Morgan's laws)* $S - \cap\mathcal{F} = \cup\{S - F \mid F \in \mathcal{F}\}$ *and* $S - \cup\mathcal{F} = \cap\{S - F \mid F \in \mathcal{F}\}$.

Relations

Many of the fundamental concepts of mathematical analysis that we rely on in modern general equilibrium theory are defined in terms of a relation that satisfies certain properties. Intuitively, when we talk about a relation, we are talking about a relation between a particular pair of objects. This consideration leads to the basic idea of representing a relation by a set of pairs of objects that are related to each other. The notions of order, function, and correspondence are based on relations.

Ordered Pairs, Relations, Inverse Relations

The *ordered pair* of x and y, with its *first coordinate* x and *second coordinate* y, is the set (x, y) defined by $(x, y) = \{\{x\}, \{x, y\}\}$. An ordered pair defined in this way has the essential property that $(x, y) = (u, v)$ if and only if $x = u$ and $y = v$. A set consisting of elements each of which is an ordered pair is called a *relation*. Given a relation $\mathcal{R}$, we sometimes write $x\ \mathcal{R}\ y$ for a pair $(x, y) \in \mathcal{R}$. The *inverse* of a relation $\mathcal{R}$ is defined by inverting every ordered pair in $\mathcal{R}$ and is denoted by $\mathcal{R}^{-1}$, that is, $\mathcal{R}^{-1} = \{(y, x) \mid (x, y) \in \mathcal{R}\}$ and $(x, y) \in \mathcal{R}$ if and only if $(y, x) \in \mathcal{R}^{-1}$. $\mathcal{R}^{-1}$ is called the *inverse relation* of $\mathcal{R}$.

Domain, Range, Image, Inferior and Superior Inverse Images, Cartesian Products

Given a relation $\mathcal{R}$, its *domain*, denoted by $\operatorname{dom}\mathcal{R}$, is the set of all first co-ordinates of elements of $\mathcal{R}$, and its *range*, denoted by $\operatorname{range}\mathcal{R}$, is the set of all their second coordinates. Thus, $\operatorname{dom}\mathcal{R} = \{x \mid (x, y) \in \mathcal{R} \text{ for some } y\}$ and $\operatorname{range}\mathcal{R} = \{y \mid (x, y) \in \mathcal{R} \text{ for some } x\}$. The *image* of $x \in \operatorname{dom}\mathcal{R}$ under the relation $\mathcal{R}$ is $\mathcal{R}(x) = \{y \mid (x, y) \in \mathcal{R}\}$. For a subset $U \subset \operatorname{dom}\mathcal{R}$ the *image* of U under $\mathcal{R}$ is the set $\mathcal{R}(U) = \{y \mid (x, y) \in \mathcal{R} \text{ for some } x \in U\}$. Let V be a subset of range $\mathcal{R}$. The *weak* or *inferior inverse image* of V under $\mathcal{R}$, denoted by $\mathcal{R}_w^{-1}(V)$, is the set $\mathcal{R}_w^{-1}(V) = \{x \mid (x, y) \in \mathcal{R} \text{ for some } y \in V\} = \{x \mid \mathcal{R}(x) \cap V \neq \emptyset\}$. The *strong* or *superior inverse image* of V under $\mathcal{R}$, denoted by $\mathcal{R}_s^{-1}(V)$, is the set $\mathcal{R}_s^{-1}(V) = \{x \mid y \in V \text{ for all } y \text{ satisfying } (x, y) \in \mathcal{R}\} = \{x \in \operatorname{dom} \mathcal{R} \mid \mathcal{R}(x) \subset V\}$.

Given two sets S, T, the *Cartesian product* of S and T, denoted by $S \times T$, is the set consisting of all ordered pairs (x, y) such that x is an element of S and y is an element of T, that is, $S \times T = \{(x, y) \mid x \in S \text{ and } y \in T\}$. Obviously, every relation is a subset of the Cartesian product of its domain and its range. If $\mathcal{R}$ is a subset of a Cartesian product $S \times T$, we say that $\mathcal{R}$ is a *relation from S to T*. If the domain of a relation $\mathcal{R}$ is a subset of a set S and its range is understood from the context, we sometimes just say that $\mathcal{R}$ is a *relation on S*.

Composition, Restriction

Given two relations $\mathcal{R}$ and $\mathcal{S}$, the *composition* of $\mathcal{R}$ and $\mathcal{S}$, denoted by $\mathcal{S} \circ \mathcal{R}$, is defined to be the set of all ordered pairs (x, z) such that $(x, y) \in \mathcal{R}$ and $(y, z) \in \mathcal{S}$ for some $y \in (\operatorname{range} \mathcal{R}) \cap (\operatorname{dom} \mathcal{S})$. Sometimes we are interested in how a given relation $\mathcal{R}$ behaves only on a subset of its domain. In this case we derive a new relation $\mathcal{R} \cap (W \times \operatorname{dom} \mathcal{S})$ whose domain is a subset W of $\operatorname{dom} \mathcal{R}$. This new relation is called the *restriction* of $\mathcal{R}$ to W and is denoted by $\mathcal{R}_W$.

Identity Relations, Diagonals of Sets

The *identity relation on a set* X or the *identity on* X, denoted by $\Delta(X)$ or Δ_X, is defined as the set of all ordered pairs (x, x) for each $x \in X$. $\Delta(X)$ is also called the *diagonal* of X, visualizing its geometric image in $X \times X$.

Relational Properties: Reflexivity, Irreflexivity, Symmetry, Asymmetry, Anti-Symmetry, Transitivity, Completeness, Equivalence

Let $\mathcal{R}$ be a relation and X be the set of elements belonging to the domain or range of $\mathcal{R}$, i.e., $X = (\operatorname{dom} \mathcal{R}) \cup (\operatorname{range} \mathcal{R})$. The following terms for a relation are commonly used to specify relational properties between or among elements of the set X. $\mathcal{R}$ is said to be

- *reflexive* if $(x, x) \in \mathcal{R}$ and $(y, y) \in \mathcal{R}$ for each $(x, y) \in \mathcal{R}$,
- *irreflexive* if $(x, x) \notin \mathcal{R}$ and $(y, y) \notin \mathcal{R}$ for each $(x, y) \in \mathcal{R}$,
- *symmetric* if $(y, x) \in \mathcal{R}$ whenever $(x, y) \in \mathcal{R}$,
- *asymmetric* if $(y, x) \notin \mathcal{R}$ whenever $(x, y) \in \mathcal{R}$,
- *antisymmetric* if $x = y$ whenever $(x, y) \in \mathcal{R}$ and $(y, x) \in \mathcal{R}$,
- *transitive* if $(x, z) \in \mathcal{R}$ whenever $(x, y) \in \mathcal{R}$ and $(y, z) \in \mathcal{R}$, and
- *complete* if $(x, y) \in \mathcal{R}$ or $(y, x) \in \mathcal{R}$ for any $x, y \in X$.

Another way to express these properties is to say that $\mathcal{R}$ is reflexive if $\Delta(X) \subset \mathcal{R}$, symmetric if $\mathcal{R} = \mathcal{R}^{-1}$, irreflexive if $\mathcal{R} \cap \Delta(X) = \emptyset$, asymmetric if $\mathcal{R} \cap \mathcal{R}^{-1} = \emptyset$, antisymmetric if $\mathcal{R} \cap \mathcal{R}^{-1} = \Delta(X)$, transitive if $\mathcal{R} \circ \mathcal{R} \subset \mathcal{R}$, and complete if $X \times X \subset \mathcal{R} \cup \mathcal{R}^{-1}$.

Equivalence Relations

A relation $\mathcal{R}$ is an *equivalence relation* if it is reflexive, symmetric, and transitive. Equivalence relations are closely related to a family of sets. Let $\mathcal{R}$ be a relation on X for which $X = \operatorname{dom} \mathcal{R} = \operatorname{range} \mathcal{R}$. A subset S of X is an *equivalence class* (with respect to $\mathcal{R}$) if there is an element x in X such that $S = \{y \in X \mid (x, y) \in \mathcal{R}\}$. The equivalence class to which the element x belongs is called the equivalence class of x and is denoted by $x/\mathcal{R}$. If $S = x/\mathcal{R}$, then $\{(x, y) \mid y \in S\} = S \times S$. The set of all equivalence classes with respect to $\mathcal{R}$ is written as $X/\mathcal{R}$ or $X \bmod \mathcal{R}$. Note that $X/\mathcal{R}$ is a partition of X, and a partition of the set X characterizes an equivalence relation on X.

Mathematical Fact 1.3 [Equivalence Relation]*A relation $\mathcal{R}$ with $X = \operatorname{dom} \mathcal{R} = \operatorname{range} \mathcal{R}$ is an equivalence relation if and only if there exists a partition $\mathscr{C}$ of X such that $\mathcal{R} = \cup\{S \times S \mid S \in \mathscr{C}\}$.*

Correspondences, Functions

The concepts of correspondences and functions are introduced as relations that satisfy certain specific properties. Let S, T be two sets and φ be a relation from S to T. Then φ is called a *correspondence* from S to T if $\operatorname{dom} \varphi = S$ and $\operatorname{range} \varphi \subset T$. It is called a *function*, (or synonymously, *map*, *mapping*, *transformation*, or *operator*), from S to T if φ is a correspondence from S to T satisfying $y = z$ whenever $(x, y), (x, z) \in \varphi$. A standard convention that will be followed is to denote correspondences by lowercase Greek letters φ, ψ, etc., and functions by lowercase Latin letters f, g, h, etc. in most cases unless there is ambiguity.

For a relation φ from S to T and for each ordered pair $(x, y) \in \varphi$, the element y belongs to the set $\{z \in T \mid (x, z) \in \varphi\}$ denoted by $\varphi(x)$. x is the *variable*, and $\varphi(x)$ is the *value* of φ at x, or the *image* of x under φ. It is a subset of T and $y \in \varphi(x)$. Sometimes the set $\{(x, y) \in S \times T \mid y \in \varphi(x)\}$ is called the *graph of* φ, but trivially $\varphi = \{(x, y) \mid (x, y) \in \varphi\} = \{(x, y) \mid y \in \varphi(x)\}$. A correspondence or function φ from S to T is often expressed in one of three ways: $\varphi : S \to T, x \mapsto \varphi(x)$, or $S \xrightarrow{\varphi} T$. The second expression emphasizes the way in which a set $\varphi(x) \subset T$ is related to the variable $x \in S$ by the "relation" or rule φ. Sometimes a correspondence is called a *set-valued function* as opposed to a function which is *single-valued*. For a function f from S to T, we sometimes say that f is defined on S taking values in T, or T-valued.

A function $f : S \to T$ is said to be *one-to-one* or *injective* or an *injection* if $x = y$ whenever $f(x) = f(y)$. It is said to be *onto* or *surjective* or a *surjection* if $\operatorname{range} f = T$. It is said to be *bijective* or a *bijection* if it is both one-to-one and onto. If f is a bijection, then for every $y \in T$ there is a unique $x \in S$ such that $f(x) = y$. Thus the inverse relation $f^{-1} : T \to S$ is a mapping and is called the *inverse (function) of* f. The function $x \mapsto x$ is the *identity mapping* and is denoted by id_S. If $S = T$, $\mathrm{id}_S = \Delta(S)$.

Indexed Families of Sets

A family of sets is sometimes represented by a function in such a way that its range carries more weight than the function itself.

Families as Indexed Sets

Let $\mathcal{F}$ be a family of sets and x be a function from a set Λ onto $\mathcal{F}$. An element of the domain of x is called an *index*, the domain Λ is the *index set*, and an element of the range of x is an *indexed set*. The function x itself can be well expressed by the family $\mathcal{F}$ consisting of the values of the function, i.e. $\mathcal{F} = \{x(\lambda) \mid \lambda \in \Lambda\}$. From this point of view, a reference to a family of sets $\{F_\lambda \mid \lambda \in \Lambda\}$ is usually understood to mean a function F, where F_λ is the value of the function F at λ. It is also natural to identify the family of sets $\mathcal{F}$ as the set $\{id_{\mathcal{F}}(F) \mid F \in \mathcal{F}\}$; that is, $\mathcal{F}$ itself could play the role of an index set of the family with the identity mapping from $\mathcal{F}$ onto $\mathcal{F}$. Then $\mathcal{F} = \{id_{\mathcal{F}}(F) \mid F \in \mathcal{F}\} = \{F \mid F \in \mathcal{F}\}$.

Given a family $\mathcal{F}$ of indexed sets $\{F_\lambda \mid \lambda \in \Lambda\}$, the standard notations for its union and its intersection are as follows:

$$\begin{aligned}\cup\mathcal{F} &= \{x \mid x \in F \text{ for some } F \in \mathcal{F}\} = \{x \mid x \in F_\lambda \text{ for some } \lambda \in \Lambda\}\\ &= \cup\{F_\lambda \mid \lambda \in \Lambda\} = \cup_{\lambda\in\Lambda}F_\lambda = \cup_\lambda F_\lambda;\\ \cap\mathcal{F} &= \{x \mid x \in F \text{ for all } F \in \mathcal{F}\} = \{x \mid x \in F_\lambda \text{ for all } \lambda \in \Lambda\}\\ &= \cap\{F_\lambda \mid \lambda \in \Lambda\} = \cap_{\lambda\in\Lambda}F_\lambda = \cap_\lambda F_\lambda.\end{aligned}$$

Cartesian Products of an Indexed Family of Sets

The notion of a Cartesian product of two sets is extended to a family of sets. Let $\mathcal{F}$ be a family of indexed sets $\{F_\lambda \mid \lambda \in \Lambda\}$. The *Cartesian product of a family of sets* $\mathcal{F} = \{F_\lambda \mid \lambda \in \Lambda\}$ is defined as the set of all maps $x : \Lambda \to \cup\mathcal{F}$ such that $x(\lambda) \in F_\lambda$ for each $\lambda \in \Lambda$. It is denoted by $\Pi\{F_\lambda \mid \lambda \in \Lambda\} = \Pi_{\lambda\in\Lambda}F_\lambda = \Pi_\lambda F_\lambda$. Usually the subscript notation is used instead of the usual functional notation, so $x(\lambda)$ is written as x_λ. The set F_λ is called the *λ-th coordinate set* or *coordinate subspace*, and the point x_λ is called the *λ-th coordinate* or *component* of the point x in the product. A mapping $\mathrm{pr}_\lambda : F \to F_\lambda$ defined by $\mathrm{pr}_\lambda(x) = x_\lambda$ for each x in the product set is surjective unless some $F_{\lambda'} = \emptyset$, $\lambda' \in \Lambda$, and is called the *λ-th projection map* or the *projection (map) onto the λ-th coordinate* set or subspace.

Binary Relations

If $\mathcal{R}$ is a relation on a set X, it is said to be a *binary relation* (on X) if $\operatorname{dom}\mathcal{R} = \operatorname{range}\mathcal{R} \subset X$.

Notational Convention for Binary Relations

Except in a few cases, it is common practice in the following chapters to express binary relations on a set by symbols such as $\prec$, $\precsim$, $<$, or $\leqq$ and related symbols, $\succ, \succsim, >, \geqq$. Let $\prec$ (or $<$) be a binary relation on a set S. Then its inverse $\prec^{-1}$ (or $<^{-1}$) is written as $\succ$ (or $>$). The negation of $x \prec y$ (or $<$) is written as $x \nprec y$ (or $x \nless y$). An associated binary relation $\precsim$ (or $\leqq$) is a relation on S defined by $x \precsim y$ (or $x \leqq y$) if and only if $x \prec y$ and $y \nprec x$ (or $x < y$ and $y \nless x$). The negation of its inverse is written similarly with a forward slash. On the other hand, if a binary relation $\precsim$ (or $\leqq$) on S is given, an associated relation $\prec$ (or $<$) on S is defined by $x \prec y$ (or $x < y$) if and only if $x \precsim y$ and $y \not\precsim x$ (or $x \leqq y$ and $y \nleqq x$). In the case of $\precsim$ on S, we will have occasion to use another related relation $\sim$ on S defined by $x \sim y$ if and only if $x \precsim y$ and $y \precsim x$. If $\precsim$ is reflexive and transitive, then the relation $\sim$ is an equivalence relation on S.

Maximal Elements, Minimal Elements, Greatest Elements, Least Elements

Let $\prec$ be a binary relation on a set S, and B be a subset of S. An element $x \in B$ is a *maximal* (resp. *minimal*) element of B if $y \prec x$ (resp. $x \prec y$) for any $y \in B$ such that $x \prec y$ (resp. $y \prec x$). If the relation $\prec$ is asymmetric, x is a maximal (resp. *minimal*) element of B if and only if $x \nprec y$ (resp. $y \nprec x$) for every $y \in B$. x is the *greatest* (resp. *least*) element of B if $y \prec x$ (resp. $x \prec y$) for every $y \in B$ with $y \neq x$. The greatest (resp. least) element is always maximal (resp. minimal), but the converse need not be true unless the relation $\prec$ is complete.

Ordering Relations, Upper Bounds, Lower Bounds, Supremum, Infimum, Linear/Total Ordering

A relation $\prec$ *transitively orders* a set S if $\prec$ is transitive on S. S is said to be a *transitively ordered set*. If $x \prec y$ for $x, y \in S$, x is said to be *less than* y or x is a *predecessor* of y. y is *greater than* x or y is a *successor* of x. For a subset B of S, an element $x \in S$ is an *upper bound* (or a *lower bound*) of B if $y \prec x$ (or $x \prec y$) or $y = x$ for any $y \in B$. There can be multiple upper and/or lower bounds of the set B. An element $x \in S$ is called the *least upper bound (l.u.b.)* or *supremum* (or *greatest lower bound (g.l.b.)* or *infimum*) if it is an upper bound that is a lower bound for the set of all upper bounds (or a lower bound that is an upper bound for the set of all lower bounds) of B. A transitively ordered set S is said to be *order complete* if every non-empty subset of S that has an upper bound has a supremum.

Mathematical Property 1.1 *A transitively ordered set S is order complete if and only if every non-empty subset that has a lower bound has an infimum.*

A transitive relation is a *partial order* (or *ordering*) if it is reflexive and antisymmetric. A partial order $\precsim$ *partially orders* a set S if it is a partial order on S. S is said to be a *partially ordered set*. A transitive relation on a set is called a *strict order* if it is asymmetric. Given a strict order $\prec$ on S, its associated order $\precsim$, defined by $x \precsim y$ if and only if $x \prec y$ or $x = y$, is a partial order on S, and the passage from $\precsim$ to its associated order, defined by $x \prec y$ if and only if $x \precsim y$ and $x \neq y$, leads back to the strict order $\prec$ on S. This assertion holds even if strict order and partial order are interchanged.

A partial order is called *linear* or *total order* (or *ordering*) if it is complete. Let $\precsim$ be a partial order on a set S and C be a subset of S. If the restriction of the relation $\precsim$ to C, $\precsim_C$, is a linear order on C, then C is called a *chain* with respect to the relation $\precsim_C$.

Given a partially ordered set S, let $\precsim$ be a partial order on S and $\prec$ its associated strict order. For $x, y \in S$ such that $x \precsim y$, each of the following subsets of S is called an *interval*:

$$[x, y] = \{z \in S \mid x \precsim z \precsim y\}, \qquad (x, y) = \{z \in S \mid x \prec z \prec y\},$$
$$[x, y) = \{z \in S \mid x \precsim z \prec y\}, \qquad (x, y] = \{z \in S \mid x \prec z \precsim y\}.$$

Axiom of Choice

The mathematical analysis begins with some postulates for constructing a consistent theory. As noted earlier, we follow the framework of "naive set theory" (see e.g. [46]), which seems to be standard practice at the level of this book. There are a number of postulates of set theory. Each of them is a formal statement and is called an *axiom*. Since we are not describing axiomatic set theory, we will only refer briefly to the Axiom of Choice among the axioms of set theory.

Axiom of Choice, Zorn's Lemma, Hausdorff Maximal Principle

In later chapters we will have occasion to refer to the *Axiom of Choice* and related conditions which are equivalent to each other. The statement of the Axiom of Choice and some of its conditions are given below.

> [Axiom of Choice] *The Cartesian product of a non-empty family of non-empty sets is non-empty.*
>
> [Hausdorff Maximal Principle] *If $\mathcal{F}$ is a family of sets and $\mathcal{D}$ is a chain consisting of sets belonging to $\mathcal{F}$, then there exists a maximal chain C containing $\mathcal{D}$.*
>
> [Zorn's Lemma] *If every chain in a partially ordered set has an upper bound, then there exists a maximal element of the set.*

An alternative way of expressing the Axiom of Choice, which gave the axiom its name, is as follows:

[Axiom of Choice] *If $\mathcal{F}$ is a non-empty family of non-empty indexed sets $\{F_\lambda \mid \lambda \in \Lambda\}$, then there exists a function $c : \Lambda \to \mathcal{F}$ (called a* choice function*) such that $c(\lambda) \in F_\lambda$ for each $\lambda \in \Lambda$.*

Mathematical Fact 1.4 (1) *The Hausdorff Maximal Principle is equivalent to the Axiom of Choice.*
(2) *Zorn's Lemma is equivalent to the Axiom of Choice.*

Fields

Binary Operations

A *binary operation* on a set S is a mapping from the Cartesian product $S \times S$ to S. Let $\triangle$ be a binary operation on S. The operation is *associative* if $(x\triangle y)\triangle z = x\triangle(y\triangle z)$ for all $x, y, z \in S$. It is *commutative* if $x\triangle y = y\triangle x$ for all $x, y \in S$. A *neutral element* or *identity* for the operation, denoted by $e(\triangle)$, satisfies $e(\triangle)\triangle x = x\triangle e(\triangle) = x$ for all $x \in S$. For $x \in S$ an *inverse element* of x, denoted by x^{-1}, satisfies $x\triangle x^{-1} = x^{-1}\triangle x$. Let $\triangle$ and $\Diamond$ be two binary operations on S. The operation $\Diamond$ is *distributive* with respect to the operation $\triangle$ if $x\Diamond(y\triangle z) = (x\Diamond y)\triangle(x\Diamond z)$ and $(x\triangle y)\Diamond z = (x\Diamond z)\triangle(y\Diamond z)$ for all $x, y, z, \in S$.

Fields, Ordered Fields

A *field* $(F, +, \cdot)$ is a triple consisting of a non-empty set F with two binary operations called *addition*, denoted by $+$, and *multiplication*, denoted by $\cdot$. The addition and multiplication operations are assumed to have the following properties:

[Additive Structure] (AS) (a) Addition is associative and commutative. (b) There is a unique neutral element of addition, $e(+)$, called the *additive identity* and denoted by 0. (c) For each $x \in F$, there is a unique inverse element of x with respect to addition, denoted by $-x$ and called the *additive inverse* of x.
[Multiplicative Structure] (MS) (a) Multiplication is associative and commutative. (b) There is a unique neutral element of multiplication, $e(\cdot)$, called the *multiplicative identity*, denoted by 1, which is distinct from 0. (c) For each $x \in F - \{0\}$, there is a unique inverse element of x with respect to multiplication, denoted by x^{-1} or $1/x$, called the *multiplicative inverse*.
[Connecting Structure] (CS) Multiplication is distributive with respect to addition.

If the set F of the field $(F, +, \cdot)$ is an ordered set, $(F, +, \cdot)$ is said to be an *ordered field*. It may be expressed by a quadruple consisting of components of a field together with an order. For $x, y \in F$, the value of the addition is called the *sum of x and y* and is denoted by $x + y$. Similarly, the value of the multiplication is called the *product of x and y* and is denoted by $x \cdot y$. The symbol $\cdot$ can be omitted entirely if no confusion is likely, and is written as xy. In cases where two binary operations of a field $(F, +, \cdot)$ are clearly understood from the context, one can simply refer to a field F instead of fully specifying two operations as in $(F, +, \cdot)$.

Numbers

Real Numbers $\mathbb{R}$*, and their Subsets* $\mathbb{N}, \mathbb{Z}, \mathbb{Q}$

A *real number system* is an ordered field $(\mathbb{R}, +, \cdot, \leqq)$ with 0 (read "*zero*") as the additive identity and 1 as the multiplicative identity, with the following properties:

(a) The order $\leqq$ is a linear (or total) order on $\mathbb{R}$ such that, for any $x, y, z \in \mathbb{R}$, $x + z \leqq y + z$ if $x \leqq y$, and $x \cdot y \geqq 0$ if $x \geqq 0$ and $y \geqq 0$.
(b) [Order Completeness (Dedekind-Completeness)] $\mathbb{R}$ is order complete with respect to the linear order $\leqq$, i.e., every non-empty subset of $\mathbb{R}$ having an upper bound (or lower bound) has a supremum (or infimum).

$\mathbb{R}$ represents the *set of all real numbers*. The set $\mathbb{R}$ is uniquely determined in the sense that any two distinct ordered fields satisfying the above properties are isomorphic; i.e., there is a one-to-one and onto mapping from one set to the other that preserves the ordering and the field operations. Two associative and commutative binary operations on $\mathbb{R}$ (addition + and multiplication $\cdot$) are accompanied by the corresponding binary operations of subtraction and division. The *subtraction*, denoted by $-$, is the mapping $(x, y) \mapsto x - y = x + (-y)$; that is, the sum of x and the additive inverse of y is assigned to an ordered pair (x, y). Its value $x - y$ is called the *difference*. The multiplication of x and the multiplicative inverse of y, i.e. $x \cdot 1/y = x \cdot y^{-1}$, may be written x/y. The binary operation $(x, y) \mapsto x/y$ is called *division* and its value the *quotient*. The additive identity 0 yields x when added to any element x. The *unique additive inverse* $-x$ of an element x, i.e., an element that yields 0 when added to x, is called a *negative*. The multiplicative identity 1, which is different from 0, yields x when multiplied by any element x. Any element x other than 0 has an *inverse* $x^{-1} = 1/x$, i.e., an element that, when multiplied by x, gives 1. Multiplication is distributive with respect to addition. For any $x \in \mathbb{R}$ one has $0 \cdot x = 0$ and $(-1) \cdot x = -x$.

Let $<$ and $>$ be the strict orders on $\mathbb{R}$ associated with the linear order $\leqq$ on $\mathbb{R}$. $x \leqq y$ (or $x < y$) reads x *less than or equal to* y (or x *less than* y). $x \leqq 0$ (or $x < 0$) reads x *non-positive* (or x *negative*). Similarly, $x \geqq y$ (or $x > y$) reads x *greater than or equal to* y (or x *greater than* y). $x \geqq 0$ (or $x > 0$) reads x *non-negative* (or x *positive*). We have $0 < 1$. A subset $\mathbb{R}_+ = \{x \in \mathbb{R} \mid x > 0\}$ (resp. $\mathbb{R}_- = \{x \in \mathbb{R} \mid x < 0\}$) of $\mathbb{R}$ is the set of all positive (resp. negative) numbers. The greatest (or least) element of a subset S of $\mathbb{R}$, if it exists, is unique; it is called the *maximum* (or *minimum*) of S and denoted by $\max S$ (or $\min S$). The *absolute value* $|x|$ of a number x is defined by $|x| = \max\{x, -x\}$. We have $|x + y| \leqq |x| + |y|$. The *sign* of a number $x \neq 0$ is defined by $\operatorname{sign} x = x/|x|$.

The elements of $\mathbb{R}$ may be called *points*. We can visualize the set $\mathbb{R}$ with a horizontal line, in which case $\mathbb{R}$ is called the *real line*. Choose two different points representing 0 and 1 on the line so that 0 is to the left of 1. Then an element x of $\mathbb{R}$ is identified with a point of the line at distance $|x|$ from 0, to the right (or left) of 0 if x is positive (or negative). The phrase *extended real numbers* refers to real numbers or one of the two symbols $+\infty$ and $-\infty$ with the property: $-\infty < x < +\infty$ for each $x \in \mathbb{R}$.

Now define a subset $\mathbb{N}$ of $\mathbb{R}$ as the minimal set among subsets S of $\mathbb{R}$ such that $0 \in S$ and $x + 1 \in S$ whenever $1 \in S$ and $x \in S$; that is,

$$\mathbb{N} = \bigcap \left\{ S \subset \mathbb{R} \,\middle|\, 0 \in S \text{ and } x + 1 \in S \text{ whenever } 1 \in S \text{ and } x \in S \right\}.$$

An element of $\mathbb{N}$ is called a *natural number*. If we write $s(x) = x+1$ for any element x in $\mathbb{R}$, the set $\mathbb{N}$ of natural numbers can be written $\{0, 1, s(1), s(s(1)), s(s(s(1))), \ldots\}$ by symbolizing its specific elements, which are commonly denoted by $0, 1, 2, \ldots,$ $n, \ldots$. Then $\mathbb{N} = \{0, 1, 2, \ldots, n, \ldots\}$. The set of positive natural numbers is denoted by $\mathbb{N}_+$ $(= \mathbb{N} - \{0\})$. The set $\mathbb{Z} \subset \mathbb{R}$ of *integers* is defined by $\mathbb{Z} = \{x \in \mathbb{R} \mid x \in \mathbb{N} \text{ or } -x \in \mathbb{N}\}$. A subset $\mathbb{Z}_+$ (or $\mathbb{Z}_-$) of $\mathbb{R}$ is the set of *positive* (or *negative*) *integers*. The set $\mathbb{Z}$ can be written $\{\ldots, -n, \ldots, -2, -1, 0, 1, 2, \ldots, n, \ldots\}$ by listing its negative and non-negative elements. The set $\mathbb{Q} \subset \mathbb{R}$ of *rational numbers* is defined by a set of real numbers of the form x/y, where $x \in \mathbb{Z}, y \in \mathbb{Z}$ and $y \neq 0$; that is, $\mathbb{Q} = \{z \in \mathbb{R} \mid z = x/y, x \in \mathbb{Z}, y \in \mathbb{Z}, y \neq 0\}$. The letters $\mathbb{R}, \mathbb{N}, \mathbb{Z}, \mathbb{Q}$ have the meanings introduced here in this book.

Cardinality, Finite Sets, Infinite Sets, Countability, Uncountability

Two sets, S and T, are said to be *equal in cardinality* or to have *equal cardinality* if there is a bijection from S to T. A set S is *finite* if it is equal in cardinality to a set of the form $\{x \mid x \in \mathbb{N} \text{ and } x < n\}$, for some n in $\mathbb{N}$. In this case, S is said to have equal cardinality with the number n. If S is not finite, it is said to be *infinite*. A set has equal cardinality with at most one natural number. The unique natural number n of a set of the form $\{x \mid x \in \mathbb{N} \text{ and } x < n\}$ with equal cardinality of a finite set S is the *number of elements* in S and is denoted by $\sharp S$. If S is an infinite set, we write $\sharp S = \infty$ using a symbol ∞ as in the extended real numbers. A set is *countable* or *denumerable* if it is finite or equal in cardinality to the set of natural numbers $\mathbb{N}$. In the latter case it is *countably infinite*. If a set is infinite but not countable, it is said to be *uncountable* or *non-denumerable*.

Mathematical Fact 1.5 [Real vs. Rational Numbers]

Given any two numbers x, y in $\mathbb{R}$ satisfying $x < y$, there is a rational number $r \in \mathbb{Q}$ such that $x < r < y$.

Mathematical Property 1.2 [Cardinality of Sets]

(1) *A subset of a countable set is countable.*
(2) *If $\mathcal{F} = \{F_\lambda \mid \lambda \in \Lambda\}$ is a countable family of countable sets, then $\cup\mathcal{F} = \cup\{F_\lambda \mid \lambda \in \Lambda\}$ is countable.*
(3) *The family of all finite subsets of a countably infinite sets is countable, but the family of all subsets is not.*
(4) *The set $\mathbb{Q}$ of all rational numbers is countable.*
(5) *The set $\mathbb{R}$ of all real numbers is uncountable.*

1.2 Linear Spaces

Informally, a linear space over a field is a set with two operations on elements of the set and the field, such that elements of the set can be added and also multiplied by members of the field. Elementary concepts and basic facts that will be used later are collected in this section.

Linear Structures

Linear Spaces, Vector Addition, Scalar Multiplication

A *linear space* or a *vector space* over a field $(F, +, \cdot)$, where elements of F are called *scalars*, is a set V of elements, called *vectors* or *points*, with two operations: a binary operation on V, called *(vector) addition*, $(x, y) \mapsto x + y$ for $x, y \in V$, and another mapping from $F \times V$ to V, called *multiplication (by scalars)*, $(t, x) \mapsto tx$ for $t \in F, x \in V$, which satisfies the following properties:

[Additive Vector Structure (AVS)] (a) Addition is associative and commutative. (b) There is a unique neutral element of addition, 0, called the *origin*. (c) For each $x \in V$, there is a unique inverse element of x with respect to addition, denoted by $-x$.

[Multiplicative Scalar Structure (MSS)] (a) Multiplication by scalars is associative, i.e., for $s, t \in F$ and $x \in V$, $s(tx) = (st)x$. (b) There is a unique (non-zero) neutral element of multiplication, 1, called the *multiplicative identity* or *one*.

[Scalar–Vector Connecting Structure (SVCS)] (a) Multiplication by scalars is distributive with respect to vector addition, that is, for $t \in F, x, y \in V, t(x + y) = tx + ty$. (b) Multiplication by vectors is distributive with respect to scalar addition, that is, for $s, t \in F, x \in V$, $(s + t)x = sx + tx$.

Here, we use the same symbols for vector addition and field addition as for scalar multiplication and field multiplication. The difference is obvious. We may also speak of a linear space V over a field F without explicitly referring to its additive and multiplicative structures. When there is no need to mention the field F, we simply express the linear space by V. Note that scalar fields of linear spaces in this book are restricted to the field of real numbers $\mathbb{R}$.

Vector Sum, Translation, Linear Dependence, Linear Combination

Given a linear space V over a field F, for $x, y \in V$, $x + y$ is called the *sum of vectors* x and y. For sets $S, T \subset V$, $S + T = \{x + y \mid x \in S \text{ and } y \in T\}$ is the *sum of sets* S and T. For an indexed set of vectors x_i, we simply speak of a set $\{x_i\}$ of vectors corresponding to each index i if it is not necessary to specify the set of indices exactly. If the set of indices under consideration is finite, we denote the sum of the corresponding vectors by $\sum_i x_i$, which is uniquely defined since addition is associative and commutative. Similarly, for a finite family $\{S_i\}$ of subsets of V, the *sum* of sets in $\{S_i\}$ is $\sum_i S_i = \{\sum_i x_i \mid x_i \in S_i \text{ for all } i\}$. For $z \in V$, the mapping $x \mapsto x + z, x \in V$, is called *z-translation* or simply *translation*. For $S \subset V$ and $z \in V$, $S + z = \{x + z \mid x \in S \text{ for all } x \in S\}$ is a *translation of set S*.

A finite set $\{x_i\}$ of vectors is said to be *linearly dependent* if there exists a corresponding set $\{t_i\}$ of scalars, not all equal to 0, such that $\sum_i t_i x_i = 0$. On the other hand, if $\sum_i t_i x_i = 0$ implies $t_i = 0$ for all i, then the set $\{x_i\}$ is said to be *linearly independent*. Whenever we have $x = \sum_i t_i x_i$ for some finite set $\{x_i\}$ of vectors with a finite set $\{t_i\}$ of scalars, we say that x is a *linear combination* of $\{x_i\}$.

Basis, Dimension

A *basis* or *coordinate system* in a linear space V is a set of linearly independent vectors such that every vector in V is a linear combination of its elements. A linear space is *finite-dimensional* if it has a finite basis. Otherwise, it is called *infinite-dimensional*. The *linear dimension* or *dimension* of a finite-dimensional linear space V, denoted by $\dim V$, is the maximum number of elements in a basis of V.

Linear Subspaces, Linear Span, Linear Mappings

A non-empty subset L of a linear space V is called a (linear) *subspace* or a *linear manifold* if for every pair of vectors x, y in L every linear combination $sx + ty$ is also in L. It is immediately obvious that a subspace L is itself a linear space. In fact, with the same definition of addition and scalar multiplication as in the linear space V, both restricted to L satisfy the properties listed above as AVS, MSS, and SVCS with V replaced by L. The intersection of any collection of subspaces is a subspace. Let S be a subset of a linear space V. The intersection of all subspaces containing S is called the linear subspace *spanned* by S or the (linear) *span* of S. The linear subspace spanned by S is the same as the set of all linear combinations of elements of S. Note that if a vector x belongs to a linear subspace, then $x - x$ also belongs to the subspace. This means that every subspace contains the origin 0. Let V, W be linear spaces over the same field F. A *linear mapping* $f : V \longrightarrow W$ is a mapping which satisfies $f(sx + ty) = sf(x) + tf(y)$ for all $x, y \in V$ and for all $s, t \in F$.

Linear Functionals, Dual Spaces, Affine Subspaces, Hyperplanes

Let V be a linear space over a field F. If the field itself is regarded as a one-dimensional linear space over itself, a linear mapping $f : V \longrightarrow F$ is called a *linear functional* on the linear space V. The set of all linear functionals on V, denoted by V^*, is called the *dual space* of V.

Let L be a subspace of V. A translation $L_x = L + x, x \in V$, is called an *affine subspace* of V. The dimension of L_x is the dimension of L. There is a special class of affine subspaces that we encounter in the context of economics. Let $p^* : V \longrightarrow F$ be a linear functional that is not identically 0. Then, for $t \in F$, the set $H(p^*, t) = \{x \in V \mid p^*(x) = t\}$ is called an *affine hyperplane* or *hyperplane* in V. A hyperplane is an example of an affine subspace. A hyperplane $H(p^*, t)$ is related to a market valuation of commodities that each individual participant faces in a market economy (see Chapter 2).

Real Vector Spaces, Convex Sets, Convex Combinations, Line Segments, Straight Lines, Convex Hulls

A linear space V over the field of real numbers $\mathbb{R}$ is called a *real vector space* or *real linear space*. As indicated earlier, linear spaces in this book are restricted to real vector spaces. Given a real vector space V, a subset S of V is said to be *convex* if for every $t \in [0,1]$ we have $tS + (1-t)S \subset S$, that is, for every $x, y \in S$ we have $tx + (1-t)y \in S$ for all $t \in [0,1]$. If a vector x in V is a linear combination of $\{x_i\}$ with a finite set $\{t_i\}$ of reals such that $x = \sum_i t_i x_i$, it is said to be a *convex combination* if $\sum_i t_i = 1$ and $t_i \in [0,1]$ for all i. A convex combination of $\{x_i\}$ is its "weighted average." For any two vectors $x, y \in V$, the set $\{t_1 x + t_2 y \mid t_1 + t_2 = 1, t_1, t_2 \geqq 0\}$ of all convex combinations of x, y is a *line segment* connecting two "endpoints" x, y. A set S is convex if the entire line segment connecting any two points in S is contained in S. The set $\{t_1 x + t_2 y \mid t_1 + t_2 = 1\}$ is a *straight line* through points x, y. The *convex hull* of a subset S of V, denoted by $\operatorname{co} S$, is the intersection of all the convex sets containing the set S. Clearly, S is a convex set if and only if $S = \operatorname{co} S$.

Mathematical Property 1.3 [Convex Hulls] *Let V be a real vector space, S a subset of V, and $\{S_i\}$ a finite family of subsets of V. Then the following properties hold for convex hulls.*

(1) *The convex hull of S is identical to the set of all convex combinations of elements in S, i.e. $\operatorname{co} S = \{x \in V \mid x = \sum_i t_i x_i$ for some finite set $\{x_i\}$ of elements in S, with a finite set of real numbers, $\{t_i\}$ satisfying $\sum_i t_i = 1$ and $t_i \in [0,1]$ for all $i\}$.*
(2) *The convex hull of a finite sum of sets is equal to the finite sum of the convex hulls of each of the sets, i.e. $\operatorname{co} \sum_i S_i = \sum_i \operatorname{co} S_i$.*

Inner Product Spaces

Scalar Products

A *scalar* or *inner product* in a real vector space V over a field F is a scalar-valued mapping from $V \times V \to F, (x, y) \mapsto x \cdot y$ such that the following properties are satisfied:

[Vector Product (VP)] The scalar-valued product is commutative, i.e., $x \cdot y = y \cdot x$ for $x, y \in V$.
[Vector Product Addition (VPA)] The scalar product is distributive with respect to vector addition, i.e., $x \cdot (y + z) = x \cdot y + x \cdot z$ for $x, y, z \in V$.
[Scalar Multiplication (SVP)] For $t \in F$ and $x, y \in V$, $t(x \cdot y) = (tx) \cdot y = x \cdot (ty)$.
[Own Non-Negativity (NVP)] $x \cdot x \geqq 0$ and $x \cdot x = 0$ if and only if $x = 0$.

Inner Product Spaces, Orthogonality, Dual Spaces

A linear space with an inner product is called an *inner product space*. In an inner product space the *norm* or *length* of a vector x, denoted by $||x||$, is defined by $||x|| = \sqrt{x \cdot x}$. A vector x is called a *unit vector* if $||x|| = 1$. In an inner product space the concept of orthogonality between two vectors plays a very important role. Let V be an inner product space. Two vectors $x, y \in V$ are said to be *orthogonal* or *perpendicular*, and are written $x \perp y$ if $x \cdot y = 0$. Given a subset $W \subset V$, the set of all elements of V that are orthogonal to all elements in W is called the *orthogonal space* of W and is denoted by $W^{\perp}$, i.e. $W^{\perp} = \{x \in V \mid x \perp z \text{ for all } z \in W\}$. Since for every pair of vectors x, y in $W^{\perp}$ every linear combination $sx + ty, s, t \in F$, is also in $W^{\perp}$, the orthogonal space of W is a linear subspace of V, even if W is not a subspace. If W is a linear subspace, $W^{\perp}$ is called the *orthogonal complement* of W, and a vector $x \in V$ is uniquely expressed as the sum $x = y + z$ of a vector y in W and a vector z in $W^{\perp}$.

In a finite-dimensional linear space V, its dual space V^* has agreeable properties such that it has the same linear dimension as the linear space V and that each linear functional p^* in V^* can be uniquely *represented by a vector* $p \in V$ so that the dual space can be identified with the given linear space.

Mathematical Fact 1.6 [Dual Linear Spaces] *Let V be a finite-dimensional real vector space over F with a scalar product.*

(1) *The dual space V^* is also finite-dimensional and* $\dim V^* = \dim V$.
(2) *For every linear functional $p^* \in V^*, p^* : V \to F$, there exists a unique element $p \in V$ such that $p^*(x) = p \cdot x$ for all $x \in V$. For each element $x \in V$, a mapping $f_x : V \to F$ defined by $f_x(y) = x \cdot y$ for all $y \in V$ is a linear functional on V.*

1.3 Topological Spaces

Topological Structures

A topological structure on a set purports to provide a way to talk about which points are close to a given point in the set even among points between two of which the concept of "distance" is not well-defined on an intuitive ground. In our physical world we can tell whether your house is closer, for example, to the train station A than to your friend's house by a comparison of the distances between your house to the station and to your friend's house. In theoretical economics, individuals' behaviors that depend upon among other things what they like to consume, buy, and sell in markets are analyzed. Thus, the likings or tastes of individuals influence their behaviors. We come across instances in which we compare tastes and likings of different individuals. Intuitively, we feel that one could say, for example, they have very similar tastes in food and fashion. A topological structure serves to help express similarity even among very abstract objects such as our likings or tastes. A *topological structure* is introduced on a set by defining a family of subsets of the set, called "open sets," satisfying a certain property.

Topology, Open Sets, Closed Sets

A *topology* on a set T is a family $\mathcal{T}$ of sets consisting of subsets of T satisfying:

(a) Any union of members of $\mathcal{T}$ belongs to $\mathcal{T}$.
(b) Any finite intersection of members of $\mathcal{T}$ belongs to $\mathcal{T}$.
(c) $\emptyset$ and T belong to $\mathcal{T}$.

A pair $(T, \mathcal{T})$ consisting of a set T and a topology $\mathcal{T}$ on T is called a *topological space*. When the pairing $\mathcal{T}$ is clearly understood from the contexts, one may simply say that "T is a topological space." Sets belonging to the topology $\mathcal{T}$ are called *open sets*. Sets F whose complements $T - F$ are open sets are called *closed sets*. Given two topologies $\mathcal{S}$ and $\mathcal{T}$ on the same set T, we say $\mathcal{S}$ is *weaker* (or *coarser*) than $\mathcal{T}$, or $\mathcal{T}$ is *stronger* (or *finer*) than $\mathcal{S}$ if $\mathcal{S} \subset \mathcal{T}$.

Closure, Interior, Boundary of Sets, Dense Sets, Separable Spaces

In a topological space $(T, \mathcal{T})$ the intersection of all the closed sets containing a set $S \subset T$ is called the *closure* of S and is denoted by $\mathrm{cl}\, S$; that is, $\mathrm{cl}\, S = \cap\{T - W \mid S \subset T - W \text{ and } W \in \mathcal{T}\}$. The *interior* of S, denoted by $\mathrm{int}\, S$, is the union of all the open sets contained in S; that is, $\mathrm{int}\, S = \cup\{W \mid W \subset S \text{ and } W \in \mathcal{T}\}$. The *boundary* ∂S of S is the set $\mathrm{cl}\, S - \mathrm{int}\, S = \mathrm{cl}\, S \cap \mathrm{cl}\,(T - S)$. A set $S \subset T$ is *dense* if $\mathrm{cl}\, S = T$. A topological space $(T, \mathcal{T})$ is said to be *separable* if there is a countable dense set $S \subset T$.

Bases and Subbases for a Topology, Neighborhoods, Neighborhood Bases

Let $(T, \mathcal{T})$ be a topological space. A subset $\mathcal{B}$ of $\mathcal{T}$ is called a *base for the topology* $\mathcal{T}$ if every (open) set $W \in \mathcal{T}$ is a union of sets in $\mathcal{B}$. Sets belonging to $\mathcal{B}$ are called *basic open sets*. A subset $\mathcal{S}$ of $\mathcal{T}$ is called a *subbase for the topology* if the sets generated by finite intersections of sets belonging to $\mathcal{S}$ form a base for the topology. If $\mathcal{S}$ is a non-empty family of subsets of T, then the family of all finite intersections of subsets in $\mathcal{S}$ is a base for some topology on T.

A *neighborhood U of a point x* in a topological space $(T, \mathcal{T})$ is a subset of T containing an open set W such that $x \in W$. A *neighborhood U of a set S* in $(T, \mathcal{T})$ is a subset of T containing an open set W such that $S \subset W \subset U$. A *neighborhood base $\mathcal{U}_x$ at x* in $(T, \mathcal{T})$ is a subset $\mathcal{U}_x$ of the set $\mathcal{N}_x$ of all neighborhoods at x such that each $U \in \mathcal{N}_x$ contains some $V \in \mathcal{U}_x$. Its elements are called *basic neighborhoods*.

Topological Separation of Points and Partial Countability of Open Sets

The definition of a topology allows topological structures that are not very practical in standard cases for analytical purposes. Let us take two extreme examples to see this point. One example is the topology $\mathcal{T}_\emptyset = \{\emptyset, T\}$, called the *trivial topology*, on a set T. The other is the topology $\mathcal{T}_\mathbb{P}$ given by the family of all subsets of T, called the *discrete topology*. For any topology $\mathcal{T}$ on T, obviously one has $\mathcal{T}_\emptyset \subset \mathcal{T} \subset \mathcal{T}_\mathbb{P}$. From a topological viewpoint, the trivial topology makes T look like a singleton set whereas the discrete topology accepts all the set-theoretic distinctions of each and every point. The other distinct points in the space cannot be "topologically close" since the discrete topology includes all the possible singleton sets of the space as open sets. For topological structures to be useful in most analytical cases, distinct points in the space need be topologically separated in some appropriate sense.

There are several separation concepts used in topological analysis. Only one among them is explicitly referred to here. The fitting levels of "abundance" of open sets are expressed in terms of countability of subsets of open sets or neighborhoods. Two types of countability are introduced. A topological space $(T, \mathcal{T})$ is said to be a *Hausdorff space* or T_2 *space* if each pair of distinct points has non-intersecting neighborhoods; that is, whenever $x, y \in T, x \neq y$, there are neighborhoods U of x and V of y such that $U \cap V = \emptyset$. The space $(T, \mathcal{T})$ is called *first countable* if there is a countable neighborhood base $\mathcal{B}_x$ at every $x \in T$. The space is said to be *second countable* if there is a base $\mathcal{B} \subset \mathcal{T}$ which is countable. *Every second-countable space is first countable.* However, the converse need not be true as any uncountable space with the discrete topology is first countable but not second countable.

Topological Subspaces, Relative Topologies

Let S be a subset of a topological space $(T, \mathcal{T})$. If a family $\mathcal{T}_S$ of subsets of S is given by $\mathcal{T}_S = \{W \cap S \mid W \in \mathcal{T}\}$, $\mathcal{T}_S$ defines a topology on S. The topological space $(S, \mathcal{T}_S)$ is called a *topological subspace* (or simply, a *subspace*) of the space $(T, \mathcal{T})$. $\mathcal{T}_S$ is called the *relative topology* for $\mathcal{T}$ or the subspace topology induced from $\mathcal{T}$ on S. Unless otherwise mentioned, whenever reference is made to the topological space $S \subset T$, it means the topological subspace.

Continuity and Convergences

Continuous Mappings, Open Maps (Closed Maps), Homeomorphisms

Given two topological spaces $(T_1, \mathcal{T}_1)$, $(T_2, \mathcal{T}_2)$ and a mapping $f : T_1 \to T_2$, f is said to be *continuous* at x if, given any (basic) neighborhood V of $f(x)$, there is a (basic) neighborhood U of x such that $f(U) \subset V$. f is said to be *continuous* if it is continuous at every $x \in T_1$.

A mapping $f : T_1 \to T_2$ is called an *open* (resp. *closed*) *map* if $f(W)$ is open (resp. closed) for any open (resp. closed) set W in T_1. A one-to-one and onto mapping, i.e. a bijection, $f : T_1 \to T_2$ is called a *homeomorphism* if both of the mapping f and its inverse f^{-1} are continuous. Thus, a bijection $f : T_1 \to T_2$ is a homeomorphism if and only if it is continuous and open (or closed). Topological spaces $(T_1, \mathcal{T}_1)$ and $(T_2, \mathcal{T}_2)$ are called *homeomorphic* if there exists a homeomorphism $f : T_1 \to T_2$. A mapping $f : T_1 \to T_2$ is called an *embedding* if T_1 is homeomorphic to $f(T_1) \subset T_2$.

Mathematical Fact 1.7 [Continuity of Mappings] *Let $(T_1, \mathcal{T}_1)$ and $(T_2, \mathcal{T}_2)$ be topological spaces and f a mapping from T_1 to T_2. The following three conditions are equivalent:*

(a) f is a continuous mapping from $(T_1, \mathcal{T}_1)$ to $(T_2, \mathcal{T}_2)$.
(b) For any open set W in $(T_2, \mathcal{T}_2)$, $f^{-1}(W)$ is open in $(T_1, \mathcal{T}_1)$.
(c) For any closed set F in $(T_2, \mathcal{T}_2)$, $f^{-1}(F)$ is closed in $(T_1, \mathcal{T}_1)$.

Nets, Sequences, Convergences, Limits

There is another way to characterize the continuity of mappings by way of the concepts of convergences of points in the domain and the range of mappings. The convergences of points are expressed in terms of a set of points having a "direction" and the nearness of some of these points to a point.

A pair $(\Gamma, \lesssim)$ composed of a non-empty (index) set Γ and a binary relation $\lesssim$ on Γ is called a *directed set* if the relation $\lesssim$ is reflexive, transitive and satisfies the condition: given any $\alpha, \beta \in \Gamma$, there exists $\lambda \in \Gamma$ such that $\alpha \lesssim \lambda$ and $\beta \lesssim \lambda$. To put it differently, the binary relation of a directed set indicates a "direction" toward which any of the two points in the set "proceed" by showing their common successor. Very often $(\Gamma, \lesssim)$ is simply written as Γ. A mapping x from a directed set Γ to a set T is called a *net* or a *directed family of points* in $S \subset T$ if $x(\Gamma) \subset S$. Usually, $x(\gamma), \gamma \in \Gamma$, is written as x_γ, and the mapping $x : \Gamma \to T$ is written as $(x_\gamma)_{\gamma\in\Gamma}$ or simply as $(x_\gamma)_\gamma$ or (x_γ). In general, the index set Γ is an uncountably infinite set. However, depending upon topological spaces and in particular, when spaces are first countable, a directed set $(\Gamma, \lesssim)$ is taken to be the set of positive integers with its usual ordering, $(\mathbb{N}_+, \leqq)$. In such cases a net $x : \mathbb{N}_+ \to T$ is called a *sequence* in T and is written as $(x_n)_{n\in\mathbb{N}_+}$ or $(x_n)_{n=!,2,\ldots}$. It is often simplified as $(x_n)_n$ or (x_n).

If x is a net from a directed set $(\Gamma, \lesssim)$ to a set T, a *subnet* of the net $(x_\gamma)_{\gamma\in\Gamma}$ is the composition of mappings $x \circ \zeta$ from a directed set $(\Theta, \prec)$ to T where ζ is a mapping $\zeta : \Theta \to \Gamma$ which satisfies the following two conditions:

(a) ζ is increasing, that is, $\zeta(\theta_1) \lesssim \zeta(\theta_2)$ if $\theta_1 \prec \theta_2$.
(b) Given $\gamma_0 \in \Gamma$, there is $\theta_0 \in \Theta$ such that $\gamma_0 \lesssim \zeta(\theta)$ if $\theta_0 \prec \theta$.

For $\theta \in \Theta$, the point $x \circ \zeta(\theta)$ is often written x_{γ_θ}. In the case where a net is a sequence $(x_n)_{n\in\mathbb{N}_+}$, a subnet $(x_{n_k})_{k\in\mathbb{N}_+}$ is called a *subsequence* of the sequence and satisfies $n_k \geqq k$ for every $k \in \mathbb{N}_+$.

Let $(x_\gamma)_{\gamma\in\Gamma}$ be a net in a topological space T and $x_0 \in T$. If for any neighborhood U of x_0 there is $\gamma_0 \in \Gamma$ such that $x_\gamma \in U$ for all γ with $\gamma_0 \lesssim \gamma$, then the net $(x_\gamma)_{\gamma\in\Gamma}$ is said to *converge* to x_0, written $(x_\gamma)_\gamma \to_\gamma x_0$ or $x_\gamma \to_\gamma x_0$, and x_0 is called a *limit* of $(x_\gamma)_{\gamma\in\Gamma}$. A limit x_0 of a net $(x_\gamma)_\gamma$ is sometimes written $\lim_\gamma x_\gamma = x_0$ especially when it is unique. The convergence of $(x_\lambda)_\lambda$ to x_0 may also be expressed by stating that it is *residually* (or *eventually*) in all neighborhoods of x_0. x_0 is called a *cluster point* of $(x_\gamma)_{\gamma\in\Gamma}$ if for any neighborhood U of x_0 and for every $\gamma_0 \in \Gamma$ there is $\gamma \in \Gamma$ with $\gamma_0 \lesssim \gamma$ such that $x_\gamma \in U$. This is sometimes put in another way by stating that x_0 is a cluster point of $(x_\gamma)_\gamma$ if it is *frequently* (or, in the case of a sequence, *infinitely often*) in all neighborhoods of x_0. Although a limit of a net or a subnet may not be unique in general, a limit is unique if for any two distinct points in a space there are disjoint neighborhoods of each point that separate the points.

Mathematical Property 1.4 [Net Convergences]

(1) In a Hausdorff space a limit of a net is unique.
(2) Let T be a topological space, $S \subset T$, and $x \in T$. Then, $x \in \mathrm{cl}\, S$ if and only if there is a net $(x_\gamma)_\gamma$ in S that converges to x.
(3) A net has x_0 as its cluster point if and only if it has a subnet that converges to x_0.

Mathematical Fact 1.8 [Net Convergences and Continuous Maps] *Let $(T_1, \mathcal{T}_1)$ and $(T_2, \mathcal{T}_2)$ be topological spaces and f a mapping from T_1 to T_2. Then:*

f is continuous at $x_0 \in T_1$ if and only if $f(x_\gamma) \to_\gamma f(x_0)$ in T_2 for any net $(x_\gamma)_{\gamma\in\Gamma}$ in T_1 such that $x_\gamma \to_\gamma x_0$.

Topological Product Spaces, Product Topologies

Let $\{(T_\lambda, \mathcal{T}_\lambda) \mid \lambda \in \Lambda\}$ be a family of topological spaces. The *product topology* or *Tychonoff topology* on the Cartesian product $T = \prod_{\lambda\in\Lambda} T_\lambda$ is defined to be the topology on T having the sets $pr_\lambda^{-1}(U_\lambda)$, $U_\lambda \in \mathcal{T}_\lambda$, $\lambda \in \Lambda$, as a subbase. Namely, it is the topology on T having as a base the sets $\prod_\lambda U_\lambda$, where $U_\lambda \in \mathcal{T}_\lambda$ and $U_\lambda = T_\lambda$ except for at most finitely many λ's. A Cartesian product $T = \prod_{\lambda\in\Lambda} T_\lambda$ with the product topology is called a *topological product space*. The topological space T_λ is called the *λ-th coordinate subspace* or *factor space*.

Some elementary facts and properties relating to product topologies are presented below.

Mathematical Fact 1.9 [Projection Maps] *Let $T = \prod_{\lambda\in\Lambda} T_\lambda$ be a Cartesian product with the product topology.*

(1) For each $\lambda \in \Lambda$ the λ-th projection map is continuous and open, but need not be closed.
(2) The product topology on T is the weakest topology on T that makes the λ-th projection map continuous for each $\lambda \in \Lambda$.

Mathematical Property 1.5 [Continuity and Convergences in Product Spaces]
Let $T = \prod_{\lambda\in\Lambda} T_\lambda$ be a Cartesian product with the product topology.

(1) Given a mapping f from a topological space S to $T = \prod_{\lambda\in\Lambda} T_\lambda$, f is continuous if and only if $\mathrm{pr}_\lambda \circ f$ is continuous for each $\lambda \in \Lambda$, that is, f is continuous on each coordinate subspace.
(2) A net $(x_\gamma)_{\gamma\in\Gamma}$ in T converges to x if and only if $pr_\lambda(x_\gamma) \to_\gamma pr_\lambda(x)$ for each $\lambda \in \Lambda$; that is, $(x_\gamma)_\gamma$ converges on each coordinate subspace.

Metric Spaces

Whenever an intuitive notion of "distances" between two points in a set is well-defined, spaces become very agreeable topological spaces. The concept of distances is specified by numerical functions named metrics.

Metrics, Metric Spaces, Metric Topologies

Given a set T, let ρ be a function from $T \times T$ to $\mathbb{R}_+$ satisfying the following properties:

(a) $\rho(x, y) = \rho(y, x)$,
(b) $\rho(x, y) + \rho(y, z) \geqq \rho(x, z)$,
(c) $\rho(x, y) = 0$ if and only if $x = y$;

then ρ is called a *metric* or a *distance function* on T. A pair (T, ρ) consisting of a set T and a metric ρ on T is a *metric space*. If a metric ρ is fixed or its specification is not needed in a particular argument, one may drop the ordered-pair notation and simply speak of a "metric space" T.

For any $\varepsilon > 0$ and $x \in T$, define a set $U_\varepsilon(x)$ by $U_\varepsilon(x) = \{z \in T \mid \rho(x, z) < \varepsilon\}$. The set $U_\varepsilon(x)$ is called the *ε-neighborhood* of x. The set of all ε-neighborhoods of x for $\varepsilon > 0$ forms a neighborhood base at x. Define a family $\mathcal{T}_\rho$ of sets by $\mathcal{T}_\rho = \{W \subset T \mid \text{for any } x \in W, \text{ there is } \varepsilon > 0 \text{ such that } U_\varepsilon(x) \subset W\}$. $\mathcal{T}_\rho$ defines a topology on T and is called the *topology induced by a metric* ρ. The space $(T, \mathcal{T}_\rho)$ is the *topological space induced by a metric space* (T, ρ). Given a topological space $(T, \mathcal{T})$, if $\mathcal{T} = \mathcal{T}_\rho$ for some metric ρ on T, then $\mathcal{T}$ is called a *metric topology* or *topology induced by a metric*. In such a case the topological space $(T, \mathcal{T})$ is said to be a *metrizable space* and any such metric ρ is said to be *compatible with the topology* $\mathcal{T}$.

Given two metrics ρ_1 and ρ_2 on the same set T, they are said to be *equivalent* if they induce the same topology on T. For any metric space (T, ρ), the metic ρ can be assumed to be bounded in the sense that there is a bounded metric which is equivalent to ρ.

Mathematical Fact 1.10 [Bounded Metrics] *Given a metric space (T, ρ), there exist equivalent bounded metrics. For example, the following metrics ρ_1 and ρ_2 are equivalent to ρ:*

$$\rho_1 = \min\{\rho(x, y), 1\},\ \rho_2 = \frac{\rho(x, y)}{\rho(x, y) + 1},\ \textit{for each } (x, y) \in T \times T.$$

The following fact is very natural in view of positive distances between any two distinct points and the denseness of rational numbers in the range of any distance functions.

Mathematical Fact 1.11 [Metrizable Spaces]

(1) A metrizable space $(T, \mathcal{T})$ is Hausdorff.
(2) A metrizable space $(T, \mathcal{T})$ is separable if and only if $(T, \mathcal{T})$ is second countable.

Although the convergence of sequences in a topological space in general might not characterize closed sets, metric spaces have the following property.

Mathematical Property 1.6 [Closed Sets in Metric Spaces] *In a metric space T a set $F \subset T$ is closed if and only if the limits of any convergent sequences in F belong to F.*

Mathematical Fact 1.12 [Metric Subspaces]

(1) Let S be a subset of a metric space (T, ρ) and ρ_S the restriction of the metric ρ to $S \times S$. Then, the topology on S induced by the metric ρ_S is identical to the relative topology induced from the metric topology on T.
(2) Any subspace of a separable metric space (T, ρ) is separable.

Mathematical Property 1.7 [Product Topologies of Metric Spaces]

Let $T = \prod_{\lambda \in \Lambda} T_\lambda$ be a Cartesian product of a finite number of topological spaces with the product topology indexed by a set $\Lambda = \{1, \ldots, n\}, n \geqq 1$.

(1) If $\{(T_\lambda, \rho_\lambda) \mid \lambda \in \Lambda\}$ is a family of metric spaces, define the following mappings from $T \times T$ to $\mathbb{R}_+$: for each $x = (x_1, \ldots, x_n), y = (y_1, \ldots, y_n) \in T$,

$$\rho_{sum}(x, y) = \sum_{\lambda \in \Lambda} \rho_\lambda(x_\lambda, y_\lambda),$$
$$\rho_{max}(x, y) = \max\{\rho_\lambda(x_\lambda, y_\lambda) \mid \lambda \in \Lambda\}, \text{ and}$$
$$\rho_{eu}(x, y) = \sqrt[n]{\sum_{\lambda \in \Lambda} \rho_\lambda(x_\lambda, y_\lambda)}.$$

Then, these three mappings are equivalent metrics on T, and the metric topology induced by each one of these metrics coincides with the product topology of the metric topologies on coordinate subspaces.
(2) If $\{(T_\lambda, \mathcal{T}_\lambda \mid \lambda \in \Lambda\}$ is a family of metrizable spaces all of which are separable, then its topological product space is also separable.

Compact Spaces

Topological spaces have very practical properties from an analytic point of view if they are topologically "close" to a space having a finite or a countable number of points. The separability that requires the space to have a countable dense subset is one such property. There is an other far-reaching concept which refers to a topological approximation by a finite number of points in a space.

Compactness and Finite Intersection Property

To substantiate an idea of topological approximation of a space with a finite number of points, let us be more specific. A family $\mathcal{S} = \{S_\lambda | \lambda \in \Lambda\}$ of subsets of a topological space $(T, \mathcal{T})$ is said to be a *covering* or *cover* of T if it satisfies $T = \cup\{S_\lambda | \lambda \in \Lambda\}$. The covering $\mathcal{S}$ is called an *open covering* or *cover* if all the sets in $\mathcal{S}$ are open sets in $(T, \mathcal{T})$. Given two coverings of T, $\mathcal{S}_1$ and $\mathcal{S}_2$, $\mathcal{S}_2$ is called a *subcovering* or *subcover* of $\mathcal{S}_1$ if $\mathcal{S}_2 \subset \mathcal{S}_1$.

A topological space $(T, \mathcal{T})$ is said to be *compact* if every open cover of T has a finite subcover. It is called *sequentially compact* if every sequence has a convergent subsequence. A subset S of T is called *compact* if S is compact as a topological subspace of $(T, \mathcal{T})$. It is called *relativly compact* if $\mathrm{cl}\, S$ is compact. A family $\mathcal{S}$ of subsets of T is said to have the *finite intersection property* if every finite number of sets S_i, $i = 1, \ldots, k$, in $\mathcal{S}$ has a non-empty intersection, i.e., $\cap_{i=1}^k S_i \neq \emptyset$.

Mathematical Property 1.8 [Compact Spaces]

(1) A topological space $(T, \mathcal{T})$ is compact if and only if one of the following conditions is satisfied:

(a) $\cap\{F_\lambda | \lambda \in \Lambda\} \neq \emptyset$ for every family $\{F_\lambda | \lambda \in \Lambda\}$ of closed subsets of T having the finite intersection property.

(b) Each net in T has a subnet which converges to some point in T.

(2) A compact set in a Hausdorff topological space is closed.

(3) A closed set in a compact topological space is compact.

Mathematical Property 1.9 [Compactness Property]

(1) [Tychonoff Theorem] *Any product space of compact spaces is compact.*

(2) If f is a continuous mapping from a compact topological space $(T_1, \mathcal{T}_1)$, to a topological space $(T_2, \mathcal{T}_2)$, the image of its domain under f, $f(T_1)$, is compact. If, in addition, T_2 is Hausdorff and f is a bijection, then f becomes a homeomorphism. In particular, by considering the identity mapping $x \mapsto x$ as f, any topology $\mathcal{T}_1'$ coarser than $\mathcal{T}_1$ coincides with the topology $\mathcal{T}_1$.

Mathematical Property 1.10 [Compact Metric Spaces]

(1) A metric space T is compact if and only if it is sequentially compact, that is, every sequence $(x_n)_n$ in T has a convergent subsequence.

(2) A compact metric space is separable.

Locally Compact Spaces, One-Point Compactification (Alexandroff Compactification)

Even when topological spaces are non-compact, Hausdorff topological spaces could be identified with compact topological spaces in cases where each point has compact neighborhoods. A topological space $(T, \mathcal{T})$ is said to be *locally compact* if each $x \in T$ has a neighborhood base consisting of compact neighborhoods. In a Hausdorff space this condition is simplified as below.

Mathematical Fact 1.13 [Local Compactness] *Let $(T, \mathcal{T})$ be a Hausdorff a topological space. Then, the space T is locally compact if and only if one of the following conditions is satisfied:*

(a) Each $x \in T$ has at least one compact neighborhood.
(b) Each $x \in T$ has an open neighborhood with compact closure.

Since compact Hausdorff spaces have nice properties for the purposes of analysis, it is of interest on various occasions to consider the process of "embedding" a given space into some compact Hausdorff space as a "sufficiently extended" subspace. This process is known as a "compactification." A compactification of T is roughly a compact space K containing T as a dense subspace. To be more precise, a *compactification* of T is an ordered pair (T, f) where $f : T \to K$ is a homeomorphism onto $f(T)$, and $f(T)$ is dense in a compact Hausdorff space K. A most simple compactification is the process of "adding a point" to compactify a given space. One might imagine the real line $\mathbb{R}$, which is "open ended" toward the directions of $+\infty$ and $-\infty$, and "put together both ends" regarding them as a single point ∞ and adding it to the space.

Every non-compact Hausdorff space $(T, \mathcal{T})$ has a *one-point compactification* or *Alexandroff compactification* $T' = T \cup \{z\}$ with $z \notin T$ (for example, $z = T$). Define a topology on T' by specifying a neighborhood base at each point. A point of T has the same neighborhood base as in T. The neighborhood base at $z \notin T$ consists of all sets $T' - B$ where B's are closed, compact and contained in T. This indeed specifies a neighborhood base at each point. Define a mapping $f : T \to T'$ by $f = id_T$. Then, f is a homeomorphism onto $f(T)$, T' is compact, and T is dense in T'. One-point compactification T' of a topological space T shall be denoted by $T_{\{\infty\}}$.

Mathematical Property 1.11 [Compactification] *Given a non-compact Hausdorff space $(T, \mathcal{T})$, its one-point compactification $T_{\{\infty\}}$ has the following properties:*

(1) The one-point compactification $T_{\{\infty\}}$ is Hausdorff if and only if the space T is Hausdorff and locally compact.
(2) If (T, ρ) is a separable and locally compact metric space, then the one-point compactification $T_{\{\infty\}}$ is metrizable.

Complete Metric Spaces

In compact spaces any net has a convergent subnet even if the net itself is not convergent (cf. Mathematical Property 1.8 (1)). A topological concept of "completeness" is related to a family of nets that are convergent. Here, we only refer to the completeness of metric spaces using nets which are sequences.

Cauchy Sequences and Completeness

A sequence $(x_n)_n$ in a metric space (T, ρ) is called a *Cauchy sequence* if for any positive number $\varepsilon > 0$ there is an integer n_0 such that $\rho(x_m, x_n) < \varepsilon$ for any $m, n \geqq$

n_0. In a metric space (T, ρ) all convergent sequences are Cauchy sequences. The space (T, ρ) is said to be *complete* if all the Cauchy sequences in T are convergent. The metric ρ is called a *complete metric*. If a topological space $(T, \mathcal{T})$ is metrizable by a complete metric ρ, it is called *completely metrizable*.

In a topological space $(T, \mathcal{T})$ a countable intersection of open sets is called a G_δ *set* and a countable union of closed sets is called an F_σ *set*. The *diameter* of a set S, $\operatorname{diam}(S)$, in a metric space (T, ρ) is defined by $\operatorname{diam}(S) = \sup\{\rho(x, y) \mid x, y \in S\}$.

Mathematical Property 1.12 [Complete Metric Spaces]

(1) [Alexandroff–Mazurkiewicz Theorem] *In a complete metric space (T, ρ), a subset S of T is completely metrizable if and only if it is a G_δ set.*
(2) A closed subspace of a complete metric space is complete.
(3) A compact metric space (T, ρ) is complete.
(4) A metric space (T, ρ) is complete if and only if, for any sequence $(S_n)_n$ of non-empty subsets of T, it is true that $\cap_{n \in \mathbb{N}_+} (\operatorname{cl} S_n) \neq \emptyset$ (and singleton) whenever one has $S_n \supset S_{n+1}, n \in \mathbb{N}_+$, and $\operatorname{diam}(S_n) \to_n 0$.
(5) The product of a countable number of complete metric spaces is complete.

Polish Spaces

A topological space $(T, \mathcal{T})$ is called a *Polish space* if it is separable and completely metrizable. Let $\mathbb{I}$ denote the closed interval $[0, 1]$ of the real line $\mathbb{R}$. The product of countable $\mathbb{I}$'s, $\mathbb{I}^\infty$, is called the *Hilbert cube*. The product space $[0, 1] \times [0, 1/2] \times [0, 1/3] \times \ldots$ is also called the Hilbert cube. Both of them are homeomorphic.

Mathematical Property 1.13 [Characterization of Polish Spaces] *A topological space $(T, \mathcal{T})$ is Polish if and only if T is homeomorphic to a G_δ set in the Hilbert cube $\mathbb{I}^\infty$.*

Mathematical Property 1.14 [Polish Spaces]

(1) A closed subspace of a Polish space is Polish.
(2) An open subspace of a Polish space is Polish.
(3) The product of a countable number of Polish spaces is Polish.

1.4 Continuity of Relations

The concepts of the continuity of relations play important analytical roles in mathematical economics as well as in some of the other areas of applied mathematics. However, generally speaking, the topic of continuity of relations is missing in standard texts of analysis or topology. For this reason, some of the proofs of properties relating to the continuity of relations are included in this section.

Concepts of Continuity of Relations

Closed Relations

The initial continuity concept is the closedness of relations. Let S, T be topological spaces and φ be a relation from S to T. φ is said to be *closed* if $\varphi \subset S \times T$ is a closed set with respect to the product topology. Similarly, φ is said to be *closed at* $x \in S$ if $(x, y) \in \varphi$ for any net $(x_\gamma, y_\gamma)_\gamma$ in φ such that $(x_\gamma, y_\gamma)_\gamma \to_\gamma (x, y)$. Clearly, φ is closed if and only if it is closed at every $x \in \operatorname{dom} S$. It is also straightforward that φ is closed if and only if $y \in \varphi(x)$ whenever $(x_\gamma)_\gamma \to_\gamma x$, $(y_\gamma)_\gamma \to_\gamma y$, and $y_\gamma \in \varphi(x_\gamma)$ for all γ. There will also be occasions to refer to the topological properties of values of relations. A relation φ from S to T is said to be *closed-valued* (resp. *compact-valued*) *at* x if $\varphi(x)$ is closed (resp. compact) in T. φ is *closed-valued* (resp. *compact-valued*) if $\varphi(x)$ is closed (resp. compact) at x for every $x \in \operatorname{dom}\varphi$. One should be aware of the difference between the closedness of a relation at a point and its closed-valuedness at that point. The former is a sufficient condition for the latter but not vice versa.

The straightforward properties of closed relations are given below.

Mathematical Fact 1.14 [Closed Relations] *Let S, T be topological spaces and φ be a closed relation from S to T.*

(1) *Each of the following conditions is equivalent to the closedness of φ in $S \times T$.*

(a) *$S \times T - \varphi$ is open in $S \times T$.*
(b) *For any $(x, y) \in S \times T$ with $y \notin \varphi(x)$ there exist neighborhoods U_x of x and V_y of y, respectively, such that $\varphi(z) \cap V_y = \emptyset$ for all $z \in U_x$.*

(2) *The relation φ is closed-valued.*

Hemi-Continuous Relations

The next continuity concept is the two types of hemi-continuity of relations. Let S, T be topological spaces and φ be a relation from S to T. φ is said to be *lower hemi-continuous* (abbreviated as *l.h.c.*) *at* $x \in \operatorname{dom}\varphi$ if for every open set W in T for which $\varphi(x) \cap W \neq \emptyset$ there exists a neighborhood V of x such that $\varphi(z) \cap W \neq \emptyset$ for every $z \in V$. φ is called *lower hemi-continuous* if it is l.h.c. at every $x \in \operatorname{dom}\varphi$. φ is said to be *upper hemi-continuous* (abbreviated as *u.h.c.*) *at* $x \in \operatorname{dom}\varphi$ if for every open set W in T for which $\varphi(x) \subset W$ there exists a neighborhood V of x such that $\varphi(z) \subset W$ for every $z \in V$. φ is called *upper hemi-continuous* if it is u.h.c. at every $x \in \operatorname{dom}\varphi$. The same abbreviations *l.h.c.* and *u.h.c.* will also be used to simplify their noun forms, lower hemi-continuity and upper hemi-continuity. It is immediate that in the case there is a neighborhood V of x such that for every $z \in V$ $\varphi(z)$ consists of a single element, i.e. "single-valued," both conditions of l.h.c. and u.h.c. coincide with that of the continuity of a function from $\operatorname{dom}\varphi \cap V$ to T at $x \in \operatorname{dom}\varphi \cap V$.

If f is a continuous function from S to T, its inverse image of an open set W under f, $f^{-1}(W)$, is open for every open set $W \subset T$, and its inverse image of a closed set F, $f^{-1}(F)$, is closed for every closed set $F \subset T$. Hemi-continuous relations do satisfy analogous "global" properties as stated below.

Mathematical Fact 1.15 [Global Characterization of Hemi-Continuities] *Let S, T be topological spaces and φ be a relation from S to T.*

(1) [Lower Hemi-Continuity]

(a) φ is l.h.c. if and only if, for every open set W in T, its inferior inverse image $\varphi_w^{-1}(W) = \{x \in S \mid (x, y) \in \varphi$ for some $y \in W\} = \{x \in S \mid \varphi(x) \cap W \neq \emptyset\}$ is open.
(b) φ is l.h.c. if and only if, for every closed set F in T, its superior inverse image $\varphi_s^{-1}(F) = \{x \in S \mid y \in F$ for all y satisfying $(x, y) \in \varphi\} = \{x \in \operatorname{dom}\varphi \mid \varphi(x) \subset F\}$ is closed.

(2) [Upper Hemi-Continuity]

(a) φ is u.h.c. if and only if, for every open set W in T, its superior inverse image $\varphi_s^{-1}(W) = \{x \in S \mid y \in W$ for all y satisfying $(x, y) \in \varphi\} = \{x \in \operatorname{dom}\varphi \mid \varphi(x) \subset W\}$ is open.
(b) φ is u.h.c. if and only if, for every closed set F in T, its inferior inverse image $\varphi_w^{-1}(F) = \{x \in S \mid (x, y) \in \varphi$ for some $y \in F\} = \{x \in S \mid \varphi(x) \cap F \neq \emptyset\}$ is closed.

The following mathematical proposition shows circumstances under which closed relations are equivalent to upper hemi-continuous relations.

Mathematical Proposition 1.1 [Closedness and Upper Hemi-Continuity] *Let φ be a closed-valued relation from a topological space S to a Hausdorff compact space T. Then, φ is closed if and only if it is upper hemi-continuous.*

Proof Let φ be closed. Given an open set W in T and $x \in \varphi_s^{-1}(W)$, we shall show there is a neighborhood U_x of x such that $\varphi(U_x) \subset W$. Set $K = T - W$. K is closed in T and compact. For each $y \in K$ there are open neighborhoods $U_y(x)$ of x and V_y of y such that $\varphi(U_y(x)) \cap V_y = \emptyset$ since φ is closed at x and $y \notin \varphi(x)$. A family $\{V_y \mid y \in K\}$ of subsets of T is an open covering of a compact set $K = T - W$. Hence, for some finite number of elements $y_1 .y_2, \ldots, y_n$, of K, we have $K \subset V_K = \cup_{k=1}^n V_{y_k}$. Defining $U_x = \cap_{k=1}^n U_{y_k}(x)$, one obtains $\varphi(U_x) \cap V_K = \emptyset$. It follows that $\varphi(U_x) \subset W$. Thus, φ is u.h.c.

Conversely, assuming φ is u.h.c., we show it is closed. Given $x \in S$ and $y \in T$ such that $y \notin \varphi(x)$, there are an open set W in T and a neighborhood V of y for which $W \cap V = \emptyset$ as $\varphi(x)$ is compact. Since φ is u.h.c., for a neighborhood U of x, one has $\varphi(U) \subset W$. Thus, for all $z \in U$, $\varphi(z) \cap V = \emptyset$, showing that φ is closed. □

Recall that a continuous function has a topological property that the image of a compact set is compact. In the case of relations the upper hemi-continuity of relations shares this property with continuous functions because of (2)(b) in Mathematical Fact 1.15 if relations are compact-valued.

Mathematical Fact 1.16 [Compact Image of U.H.C. Relations] *Let S, T be topological spaces and φ be a relation from S to T. Given a compact subset $K \subset \operatorname{dom}\varphi$, the image of K under φ, $\varphi(K)$, is compact if φ is u.h.c. and compact-valued.*

Exercise 1.1 Confirm the statement of Mathematical Fact 1.16.

Sequential Characterizations of Continuity of Relations

Seeing that there are cases where it is more convenient to use sequential expressions for relational continuities in metric spaces, let us first delineate the closedness of relations in terms of "local closedness" and "local sequences." Then, after giving the expression of two forms of topological limits of sequences of subsets, sequential characterizations of hemi-continuities are described.

Sequential Closedness

A relation φ from a metric space S to another metric space T is said to be *closed at* $x \in \operatorname{dom}\varphi$ if for every convergent sequence $(x_n)_n$ in $\operatorname{dom}\varphi$ such that $x_n \to_n x$ with $y_n \in \varphi(x_n)$ for each n and $y_n \to_n y$, one has $y \in \varphi(x)$. As noted earlier (cf. p. 35), φ being *closed-valued at* x is *not* identical with φ being *closed at* x.

Mathematical Fact 1.17 [Local Sequential Properties of Closed Relations] *Let S, T be metric spaces and φ be a relation from S to T.*

(1) *Each of the following conditions is equivalent to φ being closed at $x \in \operatorname{dom}\varphi$:*

(a) *If a sequence $(x_n, y_n)_n$ in $\varphi \subset S{\times}T$ converges to (x, y) for which $x \in \operatorname{dom}\varphi$, one has $(x, y) \in \varphi$.*
(b) *If $y \notin \varphi(x)$ for $y \in T$, there exist neighborhoods U_x of x and V_y of y, respectively, such that $\varphi(z) \cap V_y = \emptyset$ for all $z \in U_x$.*

(2) *Each of the following properties characterizes φ being closed:*

(a) *For every sequence $(x_n, y_n)_n$ in $\varphi \subset S \times T$ converging to (x, y), one has $(x, y) \in \varphi$.*
(b) *φ is closed at every $x \in \operatorname{dom}\varphi$.*

Topological Limes Inferior or Superior

Let (T, ρ) be a metric space and $(F_n)_n$ be a sequence of subsets of T. The *topological limes inferior* of $(F_n)_n$, denoted by $\mathrm{Li}\,(F_n)$, is the set of all elements $x \subset T$ such that for any neighborhood U of x one has $U \cap F_n \neq \emptyset$ *residually*, i.e., for all $n \geqq n_0$ for some integer n_0. The *topological limes superior* of $(F_n)_n$, denoted by $\mathrm{Ls}\,(F_n)$, is the set of all elements $x \in T$ such that for any neighborhood U of x there are *infinitely many* n such that F_n intersects with U, i.e., for any integer $m > 0$ there is an integer $n \geqq m$ with $U \cap F_n \neq \emptyset$. In the definition of the topological limes inferior or superior, instead of taking any neighborhoods of x it is enough to take neighborhoods belonging to a neighborhood base at x, or enough to take open sets that belong to a base for the topology $\mathcal{T}_\rho$ induced by ρ and contain x.

Exercise 1.2 Confirm each of the following statements:

1. $\mathrm{Li}\,(F_n)$ is closed.
2. $\mathrm{Ls}\,(F_n)$ is closed.
3. $\mathrm{Li}\,(F_n) \subset \mathrm{Ls}\,(F_n)$.

Sequential Hemi-Continuities

In many cases of applications it is very practical to express the two concepts of hemi-continuity of relations in terms of sequences in the domain of relations and the topological limes inferior or superior of corresponding sequences of sets in their range.

Mathematical Proposition 1.2 [Local Sequential Characterization of L.H.C.] *Let S, T be metric spaces and φ be a relation from S to T.*

(1) *φ is l.h.c. at $x \in \mathrm{dom}\,\varphi$ if and only if for every sequence $(x_n)_n$ in $\mathrm{dom}\,\varphi$ converging to x one has $\varphi(x) \subset \mathrm{Li}\ (\varphi(x_n))$.*
(2) *φ is l.h.c. at $x \in \mathrm{dom}\,\varphi$ if and only if for every sequence $(x_n)_n$ in $\mathrm{dom}\,\varphi$ converging to x one has $\varphi(x) \subset \mathrm{Ls}\ (\varphi(x_n))$.*

Proof Let $(x_n)_n$ be a convergent sequence in S with $x_n \to_n x$ and $y \in \varphi(x)$. Assuming φ is l.h.c. at $x \in S$, we show $y \in \mathrm{Li}\,(\varphi(x_n))\,(\subset \mathrm{Ls}\,(\varphi(x_n)))$. If $y \notin \mathrm{Li}\,(\varphi(x_n))$, then, there exists an open neighborhood W of y such that for any integer k we have $\varphi(x_n) \cap W = \emptyset$ for some $n \geqq k$. But since φ is l.h.c. at x and $\varphi(x) \cap W \neq \emptyset$, there exists k_0 such that for all $n \geqq k_0$ we have $\varphi(x_n) \cap W \neq \emptyset$ so that $y \in \mathrm{Li}\,(\varphi(x_n))$, a contradiction.

Conversely, assuming φ is not l.h.c. at x, we show that neither the condition in (1) nor in (2) holds. It suffices to show that (2) does not hold. If φ is not l.h.c. at x, there is an open set W with $\varphi(x) \cap W \neq \emptyset$ such that for any neighborhood U of x there exists $z \in U$ satisfying $\varphi(z) \cap W = \emptyset$. Thus, there is a sequence $(x_n)_n$ with $x_n \to_n x$ and $\varphi(x_n) \cap W = \emptyset$ for all n. Let $y \in \varphi(x) \cap W$. If $\varphi(x) \subset \mathrm{Ls}\,(x_n)$, there is a subsequence $y_{k_n} \to_n y$ with $y_{k_n} \in \varphi(x_{k_n})$ for all n. It follows that $y_{k_n} \in W$ for all n large implying $\varphi(x_{k_n}) \cap W \neq \emptyset$ for such n, contradicting φ is not l.h.c. at x. □

Mathematical Proposition 1.3 [Local Sequential Characterization of U.H.C.] *Let S, T be metric spaces and φ be a relation from S to T.*

(1) *Let φ be compact-valued. Then, φ is u.h.c. at $x \in \mathrm{dom}\,\varphi$ if and only if for every sequence $(x_n)_n$ in $\mathrm{dom}\,\varphi$ converging to x one has $\mathrm{Ls}\,(\varphi(x_n)) \neq \emptyset$ and $\mathrm{Ls}\,(\varphi(x_n)) \subset \varphi(x)$.*
(2) *Let T be a compact space and φ be closed-valued at $x \in \mathrm{dom}\,\varphi$. Then, φ is u.h.c. at $x \in \mathrm{dom}\,\varphi$ if and only if for every sequence (x_n, y_n) in φ converging to (x, y) one has $(x, y) \in \varphi$; that is, $y \in \varphi(x)$ whenever $y_n \in \varphi(x_n)$ for all n, $(x_n)_n \to_n x$, and $(y_n)_n \to_n y$.*

Proof (1) Let φ be u.h.c. at $x \in \operatorname{dom}\varphi$. Given a convergent sequence $(x_n)_n \to_n x$ in $\operatorname{dom}\varphi$, set $K = \{x, x_1, x_2, \dots\}$. Then, K is compact in $\operatorname{dom}\varphi$. By Mathematical Fact 1.16, $\varphi(K)$ is compact in T. Thus, if we choose an element from $\varphi(x_n)$ for each n, there is a convergent subsequence whose limit belongs to $\operatorname{Ls}(\varphi(x_n)) \cap \varphi(K) \neq \emptyset$. To show $\operatorname{Ls}(\varphi(x_n)) \subset \varphi(x)$, assume to the contrary that there is an element $y \in \operatorname{Ls}(\varphi(x_n)) - \varphi(x)$. Then, there is a subsequence $(x_{k_n})_n \to_n x$ and an associated sequence $(y_{k_n})_n$ for which $y_{k_n} \in \varphi(x_{k_n})$ for all n and $(y_{k_n})_n \to_n y$. Since $\varphi(x)$ is compact, it is closed and there are open neighborhoods V_y and $V_{\varphi(x)}$ of y and $\varphi(x)$, respectively, such that $V_y \cap V_{\varphi(x)} = \emptyset$. But, by the u.h.c. of φ at x, one has $\varphi(U) \subset V_{\varphi(x)}$ for an open neighborhood U of x. This is contradictory to $x_{k_n} \in U$ and $y_{k_n} \in V_y$ for n large enough. Therefore, $\operatorname{Ls}(\varphi(x_n)) \subset \varphi(x)$.

Conversely, let $(x_n)_n \to_n x$ in $\operatorname{dom}\varphi$, and assume $\operatorname{Ls}(\varphi(x_n)) \neq \emptyset$ and $\operatorname{Ls}(\varphi(x_n)) \subset \varphi(x)$. If φ is not u.h.c. at $x \in \operatorname{dom}\varphi$, then there is an open set W in T for which $\varphi(x) \subset W$ such that for a neighborhood U of x one has $\varphi(U) \cap (T - W) \neq \emptyset$. It follows that there is a subsequence $(x_{k_n})_n \to_n x$ and an associated sequence $(y_{k_n})_n$ such that $y_{k_n} \in \varphi(x_{k_n}) \cap (T - W)$ for all n and $(y_{k_n})_n \to_n y \in \operatorname{Ls}(\varphi(x_n)) \subset \varphi(x) \subset W$. Thus, $(y_{k_n})_n \in W$ for n large enough contradicting the selection of $y_{k_n} \in (T - W)$ for all n.

(2) This is a restatement of Mathematical Proposition 1.1 in metric spaces using Mathematical Fact 1.17 (2). □

Exercise 1.3 Confirm the statement of Mathematical Fact 1.17 (1-b).

Remark 1.1 [Semi-Continuities in relation to Hemi-Continuities] Economic theorists in the 1950s used the term "semi-continuity" with respect to the continuity properties of correspondences (or, set-valued functions) in place of the term hemi-continuity used in this book. In some areas of applied mathematics, the term "semi-continuity" is still often used. Semi-continuities of set-valued mappings are typically defined in terms of sequences as follows: Let S, T be metric spaces and φ be a relation from S to T.

(1) [Lower Semi-Continuity] φ is said to be *lower semi-continuous* (abbreviated as *l.s.c.) at* x if there exists a sequence $(y_n)_n$ such that $y_n \to_n y$ and $y_n \in \varphi(x_n)$ for all n whenever $x_n \to_n x$ and $y \in \varphi(x)$.
(2) [Upper Semi-Continuity] φ is said to be *upper semi-continuous* (abbreviated as *u.s.c.) at* x if $x_n \to_n x$, $y_n \in \varphi(x_n)$, and $y_n \to_n y$ imply $y \in \varphi(x)$.

It follows from the above characterizations of hemi-continuity by sequences that we have:

(1) φ is l.s.c. if and only if φ is l.h.c..
(2) φ is u.s.c. if and only if φ is closed. Thus, if φ is closed-valued and T is compact, then φ is u.s.c. if and only if φ is u.h.c..

1.5 Euclidean Spaces

Euclidean m-Space $\mathbb{R}^m$

An important special case of a Cartesian product of an indexed family of sets (cf. p. 16) is the Cartesian product with a fixed coordinate set $\mathbb{R}$. *Euclidean m-space* or *m-dimensional Euclidean space*, denoted by $\mathbb{R}^m$, is the set of all real-valued functions on a set $\{1, 2, \ldots, m\}$ consisting of integers from 1 to m. The i-th *coordinate* or *component* of an element x is x_i. An element x is usually denoted by an ordered m-tuple of real numbers $x = (x_1, \ldots, x_m)$. For any pair of ordered m-tuples of real numbers $x = (x_1, \ldots, x_m)$ and $y = (y_1, \ldots, y_m)$, we write $x = y$ if and only if $x_i = y_i$ for every $i = 1, \ldots, m$. The *i-th projection (map)* from $\mathbb{R}^m$ onto the i-th *coordinate subspace* $\mathbb{R}$ is $\text{pr}_i : x \mapsto x_i$.

Linear and Topological Structures of $\mathbb{R}^m$

$\mathbb{R}^m$ is given an inner product space structure over the scalar field of real numbers $\mathbb{R}$ by the following addition, multiplication by scalars, scalar products along with the neutral element of addition called the *zero vector* or the *origin* $\mathbf{0} = (0, \ldots, 0)$, and the neutral element of multiplication by scalars or the multiplicative identity 1:

(1) [Addition] $x + y = (x_1 + y_1, \ldots, x_m + y_m)$,
(2) [Multiplication by Scalars] $tx = (tx_1, \ldots, tx_m)$,
(3) [Scalar Product (Inner Product, Dot Product)] $x \cdot y = x_1 y_1 + \cdots + x_m y_m$,

for any $x = (x_1, \ldots, x_m), y = (y_1, \ldots, y_m) \in \mathbb{R}^m$ and any $t \in \mathbb{R}$.

$(\mathbb{R}^m)^*$ is the dual space of $\mathbb{R}^m$. For each $y \in \mathbb{R}^m$, the mapping $x \mapsto x \cdot y$ defines a linear functional on $\mathbb{R}^m$, and conversely, for each $f \in (\mathbb{R}^m)^*$ there is a unique $y_f \in \mathbb{R}^m$ such that $f(x) = x \cdot y_f$ for all $x \in \mathbb{R}^m$. Thus, $\mathbb{R}^m$ itself can be identified with the dual space $(\mathbb{R}^m)^*$.

As an inner product space the usual norm in $\mathbb{R}^m$ is called the *Euclidean norm*, that is, the usual norm of a vector x is $\|x\| = \sqrt{x \cdot x}$. In the case $m = 1$, $\|x\| = |x|$. $\mathbb{R}^m$ becomes a metric space with the *usual* (or *Euclidean*) metric given by $\rho_{eu}(x, y) = \|x - y\|$ for any two vectors $x, y \in \mathbb{R}^m$. Metrics $\rho_{sum}(x, y) = \sum_{i=1}^m |x_i - y_i|$ and $\rho_{max}(x, y) = \max\{|x_i - y_i| \mid i = 1, \ldots, m\}$, for $x = (x_1, \ldots, x_m), y = (y_1, \ldots, y_m)$, are its equivalent metrics.

The i-th *unit vector* is $\mathbf{e}_i = (0, \ldots, 0, \overset{i}{1}, 0, \ldots, 0)$. $\|\mathbf{e}_i\| = 1$. The set $\{\mathbf{e}_i \mid i = 1, \ldots, m\}$ forms a basis of $\mathbb{R}^m$ and is called the *standard basis* of $\mathbb{R}^m$. We will be using a notation $\mathbf{e}$ for the vector whose components are all 1, i.e. $\mathbf{e} = (1, \ldots, 1)$.

Inequalities between two vectors $x, y \in \mathbb{R}^m$ such as $x > y, x \geqq y, x < y, x \leqq y$ should be understood to mean corresponding inequalities between each component of vectors. Similarly, inequalities between two subsets S, T of $\mathbb{R}^m$are used to mean corresponding inequalities between each element of sets. For example, $S \leqq T$ means $x \leqq y$ for any $x \in S$ and $y \in T$. The *sum* of sets S and T is $S+T = \{x+y \mid x \in S, y \in T\}$. For $p \in \mathbb{R}^m$ and $T \subset \mathbb{R}^m$, we define $p \cdot T = \{p \cdot x \mid x \in T\}$.

The sets $\mathbb{R}^m_+ = \{\mathbb{R}^m | x \geqq 0\}, \mathbb{R}^m_{++} = \{\mathbb{R}^m | x > 0\}, \mathbb{R}^m_- = -\mathbb{R}^m_+$, and $\mathbb{R}^m_{--} = -\mathbb{R}^m_{++}$ are called, respectively, the *non-negative*, *positive*, *non-positive*, and the *negative orthant*.

Convex Sets and Hyperplanes in $\mathbb{R}^m$

Some Properties of Convex Sets in $\mathbb{R}^m$

Here are some facts about topological and linear properties of convex sets in $\mathbb{R}^m$.

Mathematical Property 1.15 [Topological Properties of Convex Sets]

(1) [Closure and Interior] *If S is a convex subset of $\mathbb{R}^m$, then* $\operatorname{cl} S$ *and* $\operatorname{int} S$ *are also convex sets.*
(2) [Compactness] *If S is compact in $\mathbb{R}^m$, then so is* $\operatorname{co} S$.

Mathematical Property 1.16 [Carathéodory's Theorem]

Let S be a subset of $\mathbb{R}^m$. Any vector belonging to the convex hull of S can be written as a convex combination of $(m + 1)$ vectors belonging to S.

For a subset P of $\mathbb{R}^m$ the set $\{z \in \mathbb{R}^m \mid p \cdot z \leqq 0 \text{ for all } p \in P\}$, denoted by P°, is called the *polar* of P.

Mathematical Fact 1.18 [Polar Cones]

If $P \subset \mathbb{R}^m$ is a closed convex cone, the polar P° of P is a closed convex cone satisfying $(P^\circ)^\circ = P$.

Hyperplanes, Half Spaces, Separation Theorem

When $\mathbb{R}^m$ is identified with its dual space $(\mathbb{R}^m)^*$, $H(p, w) = \{x \in \mathbb{R}^m \mid p \cdot x = w\}$ is an (affine) hyperplane in $\mathbb{R}^m$ for $p \in \mathbb{R}^m$ and $w \in \mathbb{R}$. Associated with a hyperplane $H(p, w)$, *closed half spaces* are defined by $H_{\leqq}(p, w) = \{x \in \mathbb{R}^m \mid p \cdot x \leqq w\}$ and $H_{\geqq}(p, w) = \{x \in \mathbb{R}^m \mid p \cdot x \geqq w\}$. *Open half spaces* $H_<(p, w)$ and $H_>(p, w)$ are defined similarly by replacing the weak inequalities $\leqq$ and $\geqq$ by the strict inequalities $<$ and $>$ respectively. Let S, T be non-empty subsets of $\mathbb{R}^m$. If the closed half spaces associated with a hyperplane $H(p, w)$ satisfy $S \subset H_{\geqq}(p, w)$ and $T \subset H_{\leqq}(p, w)$, the hyperplane $H(p, w)$ is said to *separate* the sets S and T. In such a case $H(p, w)$ is called a *separating hyperplane*.

Mathematical Property 1.17 [Minkowski's Separating Hyperplane Theorem]

Let S, T be non-empty convex subsets of $\mathbb{R}^m$. If $S \cap T = \emptyset$, then there exists a hyperplane in $\mathbb{R}^m$ which separates S and T.

Mathematical Property 1.18 [Characterization of Closed Convex Sets]

Let S be a dense set in $\mathbb{R}^m$. Then, x belongs to a closed convex set $T \subset \mathbb{R}^m$ not containing a straight line if and only if, for all $q \in S$, $q \cdot x \leqq \sup q \cdot T$ or $q \cdot x \geqq \inf q \cdot T$.

For a set S in $\mathbb{R}^m$ the interior of S relative to the smallest affine subspace containing S will be called the *linear relative interior* of S and denoted by $\text{rel.int}\, S$.

Mathematical Fact 1.19 [Linear Relative Interior of Convex Sets]

(1) *Let S be a non-empty convex set in $\mathbb{R}^m$. Then, its linear relative interior* $\text{rel.int}\, S$ *is not empty.*
(2) *Let S and T be convex sets in $\mathbb{R}^m$ with $0 \in S$. If* $\text{rel.int}\, S \cap \text{rel.int}\, T \neq \emptyset$, *then there is no separating hyperplane of S and T in the smallest linear subspace containing S and T.*

Convex Hull Relations

Let T be a subset of a linear space V and φ a relation from a set S to T. The *convex hull relation* of φ, denoted by $\text{co}\, \varphi$, is defined by $\text{co}\, \varphi = \{(x, y) \in S \times V \mid y \in \text{co}\, \varphi(x)\}$. Convex hull relations inherit the following hemi-continuity properties from their original relations.

Mathematical Property 1.19 [Hemi-Continuities of Convex Hull Relations] *Let φ be a relation from a topological space S to $\mathbb{R}^m$.*

(1) [L.H.C.] *If φ is lower hemi-continuous at $x \in \text{dom}\, \varphi$, then its convex hull relation $\text{co}\, \varphi$ is lower hemi-continuous at x.*
(2) [U.H.C.] *If φ is compact-valued and upper hemi-continuous at $x \in \text{dom}\, \varphi$, then its convex hull relation $\text{co}\, \varphi$ is compact-valued and upper hemi-continuous at x.*

Fixed Point Theorem, Continuous Selection Theorem

Let φ be a relation from a set T to itself. A point $x \in \text{dom}\, \varphi$ is called a *fixed point* of the relation φ if $x \in \varphi(x)$. The following well-known Kakutani's Fixed Point Theorem plays a fundamental role in equilibrium analysis in economics and game theory.

Mathematical Property 1.20 [Kakutani's Fixed Point Theorem] *Let K be a non-empty, compact and convex subset of $\mathbb{R}^m$.*

If φ is a convex and closed-valued upper hemi-continuous correspondence from K to K, then there exists a fixed point of φ.

Let S, T be topological spaces and φ be a relation from S to T. A continuous mapping $f : \text{dom}\, \varphi \to T$ is said to be a *continuous selection* of φ if it satisfies $f(x) \in \varphi(x)$ for all $x \in \text{dom}\, \varphi$.

Mathematical Property 1.21 [Michael's Continuous Selection Theorem]

Let φ be a convex and closed-valued lower hemi-continuous correspondence from a metric space S to $\mathbb{R}^m$. Then there exists a continuous selection of the correspondence φ.

The following version of Kakutani's fixed point theorem is sometimes used effectively in equilibrium analysis.

Mathematical Property 1.22 [Equilibrium Fixed Point Theorem] *Let $P \subset \mathbb{R}^m$ be a closed convex cone with vertex 0 containing no linear subspace, and K be a non-empty compact subset of $\mathbb{R}^m$.*

If φ is a convex and closed-valued upper hemi-continuous correspondence from a convex compact subset $PS \subset P$ to K satisfying $\varphi(p) \subset H_{\leqq}(p, 0)$ for any $p \in PS$, then there exists a point $p^* \in PS$ for which $\varphi(p^*) \cap P^\circ \neq \emptyset$.

1.6 Measurable Structures

Measurable structures have been developed in a process of refining and extending a practical framework originated in calculations of the area in a region, the volume of a solid, etc. A measurable structure on a set is given in terms of a family of subsets of the set and an extended real-valued function defined on the family of subsets satisfying specific conditions expressing properties common to those of length, area, and volume. In order for families of subsets to be able to accommodate these specific properties of the function, families themselves need to satisfy corresponding properties. Measurable spaces are pairs of a set and a family of subsets of the set that can meet these requirements.

Measurable Spaces

Measurable spaces are basic components of measurable structures. In what follows, properties to be satisfied by families of subsets of a set are delineated.

Algebra, σ-algebra / σ-field

A family $\mathcal{M}$ of subsets of a set M is called a *(Boolean) algebra* of subsets of M if it satisfies:

(a) $\emptyset \in \mathcal{M}$;
(b) $M - S \in \mathcal{M}$ for $S \in \mathcal{M}$;
(c) $S_1 \cup S_2 \in \mathcal{M}$ for $S_1, S_2 \in \mathcal{M}$.

$\mathcal{M}$ is called a *σ-algebra* or *σ-field* of subsets of M if it satisfies the above conditions (a), (b), and the following condition (c′) replacing (c):

(c′) $\cup_{n \in \mathbb{N}_+} S_n \in \mathcal{M}$ for $S_n \in \mathcal{M}$, $n \in \mathbb{N}_+$.

A pair $(M, \mathcal{M})$ consisting of a set M and a σ-field of subsets of M is called a *measurable space*. When there is no need to specify its σ-field explicitly, M is simply referred to as a measurable space. Subsets of M belonging to $\mathcal{M}$ are said to be *$\mathcal{M}$-measurable* or, simply, *measurable*.

If a σ-algebra (or a σ-field) $\mathfrak{Z}$ consists of subsets belonging to the σ-algebra $\mathcal{M}$, $\mathfrak{Z}$ is called a *σ-subalgebra* or *σ-subfield*. On the other hand, if $\mathfrak{Z}$ consists of subsets of M not necessarily belonging to the σ-algebra $\mathcal{M}$, the smallest σ-algebra (that is, the intersection of all σ-algebras) containing $\mathfrak{Z}$ is called the *σ-algebra generated by* $\mathfrak{Z}$ and is denoted by $\mathcal{B}(\mathfrak{Z})$.

Borel σ-algebra, Borel Sets

If $(T, \mathcal{T})$ is a topological space, the *Borel σ-algebra* of subsets of T is the σ-algebra $\mathcal{B}(\mathcal{T})$ generated by the topology $\mathcal{T}$ and is also denoted by $\mathcal{B}(T)$. It is identical to the σ-algebra $\mathcal{B}(\mathcal{F}(T))$ generated by the family $\mathcal{F}(T)$ of all closed subsets of T. A subset B of T is called a *Borel set* if $B \in \mathcal{B}(T)$. When a topological space $(T, \mathcal{T})$ is given, unless specifically mentioned otherwise, by the measurable space we mean the measurable space $(T, \mathcal{B}(T))$. In this respect a measurable set and a Borel set are customarily synonymous in a topological space.

Mathematical Fact 1.20 [Borel Sets in a Subspace] *Let S be a subspace of a topological space T. Then:*

(1) $\mathcal{B}(S) = \{B \cap S \mid B \in \mathcal{B}(T)\}$;
(2) $\mathcal{B}(S) = \{B \subset S \mid B \in \mathcal{B}(T)\}$ *if* $S \in \mathcal{B}(T)$.

Mathematical Fact 1.21 [Borel Sets in Locally Compact Spaces] *Let $(T, \mathcal{T})$ be a second countable Hausdorff space and $\mathcal{B}(\mathcal{K})$ be the σ-algebra generated by the family $\mathcal{K}$ of all compact subsets of T. Then,*

$\mathcal{B}(\mathcal{K}) = \mathcal{B}(T)$ *if the space T is locally compact.*

Product of Measurable Spaces

Let $(M_1, \mathcal{M}_1)$ and $(M_2, \mathcal{M}_2)$ be two measurable spaces. For any $S_1 \in \mathcal{M}_1$ and $S_2 \in \mathcal{M}_2$, $S_1 \times S_2$ is called a *measurable rectangle*. The σ-algebra generated by all the measurable rectangles in $M_1 \times M_2$ is denoted by $\mathcal{M}_1 \otimes \mathcal{M}_2$ and is called the *product σ-algebra* of $\mathcal{M}_1$ and $\mathcal{M}_2$. The measurable space $(M_1 \times M_2, \mathcal{M}_1 \otimes \mathcal{M}_2)$ is called the *product of the measurable spaces* $(M_1, \mathcal{M}_1)$ and $(M_2, \mathcal{M}_2)$.

For any subset $G \subset M_1 \times M_2$, let $G_x = \{y \in M_2 \mid (x, y) \in G\}$, $G_y = \{x \in M_1 \mid (x, y) \in G\}$. G_x is called the *x-section* of G, and G_y the *y-section* of G. We have $G_y \in \mathcal{M}_1$ and $G_x \in \mathcal{M}_2$ for every $(x, y) \in M_1 \times M_2$ if $G \in \mathcal{M}_1 \otimes \mathcal{M}_2)$. Note, however, $\mathrm{pr}_{M_1}(G)$ may not belong to $\mathcal{M}_1$ nor $\mathrm{pr}_{M_2}(G)$ need belong to $\mathcal{M}_2$. The product of a finite number of measurable spaces $(M_i, \mathcal{M}_i)$, $(i = 1, \ldots, n)$, is defined analogously and is written $(\prod_{i=1}^n M_i, \bigotimes_{i=1}^n \mathcal{M}_i)$.

Mathematical Fact 1.22 [Product of Borel σ-algebra] *Let M_1 and M_2 be two separable metric spaces and $M_1 \times M_2$ be the product metric space. Then, we have:*

$$\mathcal{B}(M_1) \otimes \mathcal{B}(M_2) = \mathcal{B}(M_1 \times M_2).$$

Measure Spaces

Measurable structures are described by measure spaces which are composed of a measurable space along with an extended real-valued function on a family of subsets. In contrast to the term *set-valued function* whose *range* is a family of sets, the term *set function* is used for a function whose *domain* is a family of sets.

Set Functions, Additivity, Measures

A *set function* is defined to be a function from a family of subsets of a set to a (real) linear space. When its range is $\mathbb{R}^m$ (resp. $\mathbb{R}$), it is at times called an $\mathbb{R}^m$-valued (resp. real-valued) set function. There are instances in which a set function is allowed to assume extended real numbers permitting at most one of the improper values ∞ and $-\infty$. If $\mathcal{M}$ is a family of subsets of a set M, a real-valued or extended real-valued set function μ defined on $\mathcal{M}$ is said to be a *finite set function* if $|\mu(M)| < \infty$ and a *positive set function* if it has non-negative values only. The set function μ is said to be *additive* or *finitely additive* if, for every disjoint subfamily $\{S_1, \ldots, S_n\}$ of sets in $\mathcal{M}$ whose union is also in $\mathcal{M}$, we have $\cup_{i=1}^{n} \mu(S_i) = \sum_{i=1}^{n} \mu(S_i)$. μ is *countably additive* if, for every disjoint countable subfamily $\{S_n \mid n \in \mathbb{N}_+\}$ of sets in $\mathcal{M}$ whose union is also in $\mathcal{M}$, we have $\cup_{n=1}^{\infty} \mu(S_n) = \sum_{n=1}^{\infty} \mu(S_n)$. If $\mathcal{M}$ is an algebra, an extended real-valued and countably additive set function μ on $\mathcal{M}$ for which $\mu(\emptyset) = 0$ is called a *signed measure*. A signed measure μ is said to be a *measure* if it is positive.

A pair of measurable space $(M, \mathcal{M})$ and a measure μ on $\mathcal{M}$, $(M, \mathcal{M}, \mu)$, is called a *measure space*. A measure μ (resp. a measure space $(M, \mathcal{M}, \mu)$) is a *finite measure* (resp. a *finite measure space*) if μ is a finite and positive signed measure. $\mu(M)$ is called the *total mass* of the measure μ (or the measure space). When the measure μ is finite and normalized by the total mass, the measure $\widehat{\mu}$ defined by $\widehat{\mu}(S) = \frac{1}{\mu(M)}\mu(S)$ for each $S \in \mathcal{M}$ is called the *normalized measure* of μ. In probability theory, when μ is equal to its normalized measure, it is said to be a *probability measure*, and $(M, \mathcal{M}, \mu)$, a *probability measure space*. If M is a countable union of sets with a finite μ measure so that $M = \cup_{n\in\mathbb{N}_+} M_n, M_n \in \mathcal{M}, \mu(M_n) < \infty$ for each $n \in \mathbb{N}_+$, the measure μ and the measure space $(M, \mathcal{M}, \mu)$ are said to be *σ-finite*. In this book all measures are meant to be σ-finite measures. When the family $\mathcal{M}$ is clearly understood from the context or unless there is a need to specify a σ-algebra and a measure, a measure space $(M, \mathcal{M}, \mu)$ may be simply referred to as a measure space M or as a measure (or signed measure) μ on M by an abuse of terminology. In the case M is a topological space, $\mathcal{M} = \mathcal{B}(M)$ if not explicitly mentioned otherwise. A measure μ on $\mathcal{B}(M)$ is called a *Borel measure*.

Limits of Sequences of Sets, Continuity of Set Functions, Supports

Given a sequence $(S_n)_n$ of subsets of a set M, the *inferior limit* of $(S_n)_n$ is defined as $\cup_{n=1}^{\infty} \cap_{k=n}^{\infty} S_k$ and denoted by $\lim_n \inf S_n$. The *superior limit* of $(S_n)_n$ is defined as $\cap_{n=1}^{\infty} \cup_{k=n}^{\infty} S_k$ and denoted by $\lim_n \sup S_n$. One always has $\lim_n \inf S_n \subset \lim_n \sup S_n$; when the both sides are equal, it is denoted by $\lim_n S_n$ and is called the *limit* of the sequence of sets $(S_n)_n$.

A sequence $(S_n)_n$ of sets is an *increasing sequence* if $S_n \subset S_{n+1}$ and a *decreasing sequence* if $S_n \supset S_{n+1}$, for every $n \in \mathbb{N}_+$. In each of these cases it is called a *monotone sequence*. If the sequence $(S_n)_n$ is a monotone sequence, then its limit $\lim_n S_n$ exists and is equal to $\cup_n S_n$ or $\cap_n S_n$ depending upon whether it is increasing or decreasing.

Let $\mathcal{M}$ be a family of subsets of M and μ an extended real-valued set function defined on $\mathcal{M}$. μ is said to be *continuous from below* at a set S (in $\mathcal{M}$) if, for every increasing sequence $(S_n)_n$ of sets in $\mathcal{M}$ for which $\lim_n S_n = S$, one has $\lim_n \mu(S_n) = \mu(S)$. Similarly, μ is *continuous from above* at S if, for every decreasing sequence $(S_n)_n$ of sets in $\mathcal{M}$ for which $|\mu(S_n)| < \infty$ for at least one value of n and for which $\lim_n S_n = S$, one has $\lim_n \mu(S_n) = \mu(S)$.

Mathematical Fact 1.23 [Continuity of Measures] *Let $(M, \mathcal{M})$ be a measurable space and μ an extended real-valued function defined on $\mathcal{M}$.*

(1) *If μ is a finitely additive measure that is either continuous from below at any $S \in \mathcal{M}$ or continuous from above at $\emptyset$, then μ is a measure on $\mathcal{M}$.*
(2) *If μ is a measure on $\mathcal{M}$, then, at any $S \in \mathcal{M}$, μ is continuous both from below and from above.*

Let T be a topological space and μ a finite measure on T. A closed subset $S \subset T$ is called the *support* of the measure μ and denoted by $\mathrm{supp}(\mu)$ if it satisfies:

(a) $\mu(S) = \mu(T)$;
(b) $S \subset F$ for any closed subset F of T such that $\mu(F) = \mu(T)$.

Mathematical Fact 1.24 [Support of Borel Measures] *In separable metric spaces the support of a finite Borel measure exists.*

Null Sets, Completion of Measure Spaces

In measure spaces with uncountably many infinite points there is a way to identify "*rare*" or "*exceptional*" cases which hinder a particular mathematical property to hold. Exceptional cases in general are regarded as "*negligible*," and measurable subsets having measure zero are identified with negligibility.

Let $(M, \mathcal{M}, \mu)$ be a measure space. A subset N of M is called *μ-null* or *μ-negligible* if there exists $S \in \mathcal{M}$ such that $N \subset S$ and $\mu(S) = 0$. References to the measure μ may be dropped in this expression. A measure space $(M, \mathcal{M}, \mu)$ is said to be *complete* if the σ-algebra $\mathcal{M}$ contains every μ-negligible subset of M. If $\mathcal{U}$ denotes the family of all negligible sets of $(M, \mathcal{M}, \mu)$, the family $\mathcal{M}_\mu$ of sets of the form $S \cup N$ with $S \in \mathcal{M}$ and $N \in \mathcal{U}$ is identical with the σ-field generated by $\mathcal{M} \cup \mathcal{U}$.

If the set function $\bar{\mu}$ is defined by $\bar{\mu}(S \cup N) = \mu(S)$, then it is a measure on $\mathscr{M}_\mu$ and the measure space $(M, \mathscr{M}_\mu, \bar{\mu})$ is complete. $(M, \mathscr{M}_\mu, \bar{\mu})$ is called the *completion* of the measure space $(M, \mathscr{M}, \mu)$. The σ-algebra $\mathscr{M}_\mu$ and the measure $\bar{\mu}$ may also be called the *completion* of the σ-algebra $\mathscr{M}$ and the measure μ, respectively.

Let a measure space $(M, \mathscr{M}, \mu)$ be given. When a proposition or a statement about points in a subset $S \subset M$ stands true except for a μ-negligible subset $N \subset S$, then the proposition or the statement is said to stand true *μ-almost everywhere* or *μ-almost every x* in S. One might skip references to the measure μ in this expression. The expressions "almost everywhere" or "almost every $x \in S$" are often abbreviated as "a.e." or "a.e. $x \in S$." For example, "for almost every $x \in S$, the proposition $P(x)$ is true" may be simply written as "a.e. $x \in S, P(x)$" or "$P(x)$, a.e. $x \in S$."

Outer Measures

Measures are positive extended real-valued set functions defined on a σ-algebra of subsets of a set, and it is well-known that measures in general cannot be extended to the set of all subsets of the set (see for example [43, pp. 67–72]). In this respect it is clear that the additivity requirement of measures needs be relaxed to some extent in order to be able to extend measures to the set of all subsets of a set. In turn, with a relaxed requirement on the additivity of set functions defined on the power set, to what extent must their domains be restricted to obtain the restrictions of set functions to be measures on restricted families of subsets of a set? Carathéodory provided an elegant answer to these questions by a milder restriction of additivity, called subadditivity, of positive extended real-valued set functions defined on the power set.

An extended real-valued set function μ^* defined on the power set $\mathbb{P}(M)$ of a set M is called a *Carathéodory outer measure* or, simply, an *outer measure* on M if it satisfies:

(a) [Non-Negativity] $0 \leqq \mu^*(S) \leqq +\infty$ for any $S \subset M$, and $\mu^*(\emptyset) = 0$;
(b) [Monotonicity] $\mu^*(S_1) \leq \mu^*(S_2)$ for $S_1 \subset S_2, S_1, S_2 \in P(M)$;
(c) [Countable Subadditivity] $\mu^*(\cup_{n\in\mathbb{N}_+} S_n) \leqq \sum_{n\in\mathbb{N}_+} \mu^*(S_n)$ for every sequence $(S_n)_{n\in\mathbb{N}_+}$ of subsets of M.

Given an outer measure μ^* on M, a subset S of M is said to be *μ^* measurable* or *measurable in the sense of Carathéodory* if, for every subset $C \subset M$, one has

$$\mu^*(C) = \mu^*(C \cap S) + \mu^*(C \cap (M - S)).$$

Mathematical Fact 1.25 [Extension of Measures] *Let $(M, \mathscr{M})$ be a measurable space and μ a measure on $\mathscr{M}$. Define an extended real-valued set function μ^* on $\mathbb{P}(M)$ by*

$$\mu^*(S) = \inf\{\sum_{n\in\mathbb{N}_+} \mu(S_n) \mid S \subset \cup_{n\in\mathbb{N}_+} S_n, S_n \in \mathscr{M}, n \in \mathbb{N}_+\}$$

for each $S \in \mathbb{P}(M)$. Then, μ^* is an extension of μ to $\mathbb{P}(M)$ and is an outer measure on M. μ^* is an outer measure induced by the measure μ. *One has:*

(1) μ^* is continuous from below at each $S \in \mathbb{P}(M)$.
(2) The completion of $\mathcal{M}$, $\mathcal{M}_\mu$, is identical to the class $\mathcal{M}^*$ of all μ^*-measurable sets in $\mathbb{P}(M)$.

Mathematical Fact 1.26 [Outer Measures and Measurable Sets] *Let μ^* be an outer measure on M and $\mathcal{M}^*$ the set of all μ^*-measurable subsets of M. Then, $\mathcal{M}^*$ is an σ-algebra, and the restriction $\bar{\mu}^*$ of μ^* to $\mathcal{M}^*$ is a complete measure on $\mathcal{M}^*$.*

Lebesgue Measures on Euclidean Spaces

Measures on a σ-algebra of subsets of a set may be viewed as a formalization and an abstract extension of the concepts of length (or a "linear measure"), areal sizes (or a "square measure") and volumes (or a "cubic measure") in our living airspace to abstract spaces. From this vantage point, specific measures that delineate length, area, volume, and measurements units in further higher dimensional linear spaces are represented by Lebesgue measures in Euclidean spaces.

Let a measure λ^m on the class $\mathcal{B}(\mathbb{R}^m)$ of all Borel sets in Euclidean m-space be defined by requiring to hold

$$\lambda^m(\{x \in \mathbb{R}^m \mid \alpha_i < x_i \leqq \beta_i, i = 1, \dots, m\}) = \prod_{i=1}^{m}(\beta_i - \alpha_i).$$

A measure λ^m is uniquely defined on $\mathcal{B}(\mathbb{R}^m)$ and is called *m-dimensional Borel–Lebesgue* (or simply, *Lebesgue*) *measure*. Its completion, also denoted by λ^m, is called the *m-dimensional Lebesgue measure*. Here are some of its basic properties.

Mathematical Fact 1.27 [Lebesgue Measures] *The m-dimensional Borel–Lebesgue measure λ^m on $\mathcal{B}(\mathbb{R}^m)$ satisfies the following properties:*

(1) [Translation Invariance] *For any $S \in \mathcal{B}(\mathbb{R}^m)$ and $x \in \mathbb{R}^m$, one has:*

$$S + x = \{z + x \in \mathbb{R}^m \mid z \in S\} \in \mathcal{B}(\mathbb{R}^m) \text{ and } \lambda^m(S) = \lambda^m(S + x).$$

(2) [Null Hyperplanes] *Every $(m-1)$-dimensional hyperplane in $\mathbb{R}^m$ has m-dimensional Lebesgue measure 0.*

Measurability of Mappings

The measurability of mappings refers to whether measurable structures in the domains and the ranges of mappings are related one to the other by way of their mappings.

Measurable Mappings

Let $(M_i, \mathscr{M}_i), i = 1, 2$, be two measurable spaces. A mapping $f : M_1 \to M_2$ is said to be *measurable* if it satisfies $f^{-1}(S) = \{x \in M_1 \mid f(x) \in S\} \in \mathscr{M}_1$ for every $S \in \mathscr{M}_2$. (This may also be written as $f^{-1}(\mathscr{M}_2) \subset \mathscr{M}_1$.) When M_i's, $i = 1, 2$, are topological spaces, the "measurability" is always with respect to the corresponding Borel σ-algebras unless otherwise stated explicitly. Some of the most elementary properties of the measurability of mappings are listed in the following.

Mathematical Fact 1.28 [Basic Measurability of Mappings]

(1) *Let $(M_i, \mathscr{M}_i)$, $i = 1, 2$, be measurable spaces and $\mathcal{S} \subset \mathscr{M}_2$ a subfamily which generates $\mathscr{M}_2$. Then, a mapping $f : M_1 \to M_2$ is measurable if $f^{-1}(\mathcal{S}) \subset \mathscr{M}_1$, that is, $f^{-1}(S) \in \mathscr{M}_1$ for all $S \in \mathcal{S}$.*
(2) *Given measurable spaces $(M_i, \mathscr{M}_i)$, $i = 1, 2, 3$, let $f : (M_1, \mathscr{M}_1) \to (M_2, \mathscr{M}_2)$ and $g(M_2, \mathscr{M}_2) \to (M_3, \mathscr{M}_3)$ be two measurable mappings. Then, their composition $g \circ f$ from $(M_1, \mathscr{M}_1)$ to $(M_3, \mathscr{M}_3)$ is also measurable.*
(3) *Let f_i be a measurable mapping from a measurable space $(M_0, \mathscr{M}_0)$ to another measurable space $(M_i, \mathscr{M}_i)$ for $i = 1, \ldots, n$. Then the mapping $x \mapsto (f_1(x), \ldots, f_n(x))$ from $(M_0, \mathscr{M}_0)$ to the product measurable space $(\prod_{i=1}^n M_i, \otimes_{i=1}^n \mathscr{M}_i)$ is measurable.*
(4) *Let f and g be measurable real-valued functions defined on a measurable space $(M, \mathscr{M})$ and $s, t \in \mathbb{R}$. Then, each of the following functions is measurable:*

(a) $f + g : x \mapsto sf(x) + tg(x)$,
(b) $f \cdot g : x \mapsto f(x) \cdot g(x)$,
(c) $\sup\{f, g\} : x \mapsto \sup\{f(x), g(x)\}$.

Further Measurability

Additional properties concerning the measurability of mappings in more involved situations are listed below.

Mathematical Property 1.23 [Further Measurability of Mappings] *Let $(M, \mathscr{M})$ be a measurable space.*

(1) *Given a sequence $(f_n)_n$ of measurable real-valued functions on M, if $\lim_n f_n(x)$ exists for every $x \in M$, then the function on M defined by $x \mapsto \lim f_n(x)$ is measurable.*
(2) *A function f from $M \times [0, 1]$ to $[0, 1]$ is $\mathscr{M} \otimes \mathcal{B}[0, 1]$-measurable if it satisfies:*

(a) *For every $t \in [0, 1]$, $f(\cdot, t) : M \to [0, 1]$ is $\mathscr{M}$-measurable;*
(b) *For every $x \in M$, $f(x, \cdot) : [0, 1] \to [0, 1]$ is right continuous at every $t_0 \in [0, 1]$. That is, given a sequence $(t_n)_{n \in \mathbb{N}_+}$ and $\varepsilon > 0$ satisfying $t_n \to t_0$ with $t_0 \leqq t_{n+1} \leqq t_n$ for all n, there is n_0 such that $t_n - t_0 < \varepsilon$ for all $n \geqq n_0$.*

Mathematical Property 1.24 [Commutative Measurability of Mappings]

Let $(M_1, \mathscr{M}_1)$ and $(M_2, \mathscr{M}_2)$ be measurable spaces. Given an $\mathbb{R}^m$-valued function f on M_1 and a measurable mapping g from M_1 to M_2, one has:

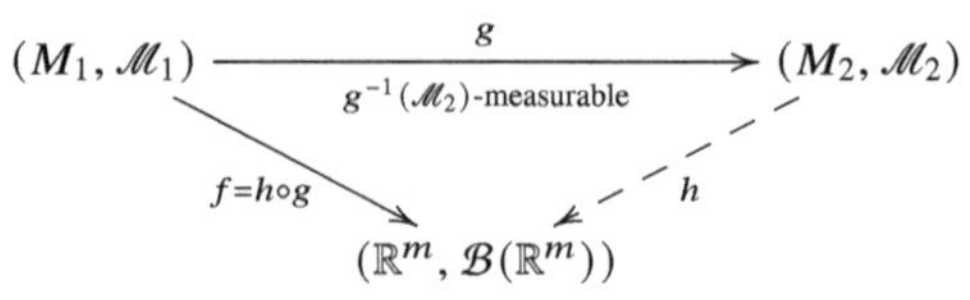

Fig. 1.1 Commutative measurability of mappings

f is measurable with respect to the σ-algebra $g^{-1}(\mathcal{M}_2)$ if and only if there exists a measurable $\mathbb{R}^m$-valued function h on M_2 such that $f = h \circ g$. (See Fig. 1.1.)

Extension of Classes of Measurable Sets

The families of subsets of a set in measurable spaces are σ-fields on which measures are defined. On the other hand, outer measures are defined on the power set of a set and satisfy countable subadditivity instead of countable additivity among the values of measures of subsets. Now, consider a set function δ_x defined on the power set of a set M such that it assigns the total mass of value 1 to each subset of M if it contains the element x of the set M and, otherwise, it assigns the value 0. Then, δ_x becomes a σ-additive set function (known as a *Dirac measure*). It is immediate that $(M, \mathbb{P}(M), \delta_x)$ is a measure space. However, it is not clear whether the σ-field of subsets of M associated with a given measure space $(M, \mathcal{M}, \mu)$ can be extended to the power set $\mathbb{P}(M)$ to enlarge the scope of measurable sets. It is well-known that it is impossible to extend the Lebesgue measure on $\mathbb{R}$ to the power set $\mathbb{P}(\mathbb{R})$ in such a way that the extended set function is still a measure (cf. [43, pp. 67–70]).

In applications of mathematical analysis it is very often the case that it is practical to enlarge families of subsets of measurable spaces beyond classes obtained by using countable elementary set-theoretic operations among measurable subsets. Let f be a mapping from a topological space T_1 to another topological space T_2, and measurable with respect to Borel σ-algebras, $\mathcal{B}(T_1)$ and $\mathcal{B}(T_2)$. Then, $f^{-1}(S_2)$ is a Borel subset of T_1 for any Borel subset S_2 of T_2. However, even if f is continuous, the set $f(S_1)$ in general need not be a Borel subset of T_2 for a Borel subset S_1 of T_1. Thus, from analytical point of view, it is convenient to broaden the scope of classes of subsets beyond families of Borel sets. Since Borel σ-fields are constructed using unions and intersections over a countable number of sets, a classical idea is to "slightly" enlarge classes of measurable subsets using unions or intersections over an uncountably infinite number of sets just large enough for continuous mappings to preserve measurability of enlarged classes of sets in the sense of making images of sets in the classes under continuous mappings to be in enlarged classes of sets in their range. For this purpose a set operation is introduced to construct sets from a given family of sets using the set of all (countably infinite) sequences of positive integers.

Souslin Extension, Souslin Sets

An index set is introduced initially so that set-theoretic operations beyond taking only countable unions or intersections among subsets of a set are well-defined. Let Λ be an index set consisting of all finite sequences of the initial k elements of infinite sequences of positive integers (or natural numbers) for any $k \in \mathbb{N}_+$. Precisely, $\Lambda = \cup_{\nu \in \mathbb{N}_+^{\mathbb{N}_+}} \Lambda_\nu$ with Λ_ν defined by $\Lambda_\nu = \{\nu_k = (n_i)_{i=1}^k \mid k \in \mathbb{N}_+\}$ for each $\nu = (n_i)_{i \in \mathbb{N}_+} \in \mathbb{N}_+^{\mathbb{N}_+}$.

Let $(M, \mathcal{M})$ be a pair of a set M and a family $\mathcal{M}$ of subsets of M with $\emptyset \in \mathcal{M}$. We call a mapping σ from Λ to $\mathcal{M}$ a *Souslin map over* $\mathcal{M}$. For $\sigma \in \mathcal{M}^\Lambda, \nu = (n_i)_{i \in \mathbb{N}_+} \in \mathbb{N}_+^{\mathbb{N}_+}, \nu_k = (n_i)_{i=1}^k, k \in \mathbb{N}_+$, one has $\sigma(\nu_k) \in \mathcal{M}$. Now, define a mapping **A** from the set $\mathcal{M}^\Lambda$ of all Souslin maps over $\mathcal{M}$ to the power set $\mathbb{P}(M)$ by $\mathbb{A}(\sigma) = \cup_{\nu \in \mathbb{N}_+^{\mathbb{N}_+}} \cap_{k \in \mathbb{N}_+} \sigma(\nu_k)$ for each $\sigma \in \mathcal{M}^\Lambda$. The mapping **A** is called the *Souslin operation* or *operation* (**A**). The family of all subsets of M obtained by Souslin operations over the set $\mathcal{M}$ indexed by Λ is denoted by $\mathcal{S}(\mathcal{M})$, that is, $\mathcal{S}(\mathcal{M}) = \{\mathbf{A}(\sigma) \mid \sigma \in \mathcal{M}^\Lambda\}$, and is called the *Souslin extension* of $\mathcal{M}$. Elements of $\mathcal{S}(\mathcal{M})$ are called $\mathcal{M}$*-Souslin* or *Souslin sets*.

Clearly, countable intersections and unions are in the images of Souslin operations. However, in general, Souslin operations involve uncountable unions since a set of all finite sequences is uncountable. Hence, one may suspect whether inverse images of Souslin sets by measurable maps could be Souslin sets in their domain without some restrictions of a given family of sets to have additional properties to being a σ-field. To prepare for a clearer view, let us proceed to a characterization of Souslin sets using projection maps.

A family of sets $\mathcal{M}$ is said to be *closed under countable (or finite) unions (resp. intersections)* if any countable (or finite) union (resp. intersection) of sets belonging to $\mathcal{M}$ belongs to it. The intersection of all the families that are closed under countable unions (resp. intersections) and contain the class $\mathcal{M}$ is denoted by $\mathcal{M}_\sigma$ (resp. $\mathcal{M}_\delta$). $(\mathcal{M}_\sigma)_\delta$ and $(\mathcal{M}_\delta)_\sigma$ may be written as $\mathcal{M}_{\sigma\delta}$ and $\mathcal{M}_{\delta\sigma}$ respectively. Given a family of sets $\mathcal{M}$ which is closed under finite unions and finite intersections, a set $S \in \mathcal{M}_\delta$ can be written as $S = \cap_n S_n$ by a decreasing sequence $(S_n)_n$ of sets in $\mathcal{M}$. Similarly, a set $S \in \mathcal{M}_\sigma$ can be written as $S = \cup_n S_n$ by an increasing sequence $(S_n)_n$ of sets in $\mathcal{M}$. A subfamily $\mathcal{S}$ of $\mathcal{M}$ is said to have the *finite intersection property* if every finite number of sets in $\mathcal{S}$ has a non-empty intersection. A family of sets $\mathcal{M}$ is said to be *countably compact* if, for any subfamily $\mathcal{S} \subset \mathcal{M}$ with countably many elements having the finite intersection property, the intersection of all the sets belonging to $\mathcal{S}$ is non-empty.

Mathematical Property 1.25 [Characterization of Souslin Sets] *Let $\mathcal{M}$ be a family of subsets of M such that $\emptyset \in \mathcal{M}$. Then, the following two conditions are equivalent:*

(a) *S is a $\mathcal{M}$-Souslin set.*
(b) *There is a countably compact family $\mathcal{K}$ of subsets of a set F with $\emptyset \in \mathcal{K}$ such that $S = \mathrm{pr}_M(T)$ for some set $T \in (\mathcal{M} \times \mathcal{K})_{\sigma\delta}$ where $\mathrm{pr}_M : M \times F \to M$ is the projection map and $\mathcal{M} \times \mathcal{K} = \{S_1 \times S_2 \subset M \times F \mid S_1 \in \mathcal{M}, S_2 \in \mathcal{K}\}$.*

Analytic Sets in Relation to Souslin Extension

The set $\mathbb{N}_+^{\mathbb{N}_+}$ of all infinite sequences has served a basic role in obtaining Souslin sets from a family of sets. Noting the fact that $\mathbb{N}_+^{\mathbb{N}_+}$ with the product topology of the discrete topology on $\mathbb{N}_+$ is a Polish space, one could infer that Polish spaces play a practical function in linking Souslin operations with topological properties. In this context the class of analytic sets is introduced as a family of sets satisfying a simple topological property. A subset C of a topological space M is called an *analytic set* if there exists a continuous mapping f from a Polish space T to M such that $C = f(T)$. The set of all analytic sets in M is denoted by $\mathcal{A}(M)$. Primary relations among Borel sets, Souslin sets, and analytic sets in Polish spaces are listed below.

Mathematical Property 1.26 [Souslin Sets, Analytic Sets, and Borel Sets in Polish Spaces] *Let $\mathcal{F}(M)$ be the family of all closed sets in a Polish space M.*

(1) *A subset of M is an analytic set if and only if it is a $\mathcal{F}(M)$-Souslin set; thus, one has $\mathcal{A}(M) = \mathcal{S}(\mathcal{F}(M))$.*
(2) *A Borel set of M is an analytic set; thus, one has $\mathcal{B}(M) \subset \mathcal{A}(M)$.*

Mathematical Property 1.27 [Preservation of Analytic Sets by Measurable Maps in Polish Spaces] *Let $M_i, i = 1, 2$, be Polish spaces. If a mapping $f : M_1 \to M_2$ is measurable with respect to $\mathcal{B}(M_1), \mathcal{B}(M_2)$, then we have:*

(a) $f^{-1}(C) \in \mathcal{A}(M_1)$ *for any* $C \in \mathcal{A}(M_2)$*;*
(b) $f(B) \in \mathcal{A}(M_2)$ *for any* $B \in \mathcal{A}(M_1)$.

Mathematical Property 1.28 [Souslin Property of Projections relating to Polish Spaces] *Let $(M, \mathcal{M})$ be a measurable space, T a Polish space, and $\mathrm{pr}_M : M \times T \to M$ the projection map. For the class $\mathcal{M} \times \mathcal{B}(T)$ of all measurable rectangles in $M \times T$, one has:*

For any $(\mathcal{M} \times \mathcal{B}(T))$-Souslin set $C \subset M \times T$, its image $\mathrm{pr}_M(C)$ in M under the projection is a $\mathcal{M}$-Souslin set.

Universal Measurability

Families of measurable sets are broadened by performing Souslin operations which allow taking unions among an uncountable number of sets in a family of sets. One might wonder to what extent each of these families is enlarged from the standpoint of the measurability of sets that are added to the families by the operations. Measure theoretically, it is not desirable for non-measurable sets to be among sets in these augmented families. It leads us to reflect whether Souslin sets in measure spaces are measurable in the sense of Catathéodory. Accordingly, we now turn to the notion of universal measurability.

Let $(M, \mathcal{M})$ be a measurable space. For a finite measure μ defined on $(M, \mathcal{M})$, let $(M, \mathcal{M}_\mu, \bar{\mu})$ be the completion of a measure space $(M, \mathcal{M}, \mu)$. Denote by $\mathcal{M}_u$ the intersection of all completions $\mathcal{M}_\mu$ for any finite measure μ on $(M, \mathcal{M})$. Then, the measurable space $(M, \mathcal{M}_u)$ is called the *universal completion* of the measurable space $(M, \mathcal{M})$. A subset $S \subset M$ is called *universally measurable* if it belongs to $\mathcal{M}_\mu$. For a topological space M, the universal completion of the measurable space $(M, \mathcal{B}(M))$ is denoted by $(M, \mathcal{B}_u(M))$.

The next Mathematical Property shows the measurability property of Souslin sets and analytic sets.

Mathematical Property 1.29 [Universal Measurability of Souslin Sets and Analytic Sets] *Let $(M, \mathcal{M})$ be a measurable space.*

(1) *$\mathcal{M}$-Souslin sets in M are universally measurable; thus, one has $\mathcal{S}(\mathcal{M}) \subset \mathcal{M}_u$.*
(2) *In particular, if M is a Polish space, then, analytic subsets of M are universally measurable; thus, $\mathcal{A}(M) \subset \mathcal{B}_u(M)$.*

The above Mathematical Properties 1.28 and 1.29 together imply the following Projection Theorem:

Mathematical Fact 1.29 [Projection Theorem] *Let $(M, \mathcal{M}, \mu)$ be a complete measure space and T a Polish space. Then, the projection $\mathrm{pr}_M(C)$ of a measurable set C in $\mathcal{M} \otimes \mathcal{B}(T)$ is measurable; thus, $\mathrm{pr}_M(C) \in \mathcal{M}$.*

Vector Measures

Let $(M, \mathcal{M})$ be a measurable space. An *m-dimensional vector measure* $\mu = (\mu_1, \ldots, \mu_m)$ on $(M, \mathcal{M})$ is an $\mathbb{R}^m$-valued set function with each coordinate μ_i being a signed measure on $(M, \mathcal{M})$. A vector measure $\mu = (\mu_1, \ldots, \mu_m)$ is said to be a *bounded vector measure* if each of its coordinates $\mu_i, i = 1, \ldots, n$, is a finite signed measure.

Mathematical Proposition 1.4 [Compactness of Vector Measure Range] *Let $\mu = (\mu_1, \ldots, \mu_m)$ be a finite vector measure on a measurable space $(M, \mathcal{M})$. Then, the range of the vector measure μ, $\{\mu(S) = (\mu_1(S), \ldots, \mu_m(S)) \mid S \in \mathcal{M}\} \subset \mathbb{R}^m$, is a compact set in $\mathbb{R}^m$.*

1.7 Integration

Generally speaking, an *integration* is an operation on a family of measurable mappings from a measure space to a linear space which associates each mapping with an element in the range in such a way that the operation becomes a specific linear mapping. The value of a measurable mapping under the integration operation is called the *integral*. Historically, the idea of an integral goes back to efforts to calculate the area of a region in a plane, the volume of a solid, etc. in ancient Greek mathematics.

Integrals of Measurable Mappings

The historical idea of calculating the area under a curve in a bounded region of a plane is roughly the following. The curve is expressed by the graph of a real-valued function

f defined on a bounded interval in $\mathbb{R}^1$, and it lies entirely above the horizontal axis which contains the domain of f. The area under the curve is approximated by a sum of a finite number of rectangular regions each of which is on or below the curve. As the number of regions approximating the area increases, approximate calculations improve and the area under the curve is defined by a limit of these approximations. With this historical background, approximations are represented by *step functions* taking only finitely many distinct values. In what follows, let us begin with integrals of real-valued functions in the form of step functions.

Step Functions, Integrals as Sums of Finite Number of Values

Let f be a function from a measure space $(M, \mathscr{M}, \mu)$ to $\mathbb{R}$. f is said to be a *step function* or a *simple function* if f takes only finitely many distinct values. Given a step function f, let $t_1, \ldots, t_k$ be all of the finitely many different values that f takes in $\mathbb{R}$, and put $S_j = \{x \in M | \ f(x) = t_j\}$ for each $j = 1, \ldots, k$. The real number $\sum_{j=1}^{k} f(S_j)\mu(S_j)$ is associated with the measurable step function f, and is called the *integral* (or the *mean*, or the *expectation*) of f. It is denoted by $\int f d\mu$ or $\int_M f d\mu$.

The next thrust in defining integrals is to extend the operation so as to assign values to non-negative measurable functions by taking a limit of approximations given by step functions.

Non-Negative Measurable Functions, Integrals as Limits of Integrals of Step Functions

Let f be a non-negative measurable function from M to $\mathbb{R}$. Then, there exists a monotone increasing sequence $(f_n)_n$ of measurable step functions converging to f at each point in M. One can show that, for any such monotone increasing sequence $(f_n)_n$, the limit of the associated sequence $\left(\int f_n d\mu\right)_n$ exists (allowing $+\infty$). It does not depend on a particular sequence $(f_n)_n$ chosen, and is, therefore, unique. If this limit is finite, the non-negative measurable function f is said to be *integrable*, and the *integral* of f is defined by $\int f d\mu = \lim_n \int f_n d\mu$.

The last step of extending the operation of defining integrals is to assign values to any measurable real-valued functions.

Integrals of Measurable Functions, Integrability

Finally, let f be any measurable function from $(M, \mathscr{M}, \mu)$ to $\mathbb{R}$. Define two functions, f^+ and f^-, from M to $\mathbb{R}$ by $f^+(x) = \max\{0, f(x)\}$ and $f^-(x) = \max\{0, -f(x)\}$ for each $x \in M$. Then, both of the functions f^+ and f^- are non-negative. f is said to be *integrable* if the two non-negative functions f^+ and f^- are integrable. The *integral* of f is defined by $\int f d\mu = \int f^+ d\mu - \int f^- d\mu$. If $|f|$ is defined by $|f|(x) = |f(x)|$, then, f is integrable if and only if the non-negative function $|f|$ is integrable. In cases where an explicit description of variables of functions being integrated helps to enhance understanding, $\int f d\mu$ may be written as $\int f(x) d\mu(x)$.

Integrals satisfy the following properties:

Mathematical Fact 1.30 [Basic Properties of Integrals] *Let f and g be integrable functions from a measure space $(M, \mathcal{M}, \mu)$ to $\mathbb{R}$.*

(1) *If $\mu(\{x \in M \mid f(x) < 0\}) = 0$, then $\int f d\mu \geqq 0$. If $f(x) \geqq g(x)$, then $\int f d\mu \geqq \int g d\mu$.*
(2) *For any $t_1, t_2 \in \mathbb{R}$, one has*

$$\int (t_1 f + t_2 g) d\mu = t_1 \int f d\mu + t_2 \int g d\mu.$$

Absolute Continuity and Radon–Nikodym Derivatives

Let $(M, \mathcal{M}, \mu)$ be a measure space. For a subset S of M, define a function $\chi_S : M \to \mathbb{R}$ by $\chi_S(x) = 1$ if $x \in S$, and $\chi_S(x) = 0$ if $x \notin S$. The function χ_S is the *indicator function* of the set S. χ_S is measurable if $S \in \mathcal{M}$. Given a measurable function f from M to $\mathbb{R}$ and a measurable set $S \in \mathcal{M}$, the function $x \mapsto \chi_S(x) f(x)$ from M to $\mathbb{R}$ is integrable if so is f. Its integral $\int \chi_S f d\mu$ is denoted by $\int_S f d\mu$.

Given a measure space $(M, \mathcal{M}, \mu)$, an extended real-valued finitely additive set function η on $\mathcal{M}$ is said to be *absolutely continuous* with respect to μ, in symbols $\eta \ll \mu$, if $\eta(S) = 0$ for every $S \in \mathcal{M}$ for which $\mu(S) = 0$. A measurable function f from M to $\mathbb{R}$ such that $\eta(S) = \int_S f d\mu$ for any $S \in \mathcal{M}$ is called a *Radon–Nikodym derivative* of η with respect to μ.

Mathematical Property 1.30 [Radon–Nikodym Theorem] *Let $(M, \mathcal{M}, \mu)$ be a measure space, and η a signed measure on $\mathcal{M}$ with finite values.*

(1) *If η is absolutely continuous with respect to μ, $\eta \ll \mu$, then there exists a Radon–Nikodym derivative of η with respect to μ.*
(2) *A Radon–Nikodym derivative of η is uniquely determined up to almost everywhere equality in the sense that if f and g are Radon–Nikodym derivatives of η, then $f(x) = g(x)$, a.e. $x \in M$.*
(3) *If η is finite, then a Radon–Nikodym derivative of η is integrable.*

Let $(M, \mathcal{M}, \mu)$ be a measure space, and $f : M \to \mathbb{R}$ an integrable function. Given a σ-subalgebra $\mathfrak{Z}$ of $\mathcal{M}$, a $\mathfrak{Z}$-measurable function $g : M \to \mathbb{R}$ satisfying $\int_S g d\mu = \int_S f d\mu$ for all $S \in \mathfrak{Z}$ is called a *conditional expectation of f given $\mathfrak{Z}$* and is denoted by $\mathbb{E}(f|\mathfrak{Z})$.

Mathematical Property 1.31 [Conditional Expectation] *Let $f : M \to \mathbb{R}$ be an integrable function on a measure space $(M, \mathcal{M}, \mu)$. If $\mathfrak{Z}$ is a σ-subalgebra of $\mathcal{M}$, then there is a conditional expectation $\mathbb{E}(f|\mathfrak{Z}) : M \to \mathbb{R}$ of f given $\mathfrak{Z}$. $\mathbb{E}(f|\mathfrak{Z})$ is unique up to almost everywhere equality.*

Convergences of Measurable / Integrable Functions

The set of all integrable functions from a measure space $(M, \mathcal{M}, \mu)$ to $\mathbb{R}$ is denoted by $\mathcal{L}_1(M, \mathcal{M}, \mu)$, or simply by $\mathcal{L}_1$ if there is no need to specify its associated measure space. By Mathematical Fact 1.30 (2), the operation of integration $\int(\cdot)d\mu$ is a linear mapping defined on $\mathcal{L}_1(M, \mathcal{M}, \mu)$ taking values in $\mathbb{R}$.

Concepts of Convergences

Let $(f_n)_n$ be a sequence of measurable functions from a measure space $(M, \mathcal{M}, \mu)$ to $\mathbb{R}$, and f a measurable function from M to $\mathbb{R}$. If the set $\{x \in M \mid \lim_n f_n(x) \neq f(x)\}$ is a μ-null set, then the sequence $(f_n)_n$ is said to *converge almost everywhere* to a measurable function f. If for every $\varepsilon > 0$ one has $\lim_n \mu\{x \in M \mid |f_n(x) - f(x)| > \varepsilon\}) = 0$, then $(f_n)_n$ is said to *converge in measure* or *converge in probability* to f. If $f, f_n \in \mathcal{L}_1(M, \mathcal{M}, \mu)$, $n \in \mathbb{N}_+$, and $\lim_n \int |f_n - f| d\mu = 0$, then, $(f_n)_n$ is said to *converge in the mean* to f.

The following mathematical properties show how the above three different concepts of convergences are related to each other, and will be used frequently.

Mathematical Fact 1.31 [Relation among Different Convergences]
Let $f, f_n \in \mathcal{L}_1(M, \mathcal{M}, \mu), n \in \mathbb{N}_+$.

(1) [Convergence Almost Everywhere and Convergence in Measure] *If a sequence* $(f_n)_n$ *of measurable functions converges almost everywhere to* f*, then it converges in measure to* f*. Conversely, if a sequence* $(f_n)_n$ *of measurable functions converges in measure, then there is a subsequence that converges almost everywhere to* f*.*
(2) [Convergence in the Mean and Convergence in Measure] *Let* $(f_n)_n$ *be a sequence of non-negative integrable functions. Then, the sequence* $(f_n)_n$ *converges in the mean to the integrable function* f *if and only if* $(f_n)_n$ *converges to* f *in measure and* $\lim_n \int f_n d\mu = \int f d\mu$.

Convergences of Integrals and Integrable Functions

In the following we list basic convergence properties related to sequences of integrable functions.

Mathematical Property 1.32 [Basic Convergence Properties of Integrals]

(1) [Monotone Convergence Theorem] *Let* $(f_n)_n$ *be an increasing (resp., decreasing) sequence in* $\mathcal{L}_1(M, \mathcal{M}, \mu)$ *for which* $\lim_n f_n(x)$ *is finite for all* $x \in M$ *and* $\lim_n \int f_n d\mu$ *is finite; then, one has* $\lim_n f_n \in \mathcal{L}_1$ *and* $\lim_n \int f_n d\mu = \int \lim_n f_n d\mu$.
(2) [Fatou's Lemma] *Let* $(f_n)_n$ *be a sequence of functions in* $\mathcal{L}_1(M, \mathcal{M}, \mu)$*. If there is an integrable function* $g \in \mathcal{L}_1(M, \mathcal{M}, \mu)$ *satisfying* $g(x) \leqq f_n(x)$ *for all* $x \in M$*,* $n \in \mathbb{N}_+$*, then* $\int \lim_n \inf f_n d\mu \leqq \lim_n \inf \int f_n d\mu$.
(3) [Lebesgue's Dominated Convergence Theorem] *Let* $(f_n)_n$ *be a sequence of functions in* $\mathcal{L}_1(M, \mathcal{M}, \mu)$*. If there is an integrable function* $g \in \mathcal{L}_1(M, \mathcal{M}, \mu)$ *satisfying* $|f_n(x)| \leqq g(x)$ *for all* $x \in M$*,* $n \in \mathbb{N}_+$*, then* $\lim_n f_n \in \mathcal{L}_1(M, \mathcal{M}, \mu)$ *and* $\int \lim_n f_n d\mu = \lim_n \int f_n d\mu$.

Image Measures

Given a measurable mapping from a measure space to a measurable space, there are instances where the values of the mapping have direct influences on objects of analysis so that points in the given measure space play the role of "parameters," so to speak. In these circumstances it is more fitted for analyses to use the measure induced by the mapping on the measurable space. It is called an *image measure*. When image measures are normalized, they are called *distributions*.

Distributions

Let f be a measurable mapping from a measurable space $(M_1, \mathscr{M}_1)$ to another measurable space $(M_2, \mathscr{M}_2)$. Given a measure μ on $\mathscr{M}_1$, define a set function η on $\mathscr{M}_2$ by $\eta(S) = \mu(f^{-1}(S)), S \in \mathscr{M}_2$. Then, η becomes a measure on $\mathscr{M}_2$. It is called the *image measure* of μ under f and is denoted by $\mu \circ f^{-1}$. When μ is a normalized measure, it is called the *distribution* of f or the *law* of f. In the remaining part of the present section, *all measures are assumed normalized* unless otherwise stated.

Given measurable spaces $(M_i, \mathscr{M}_i)$, $i = 1, 2$, and a measurable mapping f from a measure space $(M, \mathscr{M}, \mu)$ to the product measurable space $(M_1 \otimes M_2, \mathscr{M}_1 \otimes \mathscr{M}_2)$, define mappings f_i, $i = 1, 2$, by $f_i = \mathrm{pr}_i \circ f$ where $\mathrm{pr}_i : M_1 \times M_2 \to M_i$ is the projection onto the i-th coordinate subspace. We have $f = (f_1, f_2)$. The distribution of f, $\mu \circ f^{-1} = \mu \circ (f_1, f_2)^{-1}$ is called the *joint distribution* of f_1 and f_2, and the distribution of f_i, $\mu \circ f_i^{-1}$, is called the *marginal distribution* of f on M_i. In general, given a measure μ on a measurable space $(M_1 \otimes M_2, \mathscr{M}_1 \otimes \mathscr{M}_2)$, the measure $\mu \circ \mathrm{pr}_i^{-1}$, $i = 1, 2$, is called the *marginal distribution* of μ on M_i.

Let f be a measurable mapping from a measure space $(M_1, \mathscr{M}_1, \mu)$ to a measurable space M_2 and g a mapping from M_2 to another measurable space M_3. Then, the image measure of the (image) measure $\mu \circ f^{-1}$ under g is identical to the image measure of μ under $g \circ f$, that is, $(\mu \circ f^{-1}) \circ g^{-1} = \mu \circ (g \circ f)^{-1}$.

Mathematical Fact 1.32 [Image Measures and Integration/Change of Variables]

Let f be a measurable mapping from a measure space $(M_1, \mathscr{M}_1, \mu)$ to a measurable space $(M_2, \mathscr{M}_2)$, and g a measurable function from M_2 to $\mathbb{R}$. Then, g is $\mu \circ f^{-1}$-integrable if and only if $g \circ f$ is μ-integrable. In this case we have

$$\int_{M_1} g \circ f d\mu = \int_{M_2} g d(\mu \circ f^{-1}).$$

(See Fig. 1.2.)

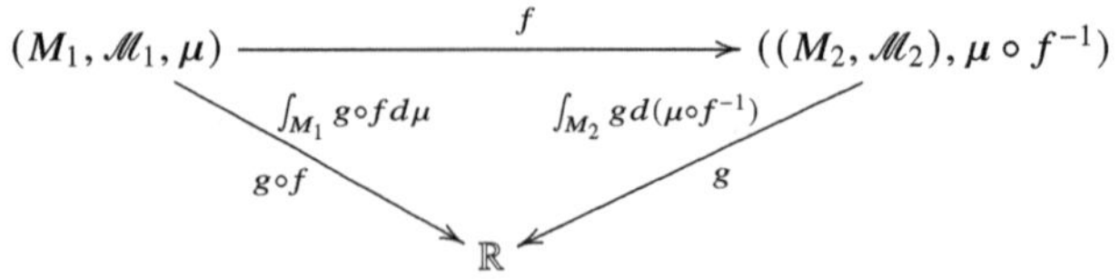

Fig. 1.2 Image measures and integration

Conditional Distributions

Mathematical Property 1.33 [Conditional Distributions] *Let* $(\Omega, \mathcal{U}), (M, \mathscr{M})$ *be measurable spaces,* T *a Polish space and* ν *a measure on* $\mathcal{U}$ *with* $\nu(T) = 1$*. Given measurable mappings* $f : (\Omega, \mathcal{U}) \to (T, \mathcal{B}(T))$ *and* $g : (\Omega, \mathcal{U}) \to (M, \mathscr{M})$*, there exists a real-valued mapping* $Q(\cdot, \cdot)$ *on* $\mathcal{B}(T) \times M$ *satisfying:*

(1) $Q(\cdot, x)$ *is a normalized (or probability) measure on* $\mathcal{B}(T)$ *for each* $x \in M$*.*
(2) $Q(S, \cdot)$ *is a measurable function on* M *for each* $S \in \mathcal{B}(T)$*.*
(3) For any $G \in \mathcal{B}(T) \times \mathscr{M}$

$$\nu \circ (f, g)^{-1}(G) = \int_M Q(G_x, x) d(\nu \circ g^{-1})(x)$$

(G_x *is the* x*-section of the set* G*).*

This normalized measure $S \mapsto Q(S, x)$ *on* $\mathcal{B}(T)$ *is called the* conditional distribution of f given that g equals x. *The conditional distribution of* f *is uniquely determined in the following sense. That is, if* Q_1, Q_2 *are two conditional distributions, there is* $N \in \mathscr{M}$ *with* $\nu \circ g^{-1}(N) = 0$ *such that one has* $Q_1(S, x) = Q_2(S, x)$ *for any* $S \in \mathscr{M}$ *and* $x \in M - N$*.*

Topological Spaces of Measures

For any given topological space T, $\mathscr{M}(T)$ will denote the set of all Borel measures defined on $\mathcal{B}(T)$. To introduce a topology on the space $\mathscr{M}(T)$, let $U_{(f,\varepsilon)}(\mu)$ be the subset of $\mathscr{M}(T)$ defined for each $\mu \in \mathscr{M}(T)$ and for each pair (f, ε) of a continuous bounded function $f : T \to \mathbb{R}$ and a positive integer ε by

$$U_{(f,\varepsilon)}(\mu) = \Big\{\eta \in \mathscr{M}(T) \mid \Big|\int f d\eta - \int f d\mu\Big| < \varepsilon \quad \text{for } \varepsilon > 0 \text{ and}$$
$$\text{a bounded continuous function } f : T \to \mathbb{R}\Big\}.$$

Then, the topology generated by the family $\mathcal{S}_T$ of subsets of $\mathscr{M}(T)$ as its subbase given by

$$\mathcal{S}_T = \{U_{(f,\varepsilon)}(\mu) \mid \text{ each } \mu \in \mathcal{M}(T), \text{ each } \varepsilon > 0, \text{ and each bounded continuous function } f : T \to \mathbb{R}\}$$

is called the *weak topology* on $\mathcal{M}(T)$ (cf. p. 64, 65). Associated with this week topology is an open neighborhood base at each $\mu \in \mathcal{M}(T)$ given by the family of all non-empty finite intersections of sets $U_{(f,\varepsilon)}(\mu) \in \mathcal{S}_T$. If not stated explicitly otherwise, the space $\mathcal{M}(T)$ is endowed with the weak topology as a topological space.

Weak Convergences of Measures

For any topological space T, it is straightforward that a net of measures $(\mu_\gamma)_{\gamma \in \Gamma}$ in $\mathcal{M}(T)$ converges to μ in the weak topology if and only if $\int f d\mu_\gamma \to_\gamma \int f d\mu$ for every continuous bounded function f from T to $\mathbb{R}$. In such a case $(\mu_\gamma)_\gamma$ is said to *converge weakly* to μ.

Mathematical Fact 1.33 [Weak Convergences I] *Let $(\mu_\gamma)_{\gamma \in \Gamma}$ be a net of measures in $\mathcal{M}(T)$ for a topological space T. Then, the following conditions are equivalent to the convergence of the net $(\mu_\gamma)_\gamma$ to μ in the weak topology.*

(a) *$(\mu_\gamma)_\gamma$ converges weakly to μ.*
(b) *$\int f d\mu_\gamma \to_\gamma \int f d\mu$ for every bounded and uniformly continuous functions $f : T \to \mathbb{R}$.*

When a topological space T is metrizable, the space of all measures $\mathcal{M}(T)$ is metrizable. Specifically, given a metric space T, define a mapping ρ_P from $\mathcal{M}(T) \times \mathcal{M}(T)$ to $\mathbb{R}$ by

$$\rho_P(\mu_1, \mu_2) = \inf\{\varepsilon > 0 \mid \mu_1(S) \leqq \mu_2(N_\varepsilon(S)) \text{ and } \mu_2(S) \leqq \mu_1(N_\varepsilon(S)) \text{ for all } S \in \mathcal{B}(T)\}$$

for any $\mu_i \in \mathcal{M}(T), i = 1, 2$; then, ρ_P is a metric on $\mathcal{M}(T)$ which is called the *Prohorov metric*.

Mathematical Fact 1.34 [Prohorov Metric Topology] *For a metrizable topological space T, the Prohorov metric on $\mathcal{M}(T)$ is compatible with the weak topology.*

Mathematical Property 1.34 [Topological Properties of $\mathcal{M}(T)$] *Let T be a topological space.*

(1) *The space $\mathcal{M}(T)$ is a compact metrizable space if and only if T is a compact metrizable space.*
(2) *The space $\mathcal{M}(T)$ is a separable metrizable space if and only if T is a separable metrizable space.*
(3) *Let T be a separable metrizable space and $S \subset T$ be a countable dense subset. Then, the set of all measures whose supports are finite subsets of S is dense in $\mathcal{M}(T)$.*

Mathematical Fact 1.35 [Weak Convergences II] *Let T be a metric space and $\mu, \mu_n, (n \in \mathbb{N}_+)$, be measures in $\mathscr{M}(T)$. Then, the following conditions are equivalent to the sequence $(\mu_n)_n$ being weakly convergent to μ.*

(a) $\lim_n \sup \mu_n(F) \leqq \mu(F)$ *for every closed subset $F \subset T$.*
(b) $\lim_n \inf \mu_n(W) \geqq \mu(W)$ *for every open subset $W \subset T$.*
(c) $\lim_n \mu_n(S) = \mu(S)$ *for every measurable set $S \in \mathcal{B}(T)$ satisfying $\mu(\partial S) = 0$, i.e., $\mu(\operatorname{int} S) = \mu(\operatorname{cl} S)$.*

Mathematical Fact 1.36 [Weak Convergences on Product Spaces] *Assume that spaces S and T are separable metric spaces.*

(1) *Given sequences $(\tau_n)_n$ and $(\mu_n)_n$ of measures in $\mathscr{M}(S)$ and $\mathscr{M}(T)$, respectively, the sequence $(\tau_n \otimes \mu_n)$ of product measures on $\mathcal{B}(S) \otimes \mathcal{B}(T)$ converges weakly to the product measure $\tau \otimes \mu$ on $\mathcal{B}(S) \otimes \mathcal{B}(T)$ if and only if the sequences (τ_n) and (μ_n) converges weakly to τ and μ respectively.*
(2) *Given measures $\mu, \mu_n, (n \in \mathbb{N}_+)$, in $\mathscr{M}(S \times T)$, let measures $\mu^S, \mu_n^S, \mu^T, \mu_n^T, n \in \mathbb{N}_+$, denote their associated marginal distributions on $\mathcal{B}(S)$ and $\mathcal{B}(T)$. Then, if the sequence $(\mu_n)_n$ of measures converges weakly to the measure μ, the sequences of marginal distributions (μ_n^S) and (μ_n^T) converge weakly to the marginal distributions μ^S and μ^T, respectively.*

Mathematical Property 1.35 [Skorokhod's Theorem] *Suppose that T is a separable metric space and $(\tau_n)_n$ is a weakly convergent sequence of measures in $\mathscr{M}(T)$ with its limit τ.*

Then, there exist measurable mappings $f, f_n, (n \in \mathbb{N}_+)$, from a common measure space $(M, \mathscr{M}, \mu)$ to T satisfying $\tau = \mu \circ f^{-1}$ and $\tau_n = \mu \circ f_n^{-1}$ such that f_n converges to f almost everywhere in M. If T is a Polish space, then the measure space $(M, \mathscr{M}, \mu)$ can be chosen to be the unit interval $[0, 1]$ with the Lebesgue measure.

Given a metric space T, every measure η in $\mathscr{M}(T)$ can be regarded as the distribution of some measurable mapping on some measure space $(M.\mathscr{M}, \mu)$. For example, take as a measure space $(M.\mathscr{M}, \eta) = (T, B(T), \mu)$ and as a measurable mapping the identity map id_M on M. Skorokhod's Theorem says that a weakly convergent sequence of measures on a family of all Borel sets of a separable metric space is a reflection of the background of almost-everywhere convergence of measurable mappings in another space. Namely, a weakly convergent sequence of measures can be identified with a convergent sequence of measurable mappings on a common measure space.

Tightness of Measures

Let T be a metric space and $\mathscr{M}(T)$ be the set of all measures on $\mathcal{B}(T)$. A family of measures $\mathscr{M} \subset \mathscr{M}(T)$ is said to be *tight* if for any given $\varepsilon > 0$ there is a compact set $K_\varepsilon \subset T$ satisfying $\mu(T - K_\varepsilon) < \varepsilon$ for all $\mu \in \mathscr{M}(T)$. A measure μ is *tight* if the family $\{\mu\}$ is tight. In view of Mathematical Fact 1.35, tight measures have a property that the values of Borel sets in the range are determined by those of compact sets in their domain.

Mathematical Fact 1.37 [Tightness of Product Measures] *Let S, T be metric spaces and $\mathcal{M}(S), \mathcal{M}(T)$ be the sets of all measures on $\mathcal{B}(S)$ and $\mathcal{B}(T)$ respectively.*

A family of measures on the product $\mathcal{B}(S) \otimes \mathcal{B}(T)$ is tight if and only if each of the families of their marginal distributions on $\mathcal{B}(S)$ and $\mathcal{B}(T)$ is tight.

Mathematical Property 1.36 [Tightness] *Let T be a metric space.*

If a family of measures $\mathcal{M} \subset \mathcal{M}(T)$ is tight, then $\mathcal{M}$ is relatively compact in the weak topology so that any sequence $(\mu_n)_n$ of measures in $\mathcal{M}$ has a weakly convergent subsequence with a limit in $\mathcal{M}(T)$ (but not necessarily in $\mathcal{M}$).

Mathematical Property 1.37 [Tightness in Polish Spaces] *Let T be a Polish space and $\mathcal{M}(T)$ be the set of all measures on $\mathcal{B}(T)$.*

(1) *Every measure in $\mathcal{M}(T)$ is tight.*
(2) *If a family of measures $\mathcal{M} \subset \mathcal{M}(T)$ is relatively compact in the weak topology, then $\mathcal{M}$ is tight.*

It follows from Mathematical Property 1.36 and 1.37 that in Polish spaces a family of Borel measures is tight if and only if it is relatively compact in the weak topology, in other words, if and only if any sequence of measures in the family has a weakly convergent subsequence. In a special case of a family of weakly convergent sequence of tight measures with its limit in the family, the family itself is tight even if its underlying metric space is only separable.

Mathematical Fact 1.38 [Convergence of Measures and Tightness]

Let T be a separable metric space and $(\mu_n)_n$ a sequence of measures on T which converge weakly to a measure μ on T. If each of the measures μ_n, $(n \in \mathbb{N}_+)$, and μ are tight, then the family $\{\mu, \mu_n, \mid n \in \mathbb{N}_+\}$ is tight.

Distributions of Measurable Mappings

Convergences in Distributions

In applications there are many instances in which one is faced with problems of the convergence in distribution of measurable mappings which may not be defined on the same measurable space but take values in an identical measure space. Specifically, let $(f_n)_n$ be a sequence of measurable mappings taking values in the same metric space T. It is said to *converge in distribution* or *converge in law* to a measurable mapping f taking values in T if the sequence $(\mu_n)_n$ consisting of the distribution of f_n, $(n \in \mathbb{N}_+)$, converges weakly to the distribution μ of f.

In the case when all the functions f_n and f are real-valued and defined on the same measure space, the convergence of the sequence $(f_n)_n$ in measure implies its convergence in distribution. However, the converse need not be true unless f is a constant function.

Mathematical Property 1.38 [Convergence in Distributions] *Let (T, ρ) be a separable metric space and f, f_n, $(n \in \mathbb{N}_+)$, be measurable mappings from an identical measure space $(M, \mathcal{M}, \mu)$ to T.*

Then, the function $x \mapsto \rho(f_n(x), f(x))$ from M to $\mathbb{R}$ is measurable, and if the sequence $(\rho(f_n(\cdot), f(\cdot)))_{n\in\mathbb{N}_+}$ of functions converges in measure to zero, then the sequence $(f_n)_n$ converges in distribution to f.

Uniform Integrability and Convergences

Let f_n be a measurable real-valued function defined on a measure space $(M_n, \mathcal{M}_n, \mu_n)$ for each $n \in \mathbb{N}_+$. The sequence of functions $(f_n)_{n\in\mathbb{N}_+}$ is said to be *uniformly integrable* if it satisfies the condition

$$\lim_{k\in\mathbb{N}_+} \left(\sup_n \int_{|f_n|>k} |f_n| d\mu_n \right) = 0.$$

If n is fixed in this expression, the above equality clearly holds. The uniform integrability means that this convergence is uniform with respect to n. More specifically, it requires for any $\varepsilon > 0$ there is $k_0 \in \mathbb{N}_+$ such that $\sup \int_{|f_n|>k} |f_n| d\mu_n < \varepsilon$ for every $n \in \mathbb{N}_+$ and for all $k \geqq k_0$. Given a sequence of real-valued functions $(f_n)_n$, suppose that there is an integrable real-valued function g defined on a measure space $(M, \mathcal{M}, \mu)$ satisfying

$$\mu_n (\{x \in M_n \mid |f_n(x)| \geqq k\}) \leqq \mu (\{x \in M \mid |g(x)| \geqq k\})$$

for each $n \in \mathbb{N}_+$ and $k \in \mathbb{N}_+$; then, the sequence is uniformly integrable. In particular, if each function f_n is defined on an identical measure space $(M, \mathcal{M}, \mu)$ and satisfies $-g(x) < f_n(x) < g(x)$ for a real-valued integrable function defined on the same measure space, then $(f_n)_n$ is uniformly integrable.

Mathematical Property 1.39 [Uniform Integrability] *Let $(f_n)_{n\in\mathbb{N}_+}$ be a sequence of measurable functions from measure spaces $(M_n, \mathcal{M}_n, \mu_n)$ to $\mathbb{R}$ for each $n \in \mathbb{N}_+$.*

(1) *The sequence $(f_n)_n$ is uniformly integrable if and only if:*

(a) $\sup_n \int |f_n| d\mu_n < \infty$;
(b) $\lim_n \int_{C_n} |f_n| d\mu_n \to 0$ *for any sequence $(C_n)_n$ of $\mathcal{M}_n$-measurable sets with $\mu_n(C_n) \to 0$.*

(2) *If the sequence $(f_n)_n$ is uniformly integrable, then the sequence $(\mu_n)_n$ of distributions of f_n is tight.*
(3) [Generalized Lebesgue's Theorem] *Assume that the sequence $(f_n)_n$ converges in distribution to a measurable function f. Then, if the sequence $(f_n)_n$ is uniformly integrable, f is integrable and an equality $\lim_n \int f_n d\mu_n = \int f d\mu$ holds. On the other hand, if f_n, $(n \in \mathbb{N}_+)$, and f are non-negative and integrable, then this equality implies that $(f_n)_n$ is uniformly integrable.*

Sample Distributions

Given a sequence $(f_n)_n$ of measurable functions from a measure space $(M, \mathscr{M}, \mu)$ to a metric space T, define for each $x \in M$ and $n \in \mathbb{N}_+$, a measure $\mu_n(x, \cdot)$ on T by

$$\mu_n(x, S) = \frac{1}{n}\sharp\{k \mid f_k(x) \in S, k = 1, \ldots, n\}, S \in \mathcal{B}(T).$$

In terms of statistical probability theory, the space T may be understood to represent a *sample space* or a *set of all observable objects* that are being analyzed. Then, elements of T are *sample points*. In this context a given sequence $(f_n)_n$ represents an *ordered sampling* and, for each x, the set $\{f_1(x), \ldots, f_n(x)\}$ is an (ordered) selection of n sample points from the sample space. It is called a *sample of size* n. $\mu_n(x, \cdot)$ for each $x \in M$ is the distribution of samples $\{f_1(x), \ldots, f_n(x)\}$ of size $n, n \in \mathbb{N}_+$. A mathematical result below in statistical probability provides an insightful and useful interpretation of economic models with measurable structures as we shall see later on.

Now, let $f_n, n \in \mathbb{N}_+$, be measurable functions from a measure space $(M, \mathscr{M}, \mu)$ to a metric space T_n and f a measurable map from M to the product space $\prod_n T_n$ defined by $x \mapsto f(x) = (f_n(x))_{n \in \mathbb{N}_+}$. Measurable functions $f_n, n \in \mathbb{N}_+$, are said to be *independent* if the distribution of f is the product measure of the distributions of f_n.

Mathematical Property 1.40 [Glivenko–Cantelli Theorem] *Let $(f_n)_n$ be a sequence of measurable functions from a measure space $(M, \mathscr{M}, \mu)$ to a separable metric space T.*

If $f_n, n \in \mathbb{N}_+$, are independent and have an identical distribution, then the sequence $(\mu_n(x, \cdot))_n$ of the distribution of samples $\{f_1(x), \ldots, f_n(x)\}$ of size $n, n \in \mathbb{N}_+$, defined for each $x \in M$ by

$$\mu_n(x, S) = \frac{1}{n}\sharp\{k \mid f_k(x) \in S,\ k = 1, \ldots, n\}, S \in \mathcal{B}(T),$$

converges weakly to the distribution of f_n for almost every $x \in M$.

1.8 Linear Topological Spaces

In this section a few basic concepts and facts about linear topological spaces are reviewed that will be applied to analysis in later chapters.

Linear Topological Spaces and Duals

Topological Dual Spaces

Let V be a vector space, and $\mathcal{T}$ a topology on V. If the basic vector space operations (i.e., mappings), $(x, y) \mapsto x + y$ $(x, y \in V)$ and $(t, x) \mapsto tx$ (t, a scalar, $x \in V$) are continuous with respect to the topology $\mathcal{T}$, then $(V, \mathcal{T})$ is called a *linear topological space* or a *topological vector space*. When the topology $\mathcal{T}$ is clearly understood from the context, V itself may be called a topological vector space for simplicity. If every neighborhood in V contains a convex neighborhood, then the space V is said to be a *locally convex space*. The topology $\mathcal{T}$ is referred to as the *original topology* on V. The set of all continuous linear functionals on V, denoted by V^*, becomes a vector space by defining vector space operations $f + g$ and tf by $(f + g)(x) = f(x) + g(x)$ and $(tf)(x) = t(f(x))$ for $f, g \in V^*$, a scalar t, and $x \in V$. It is called the *dual space* or the *conjugate space* of V. Sometimes V^* is referred to as the *topological dual space* of V in contrast to the set of all linear functionals on V, which is referred to as the *algebraic dual space*.

Dual Pairings, Weak Topologies, Weak-Star Topologies

Let V and W be linear topological spaces. A mapping on $V \times W$ is called a *bilinear functional* on $V \times W$ if it becomes a linear functional on V or on W when one of the coordinates is fixed. A *dual pairing* is an ordered pair of vector spaces (V, W) and a bilinear functional on $V \times W$. For $v \in V$ and $w \in W$, the value of the functional will be denoted by $w \cdot v$. The pairing makes each $w \in W$ a linear functional on V, and each $v \in V$ a linear functional on W.

Given a dual pairing (V, W), there is a weakest topology on V such that $W = V^*$. It is denoted by $\sigma(V, W)$. $\sigma(V, W)$ is the *topology of pointwise convergence* on W, that is, if $(v_\lambda)_\lambda$ is a net in V, then $v_\lambda \to_\lambda v$ if and only if $w \cdot v_\lambda \to_\lambda w \cdot v$ for all $w \in W$. Similarly, $\sigma(W, V)$ is the weakest topology on W such that $V = W^*$, and is the topology of pointwise convergence on V.

If V is a linear topological space, then there is a natural pairing (V, V^*). Since the topology $\sigma(V, V^*)$ is as weak or weaker than the original topology on V, $\sigma(V, V^*)$ is commonly referred to as the *weak topology*. $\sigma(V^*, V)$ is called the *weak-star topology*. When more than one topology on V is considered, a customary way to avoid a possibility of confusion is to put a prefix to topological concepts as in "$\sigma(V, V^*)$-closure" to show which topology is being considered on V.

Mathematical Property 1.41 [Convex Sets in Locally Convex Spaces]

(1) *Let S be a convex set of a locally convex space V. The closure of S with respect to the weak topology is identical to its closure with respect to the original topology.*

(2) *Let V be a metrizable locally convex space. If a sequence $(x_n)_n$ in V converges to $x \in V$ in the weak topology, then there is a sequence $(y_n)_n$ in V satisfying the following two properties:*

(a) $y_{n+k-1} \in \mathrm{co}\{x_{n+k-1} \mid k \in \mathbb{N}_+\}, (k, n \in \mathbb{N}_+)$;
(b) $y_n \to_n x$ *in the original topology.*

Mathematical Property 1.42 [Banach–Alaoglu Theorem] *Let U be a neighborhood of the vector 0 in a linear topological space V. The polar of U given by $U^\circ = \{f \in V^* | (\forall x \in U) |f(x)| \leqq 1\}$ is weak-star compact.*

Let S be a convex set of a vector space V. $x \in S$ is an *extreme point* of S if one has $y = z$ whenever x can be expressed as $x = ty + (1-t)z$ for some $y, z \in S$ and $0 < t < 1$.

Mathematical Property 1.43 [Krein–Milman Theorem] *Let V be a locally convex space. A convex and compact set S in V satisfies the following two properties:*

(a) *It has an extreme point.*
(b) *The closure of convex hull of the set of all the extreme points of S is equal to S.*

Normed Linear Spaces

Linear topological spaces with topologies induced by "norms" on linear spaces are called *normed linear spaces.*

Normed Linear Spaces, Banach Spaces, Strong Topologies

Let $x \mapsto \|x\|$ be a real-valued function defined on a real vector space V satisfying each of the following conditions:

(a) $\|0\| = 0$, $\|x\| > 0$ if $x \neq 0$;
(a) $\|x + y\| \leqq \|x\| + \|y\|$ for any $x, y \in V$;
(c) $\|tx\| \leqq |t|\|x\|$ for any scalar t and any $x \in V$.

Then, the mapping $\|\cdot\| : x \mapsto \|x\|$ is called a *norm* on V. The space $(V, \|\cdot\|)$ (or V) is called a *normed linear space* or simply a *normed space* when V is endowed with the metric topology induced by a metric given by $(x, y) \mapsto \|x - y\|$. The dual V^* of a normed linear space V becomes a normed space with the norm $\|\cdot\|$ on V^* defined by $\|v^*\| = \sup\{|v^* \cdot v| \, \|v\| \leqq 1\}$. If a normed space $(V, \|\cdot\|)$ is complete with respect to the metric induced by the norm $\|\cdot\|$, the space $(V, \|\cdot\|)$ is called a *Banach space*. In a Banach space V the original topology is usually called the *strong topology* in contrast to the weak topology $\sigma(V, V^*)$.

Spaces of Integrable Functions L_1, L_∞

Let $(M, \mathcal{M}, \mu)$ be a measure space. Recall that a subfamily of measurable functions from M to $\mathbb{R}$ that are integrable is denoted by $\mathcal{L}_1(M, \mathcal{M}, \mu)$, or simply $\mathcal{L}_1$ for short, so that $\mathcal{L}_1(M, \mathcal{M}, \mu) = \left\{f : M \to \mathbb{R} \,\middle|\, f \text{ is measurable and } \int |f| d\mu < \infty\right\}$. Another subfamily of measurable functions from M to $\mathbb{R}$ denoted by $\mathcal{L}_\infty(M, \mathcal{M}, \mu)$, or simply $\mathcal{L}_\infty$ for short, is defined by

$$\mathcal{L}_\infty(M, \mathcal{M}, \mu) = \left\{ f : M \to \mathbb{R} \;\middle|\; f \text{ is measurable and } \|f\|_\infty < \infty \right\}$$
$$\text{where } \|f\|_\infty = \sup\{t \geqq 0 \mid \mu(\{x \mid |f(x)| \geqq t\}) > 0\}.$$

Any real-valued measurable functions f, g on M for which $\mu(\{x \mid f(x) \neq g(x)\}) = 0\}$ have identical integrals if they are integrable. This being the case it is natural to define a relation $\sim$ for any pair (f, g) of real-valued measurable functions on M by $\sim = \{(f, g) \mid \mu(\{x \mid f(x) \neq g(x)\}) = 0\}$. Then, the relation $\sim$ is an equivalence relation on the set of all real-valued measurable functions defined on M. The equivalence classes of $\mathcal{L}_1$ and $\mathcal{L}_\infty$ under $\sim$ are denoted by $L_1(M, \mathcal{M}, \mu)$ and $L_\infty(M, \mathcal{M}, \mu)$, or by L_1 and L_∞ for short, respectively. Namely, when two functions f and g are regarded the same whenever $f(x) = g(x)$ almost everywhere, we write L_1 or L_∞ instead of $\mathcal{L}_1$ or $\mathcal{L}_\infty$. We have $L_1 = \mathcal{L}_1/\sim$ and $L_\infty = \mathcal{L}_\infty/\sim$.

Mathematical Fact 1.39 [L_1, L_∞ Spaces] *Let $(M, \mathcal{M}, \mu)$ be a measure space.*

(1) $\|f\|_1 = \int |f| d\mu$ *defines a norm* $\|\cdot\|_1$ *on* $L_1(M, \mathcal{M}, \mu)$. *The space* L_1 *with this norm is a Banach space.*
(2) $\|\cdot\|_\infty$ *defines a norm on* $L_\infty(M, \mathcal{M}, \mu)$. *The space* L_∞ *is a Banach space with the norm* $\|\cdot\|_\infty$. *This norm coincides with the norm of the dual space of* L_1; *that is,* $\|g\|_\infty = \sup\left\{|g \cdot f| \;\middle|\; f \in L_1, \|f\|_1 \leqq 1\right\}$ *for any* $g \in L_\infty$.
(3) *Under the dual pairing* $g \cdot f = \int g(x) f(x) d\mu(x)$ *for* $g \in L_\infty(M, \mathcal{M}, \mu)$ *and* $f \in L_1(M, \mathcal{M}, \mu)$, *we have* $L_\infty = L_1^*$.

ba, Total Variation

Let $\mathcal{M}$ be an algebra of sets consisting of subsets of M. The set of all real-valued bounded finitely additive set functions μ defined on $\mathcal{M}$ is denoted by $ba(M, \mathcal{M})$. If $\mu \in ba(M, \mathcal{M})$ satisfies $\mu(S) = \mu\left(\cup_{n \in \mathrm{N}_+} S_n\right) = \sum_{n \in \mathrm{N}_+} \mu(S_n)$ for any $S \in \mathcal{M}$ and any disjoint countable family of sets $\{S_n \in \mathcal{M} \mid n \in \mathrm{N}_+\}$ satisfying $S = \cup_{n \in \mathrm{N}_+} S_n$, then μ is said to be *countably additive in* $\mathcal{M}$. The subset of all bounded finitely additive set functions μ in $ba(M, \mathcal{M})$ that are countably additive is denoted by $ca(M, \mathcal{M})$. For two set functions τ and μ on $\mathcal{M}$, we write $\tau \leqq \mu$ if $\tau(S) \leqq \mu(S)$ for any $S \in \mathcal{M}$. In particular, $\mu \geqq 0$ if $\mu(S) \geqq 0$ for all $S \in \mathcal{M}$. μ in $ba(M, \mathcal{M})$ for which $\mu \geqq 0$ is said to be *purely finitely additive* if $\tau = 0$ for any countably additive $\tau \in ca(M, \mathcal{M})$ on $\mathcal{M}$ satisfying $0 \leqq \tau \leqq \mu$.

For a finitely additive set function $\mu \in ba(M, \mathcal{M})$ and a set $S \in \mathcal{M}$, the *total variation* of μ on S is denoted by $|\mu|(S)$ and is defined by

$$|\mu|(S) = \sup\left\{ \sum_{j=1}^{n} |\mu(S_j)| \;\middle|\; S_j \in \mathcal{M}, (j=1, \ldots, n), \text{ are disjoint and } S = \cup_{j=1}^{n} S_j \right\},$$

(that is, the least upper bound considering every possible partition of S). If $\mathscr{M}$ is a σ-algebra and μ is countably additive, then $|\mu|$ becomes a measure on $\mathscr{M}$. The integration under finitely additive set functions $\mu \in ba(M, \mathscr{M})$ for $g \in L_\infty(M, \mathscr{M}, \mu)$ is defined in a manner analogous to the Lebesgue integration. For details, see for example [34, Chapter III].

Mathematical Fact 1.40 [ba and ca Spaces] *Let $(M, \mathscr{M})$ be a measurable space.*

(1) *The total variation on M defines a norm on $ba(M, \mathscr{M})$. The space $ba(M, \mathscr{M})$ with this norm is a Banach space.*
(2) *We have $L_\infty^* = ba(M, \mathscr{M})$ under the pairing $\mu \cdot g = \int g d\mu$ for $\mu \in ba(M, \mathscr{M})$ and $g \in L_\infty(M, \mathscr{M}, \mu)$.*
(3) *Let $(M, \mathscr{M}, \mu)$ be a measure space. For each $f \in L_1(M, \mathscr{M}, \mu)$, $\eta(S) = \int_S f d\mu$ defines a countably additive set function on $\mathscr{M}$ with finite values, and η is absolutely continuous with respect to μ. Thus, by the Radon–Nikodym Theorem, there is an embedding ι from $L_1(M, \mathscr{M}, \mu)$ to $ca(M, \mathscr{M}) \subset ba(M, \mathscr{M})$ which identifies $L_1(M, \mathscr{M}, \mu)$ with the set of all bounded countably additive set functions which are absolutely continuous with respect to μ.*

Mathematical Property 1.44 [Yosida–Hewitt's Theorem] *Let $(M, \mathscr{M})$ be a measurable space. If μ is a finitely additive finite measure in $ba(M, \mathscr{M})$, then μ can be uniquely written as the sum of a countably additive measure μ_c and a purely finitely additive measure μ_f; that is, there exists a unique $\mu_c \in ca(M, \mathscr{M})$ with $\mu_c \geqq 0$ and a unique purely finitely additive $\mu_f \in ba(M, \mathscr{M})$ with $\mu_f \geqq 0$ such that $\mu = \mu_c + \mu_f$. Moreover, there exists a countable measurable partition $\{S_n | \ n = 1, 2, \ldots\}$ of M such that $\mu_f(S_n) = 0$ for every n.*

Notes 1

Notes 1.1 [Correspondences, p. 15] Debreu started to use the term "correspondences" in mathematical economics, adopting it from Bourbaki, and since then most of the works in mathematical economics and game theory use this term instead of referring to "multivalued functions" which are still used by some of researchers in applied mathematics. A basic reason for this usage is that "... the objects named by a noun and an adjective is not in the class of objects named by the noun alone ... " ([22, p. 27]). Authoritative Dieudonne [32, p. 1] is more straightforward in disclaiming it by stating that "... there is no such thing as a 'multiple-valued' function ...." Some others use the term *set-valued functions* or *mappings*.

Notes 1.2 For readers who might need introductory knowledge of some of the topics presented in Chapter 1 as a background for further reading of subsequent chapters, we give below a sample of introductory books in mathematics on these topics.

[Set Theory, p. 11] A classical and standard introductory book on set theory is [46]. Also, as an introductory text one may see [66, Chapter 0, pp. 1–36].
[Linear Spaces, p. 22] Examples of well-written introductory texts are [45] and [73].
[Topological Spaces p. 25] An excellent classical standard book is [66]. Other introductory textbooks include [33] and [125].
[Continuity of Relations, p. 34] A very useful and convenient book for those interested in mathematical economics is [12].

[Convex Sets, p. 41] To review the concepts and properties related to convex analysis, one may refer to [10], [99], etc.
[Measurable Structures, p. 43, and Integration, p. 53] An elegant introductory text is [44]. For convergences of measures and distributions, see [14] and [97].
[Extension of Classes of Measurable Sets, p. 50] One may refer to [70], [89], [61], [114], etc.
[Linear Topological Spaces, p. 63] One may refer to [105] among others.

Notes 1.3 [Section 1.5, p. 40]

[Kakutani's Fixed Point Theorem, p. 42] The original proof is in [64, Theorem 1, p. 457]. For further proofs, one can refer to [12, Chapter VIII, pp. 174–176], [95, Chapter I, Section 4, Theorem 4.4, pp. 67–69], [56, Mathematical appendix IV, pp. 201–204], etc.
[Michael's Continuous Selection Theorem, p. 43] This is originally due to [90, Theorem 3.1‴, p. 364]. For a proof see e.g. [56, Lemma AIV.2, p. 203]. His theorem holds in a more general space than the one given here.
[Equilibrium Fixed Point Theorem, p. 43] One can refer to [21], [40], or [95, Chapter V, Section 16, Theorem 16.6, pp. 265–267].

Part II
Foundations of Theoretical Economic Analysis

Chapter 2
Basic Concepts of Equilibrium Analysis

Abstract This chapter focuses on defining the consumption characteristics of individual economic agents. The central economic agents are consumers. After establishing their mathematical structures, concepts are introduced that express how individual agents respond to market transaction opportunities.

2.1 Economic Agents' Characteristics

Theoretical economic analysis is built around transactions of various commodities in markets. Accordingly, the primary mathematical space in which basic economic concepts are defined is the commodity space. What we call *economic agents* are participants in market transactions. Since the central economic agents are consumers, this chapter focuses on defining the consumption characteristics of individual economic agents. We then proceed to set up their mathematical structures. Subsequently, concepts are introduced that express how individual agents respond to market transaction opportunities.

The characteristics of economic agents are represented by their consumption characteristics, which are defined on the commodity space. Most of the analytical concepts in our analysis will be defined on the basis of this space.

Commodity Space $\mathbb{G} \subset \mathbb{R}^\ell$

A *commodity* is a good or a service that is distinguished by its *physical characteristics*, as well as by the *date* and the *location* at which it will be available. The quantity of a commodity is expressed by a real number. The set of all real numbers is denoted by the letter $\mathbb{R}$. Let ℓ be a (strictly) positive integer. We will assume that there are ℓ distinguishable commodities which are traded in markets. $\mathbb{R}^\ell$ denotes an ℓ-dimensional Euclidean space. A point $x = (x_1, \ldots, x_\ell)$ in $\mathbb{R}^\ell$ is regarded as an assortment of various quantities of ℓ commodities without presumptions as to the

A. Yamazaki, *Mathematical Economics of General Equilibrium*, Monographs in Mathematical Economics 8, https://doi.org/10.1007/978-981-96-9497-6_2

physical feasibility and the availability of their consumption. The i-th unit vector, $\mathbf{e}_i = (0, \dots, 0, \overset{i}{1}, 0, \dots, 0)$ represents one unit of the i-th commodity in $\mathbb{R}^\ell$.

Definition 2.1 (Commodity Space, Commodity Vector) The *commodity space* $\mathbb{G}$ is a non-empty closed subspace of $\mathbb{R}^\ell$ such that its convex hull is identical to Euclidean ℓ-space, i.e. $\operatorname{co}\mathbb{G} = \mathbb{R}^\ell$. An element $x = (x_1, \dots, x_\ell)$ of $\mathbb{G}$ is called a *commodity vector.*

The condition $\operatorname{co}\mathbb{G} = \mathbb{R}^\ell$ means that all ℓ types of commodities are tradable on the markets.

Definition 2.2 (Indivisible Commodity, Divisible Commodity) A commodity i is said to be *purely indivisible* if it can be consumed only in integer amounts. It is said to be *perfectly divisible* if it can be consumed in any real number of units.

It should be noted that the commodity space $\mathbb{G}$ delineates the physical feasibility of transactions of vectors in $\mathbb{R}^\ell$. In other words, a vector x in $\mathbb{R}^\ell$ that is not in $\mathbb{G}$ is physically unavailable as a combination of various quantities of ℓ commodities in transactions. To illustrate, if the first commodity is purely indivisible, then a vector $x = (3.1415926535, x_2, \dots, x_\ell)$ in $\mathbb{R}^\ell$ is not a physically possible combination of quantities of commodities, regardless of the quantities of the commodities $x_2, \dots, x_\ell$, and thus $x \notin \mathbb{G}$.

In the context of the commodity space $\mathbb{G}$, let $\mathcal{D}_\mathbb{G} \subset \{1, \dots, \ell\}$ be the set of all perfectly divisible commodities. It is possible that $\mathcal{D}_\mathbb{G} = \emptyset$. Denoting the projection map from $\mathbb{R}^\ell$ onto the i-th coordinate subspace by pr_i, i.e., $pr_i(x_1, \dots, x_i, \dots, x_\ell) = x_i$, we have $pr_i(\mathbb{G}) = \mathbb{R}$ for each $i \in \mathcal{D}_\mathbb{G}$.

Examples [Different Features of Commodities]

$\mathbb{Z}$ is the set of all integers, i.e., $\{\dots, -2, -1, 0, 1, 2, \dots\}$.

1. (Fig. 2.1) $\mathbb{G} = \mathbb{R}^2$. Both of the commodities are perfectly divisible so that $\mathcal{D}_\mathbb{G} = \{1, 2\}$.
2. (Fig. 2.2) $\mathbb{G} = \mathbb{R} \times \mathbb{Z}$. The first commodity is perfectly divisible and the second is purely indivisible so that $\mathcal{D}_\mathbb{G} = \{1\}$.
3. (Fig. 2.3) $\mathbb{G} = \mathbb{R} \times \{0\} \cup \{0\} \times \mathbb{R}$. Both of the commodities are perfectly divisible as in the first example but it is not feasible to consume both of the commodities at the same time. They are *mutually exclusive consumption goods.*

Subject to the physical feasibility of possible combinations of commodities three concepts – consumption sets, preferences, and initial endowments – to express consumption characteristics of economic agents are defined. From time to time we refer to economic agents as *consumers* whenever the consumption characteristics of agents or their consumption activities are discussed.

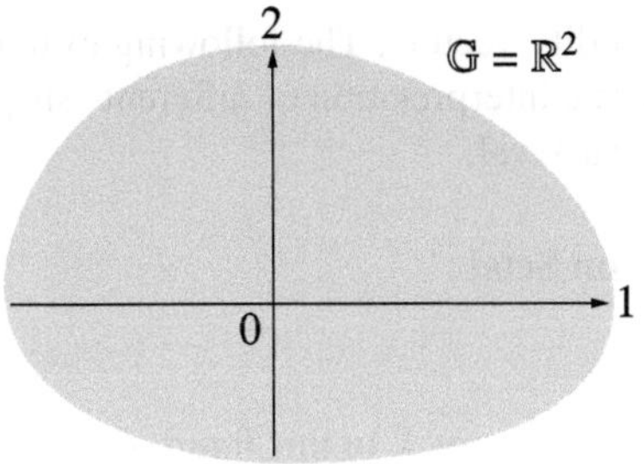

Fig. 2.1

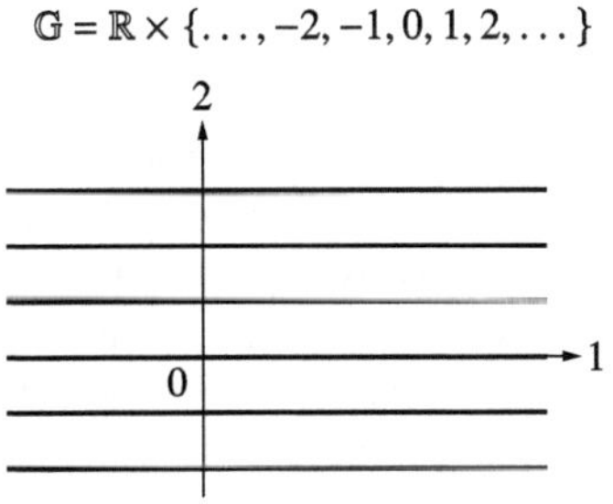

Fig. 2.2

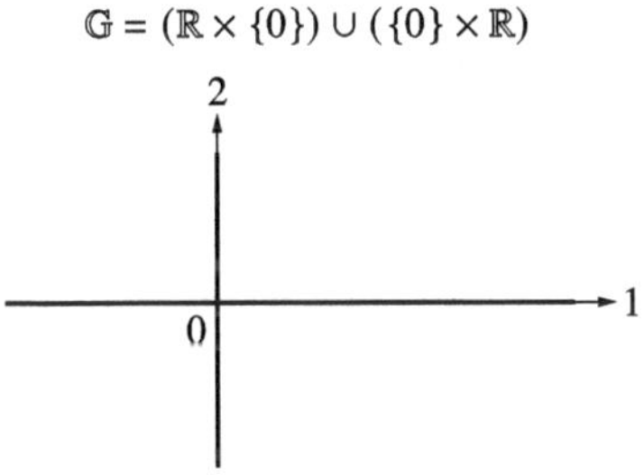

Fig. 2.3

Consumption Sets $X \subset \mathbb{G}$

Definition 2.3 (Consumption Sets, Consumption Vector) A *consumption set* X is a non-empty closed subset of the commodity space $\mathbb{G}$. A point x in X is called a *consumption vector* or *consumption plan.*

A *consumption vector* $x \in X \subset \mathbb{G}$ for an economic agent specifies the quantity of each commodity which has to be made available to the agent as well as the quantity of each commodity which is made available by the agent: a positive number represents the quantity of a commodity made available to the agent whereas a negative number represents the quantity of a commodity made available by the agent.

A consumption set $X \subset \mathbb{G}$ expresses the needs of an agent as a consumer and the feasibility of its consumption, which signal whether the consumer could *survive* by consuming given commodity vectors. The following examples will help the reader to understand the economic interpretation of different "shapes" of consumption sets in regard to consumers' survival.

Examples [Consumption Sets]

1. (Fig. 2.4) The consumption set X in this figure shows that the consumer's labor may be supplied to other agents in integer amounts only. It needs at least b_1 units of food to survive without providing any labor to others. As the consumer works longer hours it needs more food to survive. The maximum amount of labor it can provide to others is b_2 units.
2. (Fig. 2.5) A specific feature of the consumption set X in the figure is that the possibility of substitution between commodity 1 and 2 is extremely limited to reach the survival level of their consumptions. It shows that the consumer needs at least 10 units of commodity 2 to survive if it has no less than 100 units of commodity 1 but the consumer cannot reduce the consumption of commodity 1 even if it consumes commodity 2 beyond 10 units until its consumption of the commodity 2 exceeds 50 units.

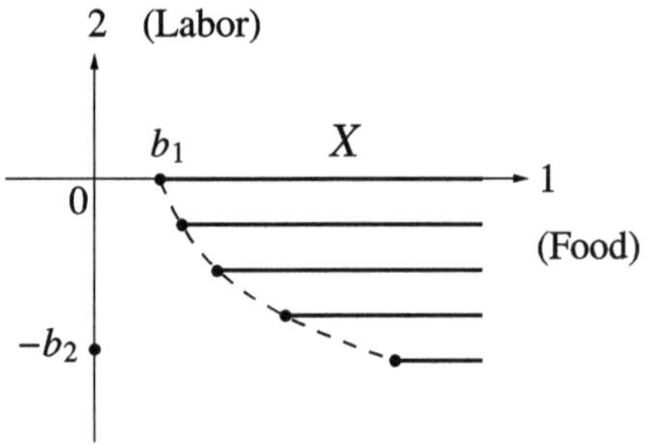

Fig. 2.4

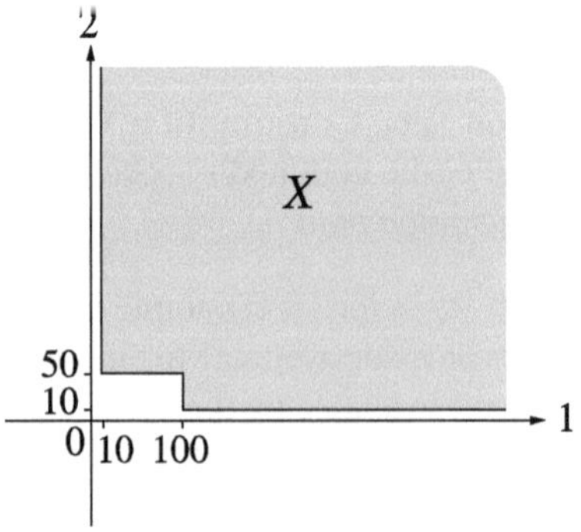

Fig. 2.5

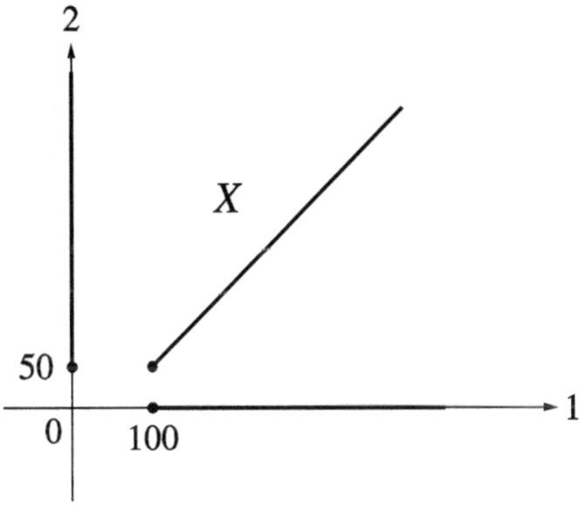

Fig. 2.6

3. (Fig. 2.6) The feasibility of consumptions shown in this figure is that consumptions are extremely restricted to very particular proportions of consumptions of two commodities; it is feasible either to consume only one of them or to consume both in the same proportion. Moreover, the survival levels of consumptions have a similar nature to the consumption set of the previous example.

Definition 2.4 (Universal Set for Consumption Sets) Let two real numbers $\underline{\beta}$ and $\overline{\beta}$ be fixed once and for all. The set of all consumption sets X satisfying the following two conditions is denoted by $\mathscr{X}$ and is called the *universal set for consumption sets*:

(a) $x_i \geqq \underline{\beta}$ for all $i = 1, \ldots, \ell$ and for all $x = (x_1, \ldots, x_\ell) \in X$;
(b) $X \cap \{x = (x_1, \ldots, x_\ell) \in \mathbb{R}^\ell \mid x_i \leqq \overline{\beta}$ for all $i = 1, \ldots, \ell\} \neq \emptyset$.

For the purpose of mathematical analysis in this book the family of all consumption sets that are considered is the universal set for consumption sets defined above. The uniform restriction (a) on the amounts of consumable commodities by an individual agent is meaningful and natural on economic grounds. In the case of commodities made available to the individual, it consumes at least zero or possibly some positive amounts for the commodities whose consumptions are vital to maintain its life, such as common salt. Meanwhile, commodities made available by the individual such as various types of labor and services must be within the amounts that could be made available in 24 hours per day. The condition (b) means that there is a uniform upper bound on the minimum amounts of commodities essential for maintaining lives of individual consumers.

Preferences $\succ \subset X \times X$

Choices made by an economic agent of a consumption vector depend upon its tastes or preferences over possible consumption vectors. Given that its consumption set is $X \in \mathscr{X}$, its preferences are represented by a binary relation on X (i.e., a subset of the Cartesian product $X \times X$).

Definition 2.5 (Preference Relations) A *preference relation* is an ordered pair $(X, \succ)$ composed of a consumption set $X \in \mathscr{X}$, and a non-empty irreflexive and asymmetric binary relation $\succ \subset X \times X$ such that $\succ$ is *continuous*, i.e., *open in* $X \times X$. A preference relation $(X, \succ)$ is sometimes written $\succ$ for short. It may be simply referred to as *preferences* (on X) of the agent.

$(x, y) \in \succ$ is read "x is *preferred to* y." $(x, y) \in \succ$ is also written as $x \succ y$. $x \prec y$ means $y \succ x$. A preference relation $(X, \succ)$ induces two other binary relations, $\precsim$ and $\sim$, on X defined as follows:

$$\precsim = \{(x, y) \in X \times X \mid (x, y) \notin \succ\}, \text{ and}$$
$$\sim = \{(x, y) \in X \times X \mid (x, y) \notin \succ \text{ and } (y, x) \notin \succ\}.$$

$(X, \precsim)$ expresses *"weak preferences"* and $(X, \sim)$ describes *"indifferences"* between two consumption vectors. $x \precsim y$ and $x \sim y$ mean $(x, y) \in \precsim$ and $(x, y) \in \sim$ respectively. $x \sim y$ is read "x is *indifferent to* y." $x \succsim y$ stands for $y \precsim x$. It is read "x is *preferred or indifferent to* y." $(x, y) \notin \succ$ may be written $x \nsucc y$. $x \not\precsim y$ and $x \nsim y$ are defined similarly.

Given $(X, \succ)$ and $x \in X$, two sets $\succ(x)$ and $\succsim(x)$ are defined as follows:

$$\succ(x) = \{y \in X \mid y \succ x\} \text{ and } \succsim(x) = \{y \in X \mid y \succsim x\}.$$

$\succ(x)$ denotes the set of all consumption vectors that are preferred to x, and $\succsim(x)$ the set of all consumption vectors that are preferred or indifferent to x.

Initial Endowments $e \in \mathbb{G}$

Each economic agent might possess some commodities prior to its participation in market transactions. They are expressed by a vector $e \in \mathbb{G}$ and will be called *initial endowments* of agents. Initial endowments e must belong to the commodity space $\mathbb{G}$ since they must conform to the physical characteristics of various commodities. One may note, nonetheless, that initial endowments need not belong to consumption sets of individual economic agents. It is because some of the commodities which each of economic agents possess initially before engaging in market transactions might not be enough to sustain their lives. In view of the valuation of commodities taking place in market economies through transactions, the endowments of each agent might be conceived as "assets " and are expressed by their value. This value is referred to as the *wealth level* of the agent.

Definition 2.6 (Initial Endowments, Wealth, and Wealth Space) Given that each economic agent is endowed with a commodity vector $e \in \mathbb{G}$ prior to market trading, e is called its *initial endowments*, or simply *endowments*. A real number $w \in \mathbb{R}$ may indicate its endowments instead of a commodity vector $e \in \mathbb{G}$, in which case w is called its *wealth* or *wealth level*, and the space $\mathbb{R}$ is called the *wealth space*.

Consumption Characteristics

The consumption characteristics of economic agents will be depicted by way of economic concepts introduced thus far.

Definition 2.7 (Consumption Characteristics) A pair $((X, \succ), e)$ composed of a preference relation $(X, \succ)$ and initial endowments e of an economic agent is called the *consumption characteristics*, or simply *characteristics* of the agent. The consumption characteristics of an agent are sometimes defined as $((X, \succ), w)$ by specifying its wealth level w instead of initial endowments e.

Exercise 2.1 Let Ψ be a binary relation on S satisfying $(y, x) \notin \Psi$ for any $(x, y) \in \Psi$. Let $B \subset S$. Confirm that

x is the greatest element in B if and only if $(x, z) \notin \Psi$ for every $z \in B$.

Exercise 2.2 Let Ψ be a binary relation on S satisfying $(y, x) \notin \Psi$ for any $(x, y) \in \Psi$.

1. Show that the greatest element is always maximal.
2. Construct an example showing that the converse is false.
3. Assume that for all x, y in S with $x \neq y$ we have $(x, y) \in \Psi$ or $(y, x) \in \Psi$. Prove that a maximal element is the greatest element.

Exercise 2.3 Let $(X, \succ)$ be a preference relation. Let Δ_X denote the diagonal in $X \times X$, i.e., $\Delta_X = \{(x, x) | x \in X\}$.

1. Show that $\precsim$ is closed in $\mathbb{G} \times \mathbb{G}$.
2. Show that the set $X_{\precsim}$ defined by $X_{\precsim} = \{x \in \mathbb{G} \mid (x, y) \in \precsim \text{ for some } y \in \mathbb{G}\}$ is identical to X.
3. Show that we have $\Delta_X \subset \precsim$.

Exercise 2.4 Let $\precsim$ be a binary relation on $\mathbb{G}$ such that $\precsim$ is a closed and proper subset of $\mathbb{G} \times \mathbb{G}$. Define $X_{\succ} = \{x \in \mathbb{G} \mid (x, x) \in \precsim\}$ and $G_{\precsim} = X_{\succ} \times X_{\succ} - \precsim$. Assume that we have $\Delta_X \subset \precsim$.

1. Show that $(X_{\succ}, G_{\precsim})$ is a preference relation.
2. Given a preference relation $(X, \succ)$, let $\precsim$ be the induced binary relation on X. Then, defining $(X_{\succ}, G_{\precsim})$ from this $\precsim$ as above, confirm that we have $(X_{\succ}, G_{\precsim}) = (X, \succ)$.

2.2 Markets and Individual Behavior

Following the introduction of concepts specifying characteristics of economic agents, several concepts related to the behavior of agents in markets and their individual market transactions are introduced in this section.

Prices $p \in \mathbb{R}^\ell$

It is assumed that all the ℓ commodities are traded in markets. The basic working of markets is symbolized by a price system under which commodities are traded in markets. Accordingly, a price system in the markets shows the value in exchange of one unit of each of the commodities traded in the markets. Thus, from a functional point of view, a price system assigns any one of commodity vectors its market value in an arithmetically consistent way. Specifically, let x, y be commodity vectors, and $p(x), p(y)$ be their market values. For any real numbers s, t, consider the combined commodity vector, $sx + ty$, composed of s times the vector x and t times the vector y. Then, its market value, $p(sx + ty)$, will be equal to s times the market value of x, $sp(x)$, plus t times the market value of y, $tp(y)$. As a consequence, it is natural to give the following definition of a price system.

Definition 2.8 (Price System, Price Space, Price Vectors, Market Values) A *price system* p is a linear functional defined on $\mathbb{R}^\ell$. A price system $p : x \mapsto p(x) \in \mathbb{R}$ belongs to the dual space of $\mathbb{R}^\ell$. By identifying $\mathbb{R}^\ell$ with its dual, p can be regarded as a vector $p \in \mathbb{R}^\ell$ in which case $\mathbb{R}^\ell$ is called the *price space* and p a *price vector* or *prices*. One has $p : x \mapsto p \cdot x = p_1x_1 + \cdots + p_\ell x_\ell$. $p \cdot x$ will be called the *market valuation* of x by p or simply the *market value* of x.

Exercise 2.5 1. Let commodities 1 and 2 represent two physically different commodities which are available at the same location and on the same date. Give an interpretation of p_1/p_2.

2. Let commodities 1 and 2 represent two commodities with the same physical characteristics, available at the same location but at different time periods t_1 and t_2. Give an interpretation of p_1/p_2.
3. Let commodities 1 and 2 represent two commodities with the same physical characteristics, available on the same date but at different locations. Give an interpretation of p_1/p_2 or $p_1 - p_2$.

Individual Market Opportunity

For an economic agent with its consumption set X, its choice of a commodity vector in consumption must be inherently restricted to those in its consumption set X. Otherwise, the agent cannot maintain its life for survival. In addition to this reservation there is a restriction coming from market transactions. Given a price vector p and the wealth level w of the agent, the market value of the choice of a consumption vector x must not exceed the bound of its wealth level w. Within a framework in which production sector is abstracted and its profit distribution among economic agents is not taken account, the market values of their initial endowments that they possess delimit the wealth levels of agents. One may note that their labor incomes are not accounted in their wealth levels but are included in the market values of consumption vectors that they choose.

It will be convenient to introduce some mathematical notations relating to the market valuation of commodity vectors by a price vector p. They help to describe constraints imposed on choices of commodity vectors by the wealth levels of agents. Let $H(p,w) = \{x \in \mathbb{R}^\ell \mid p \cdot x = w\}$ be a *hyperplane* in $\mathbb{R}^\ell$. Define the (associated) *lower closed half space* $H_{\leqq}(p,w) = \{x \in R^\ell \mid p \cdot x \leqq w\}$ and *upper closed half space* $H_{\geqq}(p,w) = \{x \in \mathbb{R}^\ell \mid p \cdot x \geqq w\}$. (Associated) *open half spaces* $H_<(p,w), H_>(p,w)$ are defined similarly. (See Fig. 2.7.) Vectors on $H(p,w)$ are exactly those whose market values are just equal to the wealth level w. The market values of vectors belonging to the closed half space $H_{\leqq}(p,w)$ are within the wealth constraint w whereas those of vectors in the open half space $H_>(p,w)$ are not tradable or "purchasable" under the wealth constraint w.

Definition 2.9 (Budget Sets, Budget Hyperplanes, Budget Set Relation) The *budget set* $B(X,w,p)$ for X, w, and p, is defined by $B(X,w,p) = X \cap H_{\leqq}(p,w) = \{x \in X \mid p \cdot x \leqq w\}$. $X \cap H(p,w)$ is termed a *budget hyperplane*. The relation $B \subset \mathscr{X} \times \mathbb{R} \times \mathbb{R}^\ell \times \mathbb{G}$ from $\mathscr{X} \times \mathbb{R} \times \mathbb{R}^\ell$ to $\mathbb{G}$ defined by $(X,w,p) \mapsto B(X,w,p)$, that is,

$$B = \{(X,w,p,x) \in \mathscr{X} \times \mathbb{R} \times \mathbb{R}^\ell \times \mathbb{G} \mid x \in B(X,w,p)\}$$

is called the *budget set relation*.

The budget set $B(X,w,p)$ gives expression to the scope of market transactions that an agent with a preference relation $(X,\succ)$ and wealth w faces when market transactions are to be carried out under market prices p.

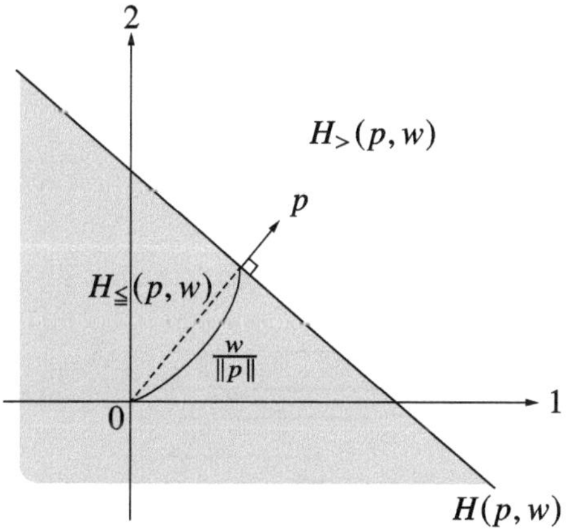

Fig. 2.7 A price vector p and $H_{\leqq}(p,w)$, $H(p,w)$, $H_>(p,w)$

Examples [Budget Sets]

Two examples of budget sets are given below for the case of $\ell = 2$.

1. In Fig. 2.8 two commodities are perfectly divisible. The non-convexity of the budget sets $B(X, w, p)$ is due to the nature of consumption feasibility attributable to the consumption set X.
2. The budget set $B(X, w, p)$ in Fig. 2.9 arises from the consumption set in Fig. 2.4. Commodity 1 is perfectly divisible but commodity 2 is purely indivisible. The budget set is not even connected.

Demand Set $D(X, \succ, w, p) \subset \mathbb{G}$

Market transactions of economic agents as consumers are conceived to be consequences of their optimal choices made with regard to consumption vectors in their budget sets. A key economic concept to describe the optimal consumption choices of economic agents is the notion of "*demands*."

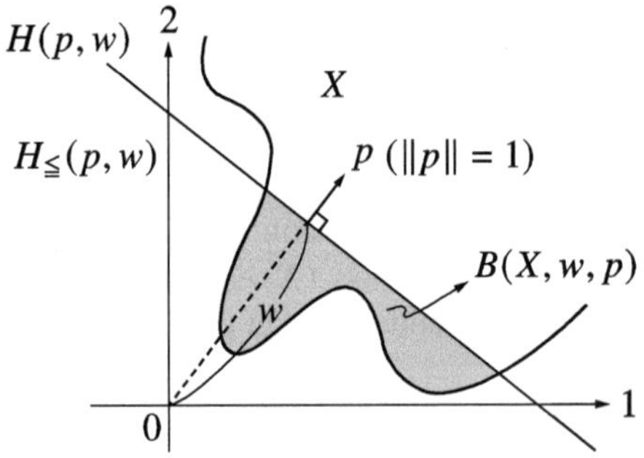

Fig. 2.8 $B(X, w, p)$

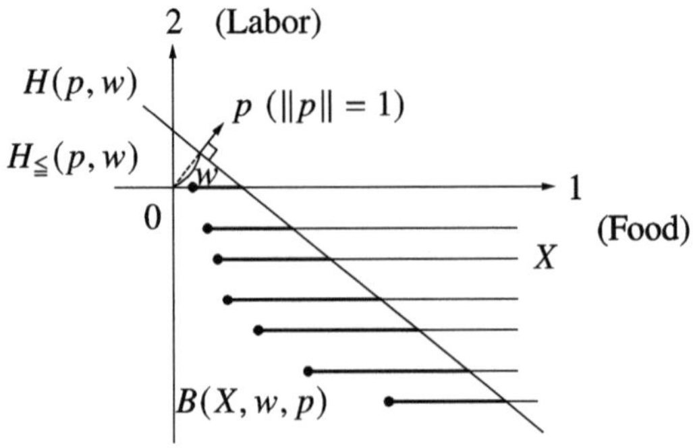

Fig. 2.9 $B(X, w, p)$

Definition 2.10 (Demand Sets, Demand Vectors, and Demand Relation)
Let $(X, \succ)$ be a preference relation, $w \in \mathbb{R}$, and $p \in \mathbb{R}^{\ell}$. The *demand set* for $(X, \succ)$ at (w, p), denoted by $D(X, \succ, w, p)$, consists of maximal elements (with respect to the preference relation $(X, \succ)$) in the budget set $B(X, w, p)$; that is,

$$\begin{aligned} D(X, \succ, w, p) &= \{x \in B(X, w, p) \mid z \not\succ x \text{ for all } z \in B(X, w, p)\} \\ &(= \{x \in B(X, w, p) \mid p \cdot z > w \text{ for all } z \in X \text{ such that } z \succ x\} \\ &= \{x \in B(X, w, p) \mid B(X, w, p) \cap \succ(x) = \emptyset\}). \end{aligned}$$

The relation D defined by $(X, \succ, w, p) \mapsto D(X, \succ, w, p)$, that is, $D = \{(X, \succ, w, p, x) \mid x \in D(X, \succ, w, p)\}$, is called the *demand relation*. A consumption vector in the demand set is called a *demand vector*.

Examples [Demand Sets]

Two examples of demand sets given below may arise from the budget sets shown in Fig. 2.8 and Fig. 2.9 of the previous example.

1. In Fig. 2.10 the demand set $D(X, \succ, w, p)$ is given by $\{x, y\}$. The dotted line shows the boundary of the preferred set $\succ(x) = \succ(y)$.
2. In Fig. 2.11 the demand set $D(X, \succ, w, p)$ is also given by $\{x, y\}$. The intersection of the dotted line passing through x or y with the consumption set X depicts the boundary of its corresponding preferred set $\succ(x)$ or $\succ(y)$.

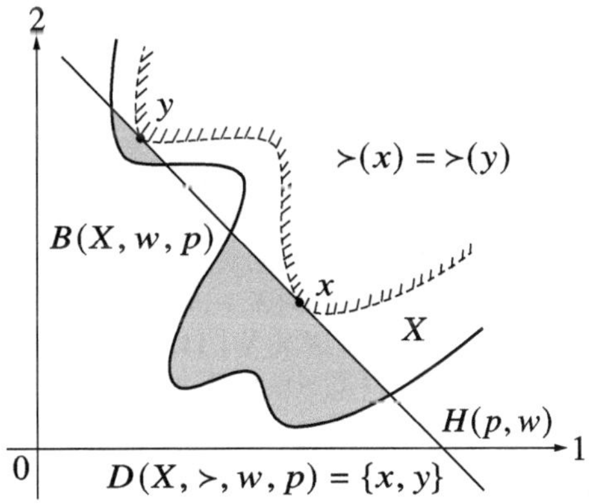

Fig. 2.10 $D(X, \succ, w, p)$

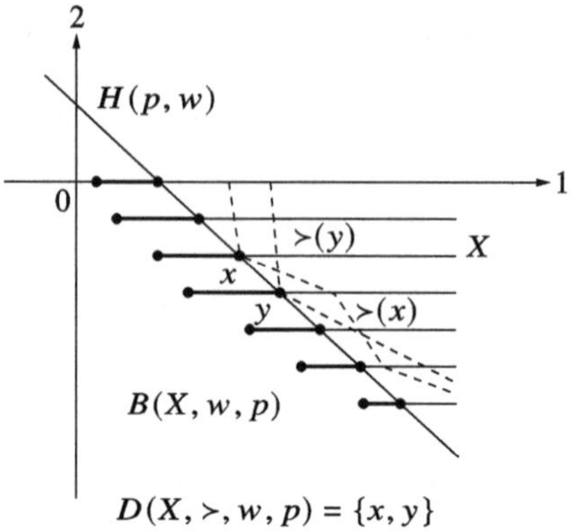

Fig. 2.11 $D(X, \succ, w, p)$

2.3 Spaces of Preference Relations

The property of preference relations postulated in Definition 2.5 encompasses a range of preferences that is much too extensive. Indeed, some of the preference relations satisfying the postulate may not admit any maximal elements even in budget sets which are non-empty and compact. From a theoretical point of view it is essential to abstain from such pathological cases since individual agents' market behaviors are not well-defined unless their demand sets are non-empty.

The following examples illustrate some of pathological preference relations that would induce empty demand sets.

Examples [Pathological Preferences]

1. Let the commodity space $\mathbb{G} \subset \mathbb{R}^\ell$, a wealth level $w \in \mathbb{R}$, and a price vector $p \in \mathbb{R}^\ell_{++}$ be given. Define a preference relation $(X, \succ)$ by $X = \{x, y, z\}$ for three mutually distinctive commodity vectors x, y, z for which $p \cdot x = p \cdot y = p \cdot z = w$, and $\succ = \{(x, y), (y, z), (z, x)\}$. Then, $D(X, \succ, w, p) = \emptyset$.
2. Let $\mathbb{G} = \mathbb{R}^2$, $X = \mathbb{R}^2_+, \succ = \{(x, y) \in X \times X \mid [x_1 > y_1 - 1 \text{ and } x_2 > y_2] \text{ or } [x_1 > y_1 \text{ and } x_2 > y_2 - 1]\}$. For this $(X, \succ)$, $D(X, \succ, w, p) = \emptyset$ for any $w > 0$ and $p \in \mathbb{R}^2_{++}$. In Fig. 2.12 the shaded areas of two preferred sets $\succ(x)$ and $\succ(y)$ for $x, y \in \mathbb{G}, x \neq y$, are depicted along with their boundaries shown by dotted lines through x and y respectively.

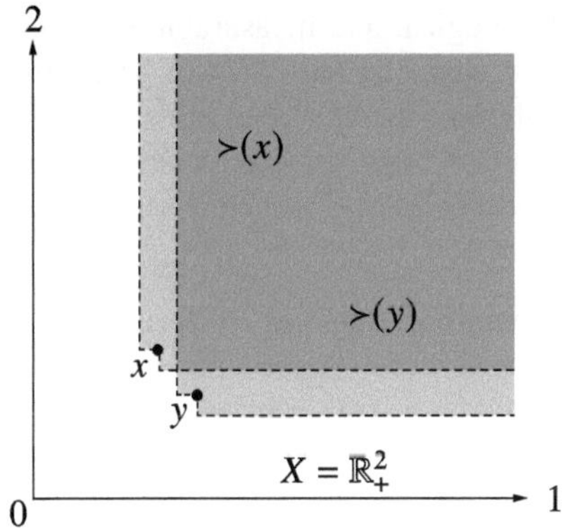

Fig. 2.12

Universal Preference Space $\mathcal{P}$

In view of the above examples, additional qualifications for preference relations are imperative beyond the basic properties of irreflexivity, asymmetry, and continuity. The minimal requirement on preference relations for the purpose of general equilibrium analysis is that they induce well-defined market behaviors of economic agents in terms of their demand sets. For this reason we like to fix as the universal set of preferences a class of preference relations which are *observable in markets*. It surely does not mean that preference relations themselves are directly observable but means that they are "indirectly observable" through the market behaviors of agents which are consequences of their demands of various commodities. The universal set defined in this manner will be most general from an analytic point of view.

Definition 2.11 (Universal Preference Space) Define a class $\mathcal{P}$ of preference relations by

$$\mathcal{P} = \{(X, \succ) \mid (X, \succ) \text{ is a preference relation such that for any } (w, p) \in \mathbb{R} \times \mathbb{R}^{\ell}, \; D(X, \succ, w, p) \neq \emptyset \text{ whenever } B(X, w, p) \text{ is non-empty and compact}\}.$$

$\mathcal{P}$ will be called the *universal preference space.*

On the basis of the universal preference space the space of economic agents' characteristics is introduced.

Definition 2.12 (Space of Economic Agents' Characteristics) The product space of the universal preference space and the commodity space, $\mathcal{P} \times \mathbb{G}$, will be called the *space of economic agents' characteristics* or the *space of consumption characteristics*. When wealth levels are studied directly in place of their sources specified by initial endowments, $\mathcal{P} \times \mathbb{R}$ will be taken as the space of economic agents' characteristics.

In the next chapter topological and measurable structures are introduced in the space of economic agents' characteristics $\mathcal{P} \times \mathbb{G}$. The remainder of this section is devoted to a characterization of the universal preference space $\mathcal{P}$ and to clarifying some of its properties. Let us begin by particularizing some of the properties of preference relations that will help us to grasp the aspects of the universal preference space.

Definition 2.13 (Finite Dominatedness of Preferences) Given a preference relation $(X, \succ)$, let $\pi_p(Q)$ be a subset of $\mathbb{R}$ for a price vector $p \in \mathbb{R}^\ell$ and a non-empty finite subset $Q \subset X$, defined by $\pi_p(Q) = \max\{p \cdot x \mid x \in Q\}$. The preference relation $(X, \succ)$ is said to satisfy the *finite dominatedness* (or, for simplicity, the property (FD)) if it satisfies the following:

(FD) *Given any $p \in \mathbb{R}^\ell$ and any non-empty finite subset $Q \subset X$, there is a consumption vector $x^* \in B(X, \pi_p(Q), p)$ such that $z \not\succ x^*$ for any $z \in Q$ whenever $B(X, \pi_p(Q), p)$ is non-empty and compact.* (See Fig. 2.13.)

The set of all preference relations that satisfy the finite dominatedness is denoted by $\mathcal{P}_{FD}$.

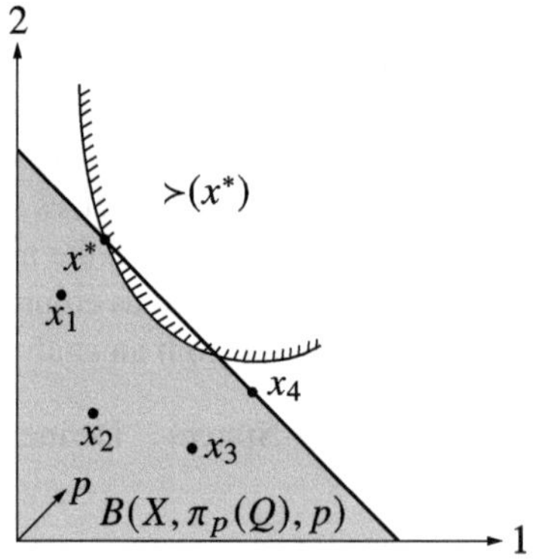

Fig. 2.13 Finite dominatedness of preferences: $X = \mathbb{R}^2_+$, $Q = \{x_1, x_2, x_3, x_4\}$

Subspaces of Universal Preference Space

Among the properties of binary relations on consumption sets $X \in \mathcal{X}$ to which the property (FD) is related, typical ones are the transitivity, acyclicity, and convexity. Except for transitivity of binary relations, these properties are not explained in Section 1.1, p. 14. Allowing for a repetition of transitivity, these three properties are as defined below.

Definition 2.14 (Transitivity of Preferences) A preference relation $(X, \succ)$ is said to be *transitive* if it satisfies the property that $x \succ y$ and $y \succ z$ imply $x \succ z$. The set of all preference relations satisfying the transitivity is denoted by $\mathcal{P}_t$.

Definition 2.15 (Acyclicity of Preferences) A preference relation $(X, \succ)$ is said to be *acyclic* if, for any finite subset $\{x_1, \ldots, x_k\}$ of X, it satisfies the property that $[x_1 \succ x_2, \ldots, x_{k-1} \succ x_k]$ implies $x_k \not\succ x_1$. The set of all preference relations satisfying the acyclicity is denoted by $\mathcal{P}_{ac}$.

The property of the convexity of preferences being defined next makes sense only when underlying commodity space is a convex subset of $\mathbb{R}^\ell$. Although the main thrust of this book is to present an analytical framework with non-convex commodity space, for the sake of showing an aspect of the property (FD), we shall include the case of a convex commodity space when we refer to convex preferences.

Definition 2.16 (Convexity of Preferences) In the case where the commodity space $\mathbb{G} \subset \mathbb{R}^\ell$ is a convex set, a preference relation $(X, \succ)$ is said to be *convex* if the consumption set X is convex and the set $\succ(x)$ of all consumption vectors that are preferred to x is convex for any $x \in X$. The set of all convex preference relations is denoted by $\mathcal{P}_{co}$.

The following proposition shows that the universal preference set $\mathcal{P}$ contains all of the sets of preference relations introduced above, $\mathcal{P}_{FD}, \mathcal{P}_t, \mathcal{P}_{ac}, \mathcal{P}_{co}$, and is characterized by the class $\mathcal{P}_{FD}$ of preferences satisfying the finite dominatedness.

Proposition 2.1 (Characterization of the Universal Preference Set)

The universal preference set $\mathcal{P}$ satisfies each of the following properties with respect to the sets of preference relations, $\mathcal{P}_{FD}, \mathcal{P}_t, \mathcal{P}_{ac}, \mathcal{P}_{co}$:

(1) *The universal preference set $\mathcal{P}$ is identical with the class $\mathcal{P}_{FD}$ of preference relations satisfying the finite dominatedness, i.e. $\mathcal{P} = \mathcal{P}_{FD}$.*
(2) *The universal preference set $\mathcal{P}$ contains the set of all preference relations satisfying the transitivity, i.e. $\mathcal{P}_t \subset \mathcal{P}$.*
(3) *The universal preference set $\mathcal{P}$ contains the set of all preference relations satisfying the acyclicity, i.e. $\mathcal{P}_{ac} \subset \mathcal{P}$.*
(4) *In the case where the commodity space $\mathbb{G} \subset \mathbb{R}^\ell$ is convex, the universal preference set $\mathcal{P}$ contains the set of all preference relations satisfying the convexity, i.e. $\mathcal{P}_{co} \subset \mathcal{P}$.*

Proof (1) Let $(X, \succ) \in \mathcal{P}$. Given $p \in \mathbb{R}^\ell$ and a non-empty finite subset $Q \subset X$, if $B(X, \pi_p(Q), p)$ is non-empty and compact, there is $x^* \in B(X, \pi_p(Q), p)$ such that one has $z \not\succ x^*$ for any $z \in B(X, \pi_p(Q), p)$. Since $Q \subset B(X, \pi_p(Q), p)$, one has $(X, \succ) \in \mathcal{P}_{FD}$. Hence, $\mathcal{P} \subset \mathcal{P}_{FD}$.

Conversely, let $(X, \succ) \in \mathcal{P}_{FD}$. Given $(w, p) \in \mathbb{R} \times \mathbb{R}^\ell$, assume $B(X, w, p)$ is non-empty and compact, but $D(X, \succ, w, p) = \emptyset$. Consider an open covering of $B(X, w, p)$ defined by the sets $\prec(x) = \{z \in B(X, w, p) \mid x \succ z\}, x \in B(X, w, p)$. Indeed, each $\prec(x)$ is open in $B(X, w, p)$ by continuity of $(X, \succ)$. One also has $B(X, w, p) \subset \cup_{x \in B(X,w,p)} \prec(x)$, as for any $y \in B(X, w, p)$, there is $x \in B(X, w, p)$ such that $y \in \prec(x)$ because if for every $x \in B(X, w, p)$ one had $x \not\succ y$, one would have $y \in D(X, \succ, w, p)$, which is a contradiction. Since $B(X, w, p)$ is compact, there is a non-empty finite subset $Q \subset B(X, w, p)$ such that $\cup_{x \in Q} \prec(x)$ is an open subcovering of $B(X, w, p)$. By the property (FD) there is $x^* \in B(X, \pi_p(Q), p) \subset B(X, w, p)$ such that $x \not\succ x^\star$ for any $x \in Q$. It follows that $x^* \notin \prec(x)$ for every $x \in Q$. This contradicts the fact that $\cup_{x \in Q} \prec(x)$ is a covering of $B(X, w, p)$. Therefore, one must have $D(X, \succ, w, p) \neq \emptyset$, establishing $(X, \succ) \in \mathcal{P}$. Consequently, $\mathcal{P}_{FD} \subset \mathcal{P}$.

(2) , (3) Let $(X, \succ) \in \mathcal{P}_t \cup \mathcal{P}_{ac}$. Given $(w, p) \in \mathbb{R} \times \mathbb{R}^\ell$, if $D(X, \succ, w, p) = \emptyset$ when $B(X, w, p)$ is non-empty and compact, it follows from the argument in (1) that there is a finite open covering $\prec(x_1), \ldots, \prec(x_k)$ of $B(X, w, p)$. But, the transitivity or the acyclicity, either way, implies that there is a maximal element x^* in the finite subset $\{x_1, \ldots, x_k\} \subset B(X, w, p)$, that is, $x_j \not\succ x^*$ for any $j = 1, \ldots, k$. It means $x^* \notin \prec(x_j)$ for all $j = 1, \ldots, k$, contradicting the fact that $\prec(x_1), \ldots, \prec(x_k)$ is a covering of $B(X, w, p)$. Hence, one must have $D(X, \succ, w, p) \neq \emptyset$, establishing $\mathcal{P}_t \cup \mathcal{P}_{ac} \subset \mathcal{P}$.

(4) Although proofs of (1)–(3) are straightforward, a proof of this fact requires a fixed point argument and is moderately involved. Before proceeding to prove (4), let us affirm the following lemma:

Lemma 2.1 *Let $\succ$ be a non-empty relation from a non-empty closed convex subset $X \subset \mathbb{R}^\ell$ to itself. Assume that it is l.h.c. and $x \notin \mathrm{co} \succ (x)$ for all $x \in X$. Then, $D(X, \succ, w, p) \neq \emptyset$ for every $(w, p) \in \mathbb{R} \times \mathbb{R}^\ell$ whenever $B(X, w, p)$ is non-empty and compact.* □

Proof of the Lemma

Let $B(X, w, p)$ be non-empty and compact for $(w, p) \in \mathbb{R} \times \mathbb{R}^\ell$. Since X is convex, so is $B(X, w, p)$. Define a binary relation $\succ_B$ by

$$\succ_B = \{(x, y) \in B(X, w, p) \times B(X, w, p) \mid y \in \mathrm{co} \succ (x)\}.$$

Then, $\succ_B$ is l.h.c. as it is a convex hull relation of the restriction to a closed set in $\mathbb{R}^\ell$ of the l.h.c. relation $\prec$ (cf. Mathematical Property 1.19 (1), p. 42). Accordingly, there is a continuous mapping f from dom $\succ_B$ to $B(X, w, p)$ such that one has $f \subset \succ_B$ by Michael's Continuous Selection Theorem (Mathematical Property 1.21, p. 43)). Define a correspondence φ from $B(X, w, p)$ to itself using the mapping f as follows:

$$\varphi(x) = \begin{cases} \{f(x)\} & \text{if } x \in \mathrm{dom} \succ_B, \\ B(X, w, p) & \text{if } x \in B(X, w, p) - \mathrm{dom} \succ_B \end{cases}$$

$\varphi(x)$ is a non-empty closed compact set for each $x \in B(X, w, p)$. To show $\varphi \subset B(X, w, p) \times B(X, w, p)$ is a closed set, let $(x_n, y_n)_n$ be a sequence in φ converging to (x, y). $y \in \varphi(x)$ if $x \notin$ dom $\succ_B$. So, let $x \in$ dom $\succ_B$. By l.h.c. of $\succ_B$, for any open neighborhood W of $f(x)$ the set $\{z \mid \{z\} \times W \cap \succ_B \neq \emptyset\}$ is open. Thus, for all n large enough, $\{x_n\} \times W \cap \succ_B \neq \emptyset$, that is, $x_n \in$ dom $\succ_B$ and $y_n = f(x_n)$. Hence, it follows from the continuity of f that one has $y = f(x)$, meaning $y \in \varphi(x)$. This shows that φ is closed in $B(X, w, p) \times B(X, w, p)$.

Now, by Kakutani's Fixed Point Theorem, there exists a fixed point x^* of φ, $x^* \in \varphi(x^*)$. If $x^* = f(x^*)$, then $x^* \in \mathrm{co} \succ (x^*)$, contradicting the assumption of the lemma. Thus, $x^* \notin \text{dom} \succ_B$ and $z \not\succ x^\star$ for every $z \in B(X, w, p)$. Therefore, $x^* \in D(X, \succ, w, p)$. This proves the lemma. □

We are now ready to prove (4) of the proposition. Let $(X, \succ) \in \mathcal{P}_{co}$; then, by the irreflexivity, $x \notin \text{co} \succ(x)$ for every $x \in X$. If $\succ \subset X \times X$ is shown to be l.h.c., it follows from Lemma 2.1 that we have $(X, \succ) \in \mathcal{P}$. To confirm that the relation $\succ$ is l.h.c., given any $x \in X$ and an open set $W \subset X$ such that $\{x\} \times W \cap \succ \neq \emptyset$, it needs be shown that for some neighborhood U of x one has $\{y\} \times W \cap \succ \neq \emptyset$ for any $y \in U$. Since $x \succ z$ for some $z \in W$, it follows from the continuity of the relation $\succ$ that there are a neighborhood U of x and a neighborhood V of z satisfying $U \times V \subset \succ$. This completes the proof of (4). □

Some comments on Proposition 2.1 may be in order. (2)–(4) of the assertion give sufficient conditions for individual demand sets to be non-empty. (2) and (3) say that any compact budget sets have maximal elements with respect to an individual preference relation provided it has the property of transitivity or acyclicity. In other words, the *rationality* in the sense of absence of inconsistencies in consumers' tastes or "preference patterns" ensures the existence of individual demand vectors. In this respect, (4) might appear "odd" since at a first glance it does not seem to require rational preference relations at all for demand sets to be non-empty. Although one can ask whether convex preference relations have qualities of rational preferences, instead one might understand the convexity property of preference relations to connote a minimal rationality required in the analysis of consumers' market behavior under consumption sets which satisfy connectedness in mathematical sense. However, the condition in (4) is taking a retrograde step as compared to (2) or (3) with regard to a presumption of the convexity of consumption sets. It has been left unsolved whether consumers' behavior in terms of demand sets may reasonably be analyzed under the rationality weaker than the acyclicity or the transitivity of preference relations without the precondition of the convexity of consumption sets. (1) admittedly shows that (FD) gives a necessary and sufficient condition that allows one to express consumers' market behavior by way of demand sets based on budget sets and preference relations. Nevertheless, it does not seem to give us an effectual insight into answering the question of to what extent consumers' rationality is required in general when we talk about consumers' demands on the basis of their preference relations.

Finally, we like to bring in an additional property of preference relations which is commonly encountered in general equilibrium analysis.

Definition 2.17 (Local Non-Satiation of Preferences) A preference relation $(X, \succ)$ is *locally non-satiated* if it satisfies the property that, for any $x \in X$ and for any neighborhood U of x, there is a consumption vector $y \in X \cap U$ such that $y \succ x$. The subspace of all preference relations in the universal preference space $\mathcal{P}$ satisfying the local non-satiation is denoted by $\mathcal{P}_{lns}$.

The local non-satiation property of preference relations has occasions to be appealed to in the analysis of later chapters. Its basic role for each consumer is the "full use of budget" as expressed in the following proposition.

Proposition 2.2 (Full Use of Budget)

Let $(X, \succ)$ *be a locally non-satiated preference relation in* $\mathcal{P}$. *Then, we have* $p \cdot x = w$ *for any* $x \in D(X, \succ, w, p)$ *whenever* $B(X, w, p) \neq \emptyset$ *at* $(w, p) \in \mathbb{R} \times \mathbb{R}^{\ell}$.

Proof Given $(X, \succ) \in \mathcal{P}_{lns}$ and $(w, p) \in \mathbb{R} \times \mathbb{R}^{\ell}$ with $B(X, w, p) \neq \emptyset$, let $x \in D(X, \succ, w, p)$. Since $x \in B(X, w, p)$, $p \cdot x \leqq w$. Suppose we had $p \cdot x < w$. Take a neighborhood U of x such that $p \cdot z < w$ for every $z \in X \cap U$. As $(X, \succ)$ is locally non-satiated, there would be a consumption vector $y \in X \cap U$ for which $y \succ x$, in contradiction to $x \in D(X, \succ, w, p)$. Therefore, we must have $p \cdot x = w$. □

NOTES 2

NOTES 2.1 [Section 2.1 , p. 71]

[Indivisibility and divisibility of commodities, p. 72] In the context of economic theory, the terms "divisible goods" and "nondivisible goods" are typically used to describe two distinct categories of commodities. These correspond in Definition 2.2 to *perfectly divisible commodities* and *purely indivisible commodities*. While the role of perfect divisibility is often emphasized in economic theory, the role of pure indivisibility should also be recognized. A notable example of this is [112].

[Commodity space and commodity vector, p. 72, Consumption set and consumption vector, p. 73, Initial endowments and wealth, p. 77] The theoretical concepts of the *commodity space, commodity vector, consumption set, consumption vector, initial endowments* and *wealth* were established by classical general equilibrium models as in [3], [84, 86], [40], [94], and [22]. In particular, a convenient and systematic account is provided in [22, Chapter 2 and Chapter 4]. The measure-theoretic general equilibrium models of [6, 8] and [55] represent an extension of the classical models, but they do not differ from the classical models in that they consider only perfectly divisible commodities.

It should be noted that the definition in the literature postulates that the commodity space is the entire set of a finite-dimensional Euclidean space. However, in this book, the commodity space is a non-empty closed subset of this space. In essence, the assumption of perfect divisibility of commodities is not made in this book. In contrast to the classical general equilibrium models, the definitions of the commodity space, the consumption sets, and so forth are based on those presented in [127].

[Example Fig. 2.3, p. 73] An example of *mutual exclusivity in consumption* is attributed to [77, Fig. 6, p. 23].

[Preference relation, p. 76] The definition of preference relations in Definition 2.5 differs from the usual definitions, e.g., [22, pp. 54–55] and [55, p. 86], in that it does not assume their transitivity (see the arguments in (4) of Proposition 2.1, p. 85). This was inspired by [116] and [79].

[The number of distinguishable commodities, p. 71] Although the general equilibrium analysis in this book is based on a commodity space with a finite number of commodities, there are contributions based on an infinite number of commodities. Let us list some of these contributions for the reader's reference. [13] introduced an infinite-dimensional commodity space given by L_∞ with the Macky topology $\tau(L_\infty, L_1)$ for its general equilibrium analysis with a finite population of economic agents (note that given a dual pairing (V, W) of linear topological spaces, the weakest topology on V that makes $W = V^*$ is called the Macky topology on V and is denoted by $\tau(V, W)$). [80] analyzed Walras' equilibrium in an infinite economy by considering a compact space K that expresses the characteristics of commodities, and a commodity space given by the set of all measures, ca(K), defined on K is given a weak star topology. Attempts to introduce a more general infinite-dimensional commodity space than these two different types of commodity spaces can be found in [1], [39], [63], and [82], etc.

Notes 2.2 [Section 2.2, p. 78]

[Price system, p. 78] Although it is not necessary, the need to recognize the price system of a competitive market as a linear functional defined on a commodity pace, as in this definition, arises especially when dealing with infinite-dimensional commodity spaces. On this point, see for example [13].

Notes 2.3 [Section 2.3, p. 82]

[Characterization of the universal preference set, p. 85] In Proposition 2.1, (1) is due to [128] and (4) to [116]. It should be noted that (2) and (3) are standard, but *the completeness of the preference relation is not assumed*. An alternative proof of (4), which does not rely on Michael's Continuous Selection Theorem, is presented by [2].

Chapter 3
Mathematical Structures on the Space of Agents' Characteristics

Abstract The concept of demands introduced in Chapter 2 may be regarded as relations defined from the space of agents' characteristics to the commodity space. From this perspective it is natural to be concerned with the mathematical properties of the relation including their continuity and measurability properties. Correspondingly, topological as well as measure theoretical structures in the universal preference set $\mathcal{P}$ will be introduced in this chapter. It allows the space of agents' characteristics, $\mathcal{P} \times \mathbb{G}$, to be endowed with these structures. We then proceed to verify topological and measurable properties of demand relations.

3.1 The Space of Subsets of a Metric Space

Let us begin by acquainting ourselves with some topological concepts that are very convenient in describing continuity concepts in spaces composed of subsets of a given metric space which is related to the mathematical structures of the space of agents' characteristics.

Topology of Closed Convergence

Closed Limits

Definition 3.1 (Closed Limits and Closed Convergence) Given a sequence $(S_n)_n$ of subsets in a metric space T, a subset $S \subset T$ is called the *closed limit* of $(S_n)_n$ if $\mathrm{Li}\,(S_n) = S = \mathrm{Ls}\,(S_n)$. A sequence $(S_n)_n$ is said to be *closed convergent* if it has a closed limit.

Mathematical Proposition 3.1 [Closed Convergence] *A sequence $(T_n)_n$ of subsets in a separable metric space (T, ρ) has a closed convergent subsequence whose closed limit may be the empty set.*

A. Yamazaki, *Mathematical Economics of General Equilibrium*, Monographs in Mathematical Economics 8, https://doi.org/10.1007/978-981-96-9497-6_3

Proof Let $\{G_i \mid i \in \mathbb{N}_+\}$ be a countable base of open sets in the separable space T. Define subsets T_n^i for each $i, n \in \mathbb{N}_+$ using the terms of the sequence $(T_n)_n$ as follows:

$$T_n^i = \begin{cases} (a)\, T_n & \text{for } i = 1 \text{ and all } n \\ (b)\, T_{j_n}^{i-1} & \text{for } i \geqq 2 \text{ and if there is a subsequence } (T_{j_n}^{i-1})_n \text{ such that} \\ & G_i \cap \mathrm{Ls}\,(T_{j_n}^{i-1}) = \emptyset \\ (c)\, T_n^{i-1} & \text{for } i \geqq 2 \text{ and if there are no such subsequences.} \end{cases}$$

Set $S_n = T_n^n$ for every n. We shall show that the sequence $(S_n)_n$ is a closed convergent subsequence of $(T_n)_n$. Suppose $(S_n)_n$ is not closed convergent. Then, $\mathrm{Li}\,(S_n) \neq \mathrm{Ls}\,(S_n)$. Since $\mathrm{Li}\,(S_n) = \cap \mathrm{Ls}\,(S_{j_n}) \subset \cup \mathrm{Li}\,(S_{j_n}) = \mathrm{Ls}\,(S_n)$ where $\cap$ and $\cup$ are taken over all subsequences, there must be a subsequence $(S_{k_n})_n$ such that the two closed sets $\mathrm{Ls}\,(S_{k_n})$ and $\mathrm{Ls}\,(S_n)$ are not equal. It follows that there is i_0 for which $G_{i_0} \cap \mathrm{Ls}\,(S_{k_n}) = \emptyset$ and $G_{i_0} \cap \mathrm{Ls}\,(S_n) \neq \emptyset$.

Now, $(S_{k_n})_n$ is a subsequence of $(S_n)_n$ whose terms from the $(i_0 - 1)$-th term onward is a subsequence of $(T_n^{i_0-1})_n$. Moreover, $G_{i_0} \cap \mathrm{Ls}\,(S_{k_n}) = \emptyset$ implies that this is the case (b) with i replaced by i_0 in the definition of T_n^i. Since $(S_n)_n$ is a subsequence of $(T_n^{i_0})_n$ possibly except for the first i_0 terms (that are in any case part of the sequence $(T_n)_n$), it follows that $G_{i_0} \cap \mathrm{Ls}\,(S_n) = \emptyset$, a contradiction. Therefore, $(S_n)_n$ is a closed convergent subsequence of the given sequence $(T_n)_n$. □

Distance between Two Sets

Definition 3.2 (Distance from a Point to a Set and ε-Neighborhood of a Set) Let x be a point in a metric space (T, ρ) and S a non-empty subset of T. The *distance from a point x to a set S* denoted by $\mathrm{dist}\,(x, S)$ is defined by

$$\mathrm{dist}\,(x, S) = \inf\{\rho(x, y) \mid y \in S\}.$$

The *ε-neighborhood of a set S* in (T, ρ) for $\varepsilon > 0$ denoted by $N_\varepsilon(S)$ is defined by

$$N_\varepsilon(S) = \{x \subset T \mid \mathrm{dist}\,(x, S) < \varepsilon\}.$$

Definition 3.3 (Hausdorff Distance) Let E and F be two non-empty subsets of a metric space (T, ρ). The *Hausdorff distance* $\delta(E, F)$ between the two sets E and F (with respect to the metric ρ) is defined by

$$\delta(E, F) = \inf\{\varepsilon \in]0, \infty] \mid E \subset N_\varepsilon(F) \text{ and } F \subset N_\varepsilon(E)\}.$$

Examples [Hausdorff Distances]

1. Let $T = \mathbb{R}$, $E =]0, 1[$, and $F = [1, 2]$. Then, $N_\varepsilon(E) =]-\varepsilon, 1+\varepsilon[$, and $N_\varepsilon(F) =]1-\varepsilon, 2+\varepsilon[$. We have

$$\begin{aligned}\delta(E, F) &= \inf\{\varepsilon \in]0, \infty] \mid E \subset N_\varepsilon(F) \text{ and } F \subset N_\varepsilon(E)\}\\ &= \inf\{\varepsilon \in]0, \infty] \mid \varepsilon \geq 1 \text{ and } \varepsilon > 1\}\\ &= 1.\end{aligned}$$

2. Let $T = \mathbb{R}$, $E = [0, 1]$, and $F = [2, \infty]$. Then, $N_\varepsilon(E) =]-\varepsilon, 1+\varepsilon[$, and $N_\varepsilon(F) =]2-\varepsilon, \infty]$. One has

$$\begin{aligned}\delta(E, F) &= \inf\{\varepsilon \in]0, \infty] \mid E \subset N_\varepsilon(F) \text{ and } F \subset N_\varepsilon(E)\}\\ &= \inf\{\varepsilon \in]0, \infty] \mid \varepsilon > 2 \text{ and } \varepsilon = \infty\}\\ &= \infty.\end{aligned}$$

Lemma 3.1 *Let E, F be any non-empty subsets of a metric space (T, ρ). The Hausdorff distance δ between two sets satisfies the following properties:*

(1) $\delta(E, F) \geqq 0$.
(2) $\delta(E, F) = \delta(F, E)$.
(3) *If E, F are closed, $\delta(E, F) = 0$ if and only if $E = F$.*
(4) $\delta(E, F) \leqq \delta(E, G) + \delta(G, F)$ *for any non-empty set $G \subset T$.*

Proof (1) This is immediate from the non-negativity of the metric ρ.

(2) This trivially follows from the definition of the Hausdorff distance δ.

(3) $E = F$ immediately implies $\delta(E, F) = 0$. If E and F are closed, $\delta(E, F) = 0$ implies $E \subset F$ and $F \subset E$. Thus, $E = F$.

(4) Let $\bar{\rho}(E, F) = \sup\{\text{dist}\,(x, F) \mid x \in E\}$. Then, we have $\delta(E, F) = \max\{\bar{\rho}((E, F), \bar{\rho}((F, E)\}$. If $E \subset N_\varepsilon(G)$ and $G \subset N_\eta(F)$, then, it follows from the triangle inequality with respect to the metric ρ that one has $E \subset N_{\varepsilon+\eta}(F)$. (Show why.) Thus, one obtains $\bar{\rho}(E, F) \leqq \bar{\rho}(E, G) + \bar{\rho}(G, F)$. Hence,

$$\begin{aligned}\delta(E, F) &= \max\{\bar{\rho}(E, F), \bar{\rho}(F, E)\}\\ &\leqq \max\{\bar{\rho}(E, G) + \bar{\rho}(G, F), \bar{\rho}(F, G) + \bar{\rho}(G, E)\}\\ &\leqq \max\{\delta(E, G) + \delta(G, F), \delta(F, G) + \delta(G, E)\}\\ &= \delta(E, G) + \delta(G, F).\end{aligned}$$

□

Definition 3.4 (Spaces of Closed Subsets $\mathcal{F}(T)$ and $\mathcal{F}_+(T)$) Given a topological space $(T, \mathcal{T})$, $\mathcal{F}(T)$ or $\mathcal{F}$ (when no ambiguity arises) denotes the set of all closed subsets of T. The set of all non-empty closed subsets of T is denoted by $\mathcal{F}_+(T)$ or $\mathcal{F}_+$ for short. Since the topology $\mathcal{T}$ consists of all the open subsets, one has

$$\mathcal{F}(T) = \{F \subset T \mid F = T - G \text{ for } G \in \mathcal{T}\} \text{ and}$$
$$\mathcal{F}_+(T) = \{F \in \mathcal{F}(T) \mid F \neq \emptyset\}.$$

As an immediate corollary to Lemma 3.1, we have:

Mathematical Proposition 3.2 [Hausdorff Distance Function] *In a compact metric space (T, ρ) the Hausdorff distance δ as a function from $\mathcal{F}_+(T) \times \mathcal{F}_+(T)$ to $\mathbb{R}$ is a metric on the space $\mathcal{F}_+(T)$. (δ may also be called the* Hausdorff metric *on $\mathcal{F}_+(T)$.)*

Lemma 3.2 *Let (T, ρ) be a compact metric space and $F, F_n \in \mathcal{F}_+(T)$ for each $n \in \mathbb{N}_+$. Then,*

(1) *$F \subset \mathrm{Li}\,(F_n)$ if and only if $F \subset N_\varepsilon(F_n)$ for n sufficiently large for any $\varepsilon > 0$.*
(2) *$\mathrm{Ls}\,(F_n) \subset F$ if and only if $F_n \subset N_\varepsilon(F)$ for n sufficiently large for any $\varepsilon > 0$.*

Proof (1) Let $\varepsilon > 0$. For any $x \in F \subset \mathrm{Li}\,(F_n)$ there is $n_0(x)$ such that for all $n \geqq n_0(x)$, the intersection of $(\varepsilon/2)$-neighborhood of x with F_n is non-empty, i.e., $U_{\varepsilon/2}(x) \cap F_n \neq \emptyset$. So, for all $n \geqq n_0(x)$, $\mathrm{dist}\,(x, F_n) < \varepsilon/2$. Take a finite subcover $\{U_{\varepsilon/2}(x_j) | \, j = 1, \ldots, k\}$ of a cover $\{U_{\varepsilon/2}(x) | \, x \in F\}$ of the compact set F. Set $n_0 = \max\{n_0(x_j) | \, j = 1, \ldots, k\}$. Then, for any $x \in F$, there is x_j such that $x \in U_{\varepsilon/2}(x_j)$. It follows that for all $n \geqq n_0$ we have $\mathrm{dist}\,(x, F_n) \leqq \rho(x, x_j) + \mathrm{dist}\,(x_j, F_n) < \varepsilon$. Hence, for any $x \in F$ and for every $n \geqq n_0$ we have $x \in N_\varepsilon(F_n)$, showing $F \subset N_\varepsilon(F_n)$.

Conversely, let $\varepsilon > 0$ be given and assume $x \in F$. Then, $\mathrm{dist}\,(x, F_n) < \varepsilon$ for all n sufficiently large. Thus we must have $U_\varepsilon(x) \cap F_n \neq \emptyset$ for all n sufficiently large. It follows that $F \subset \mathrm{Li}\,(F_n)$.

(2) Assuming $\mathrm{Ls}\,(F_n) \subset F$, let $\varepsilon > 0$ be given. Suppose there is a subsequence $(F_{k_n})_n$ of $(F_n)_n$ and $x_{k_n} \in F_{k_n} - N_\varepsilon(F)$ for n large enough. By compactness of (T, ρ) one can assume without loss of generality that $(x_{k_n})_n$ is convergent. Let $x_{k_n} \to_n x$. Since $\mathrm{Ls}\,(F_{k_n}) \subset F$, one must have $x \in F$. It follows that $x_{k_n} \in N_\varepsilon(F)$ for n large enough, contradictory to $x_{k_n} \in F_{k_n} - N_\varepsilon(F)$. Therefore, $F_n \subset N_\varepsilon(F)$ for n sufficiently large.

Conversely, assume $F_n \subset N_\varepsilon(F)$ for n sufficiently large for any $\varepsilon > 0$, and let $x \in \mathrm{Ls}\,(F_n)$. Then, there exists a subsequence $(F_{k_n})_n$ of $(F_n)_n$ with $x_{k_n} \in F_{k_n}$ such that $x_{k_n} \to_n x$ and $\mathrm{dist}\,(x_{k_n}, F) \to_n 0$. If $x \notin F$, then $\mathrm{dist}\,(x, F) > 0$ as the set F is closed. But we have

$$\begin{aligned}
\rho(x, x_{k_n}) + \mathrm{dist}\,(x_{k_n}, F) &= \rho(x, x_{k_n}) + \inf\{\rho(x_{k_n}, z) \mid z \in F\} \\
&= \inf\{\rho(x, x_{k_n}) + \rho(x_{k_n}, z) \mid z \in F\} \\
&\geqq \inf\{\rho(x, z) \mid z \in F\} = \mathrm{dist}\,(x, F) > 0.
\end{aligned}$$

This is contrary to $\mathrm{dist}\,(x_{k_n}, F) \to_n 0$ as we have $\rho(x, x_{k_n}) \to_n 0$. Thus, $x \in F$, and hence, $\mathrm{Ls}\,(F_n) \subset F$. □

Mathematical Proposition 3.3 [Space of Non-empty Closed Sets] *For a compact metric space (T, ρ), the metric space of non-empty closed sets in T, $(\mathcal{F}_+(T), \delta)$, satisfies the following properties:*

(1) *$(\mathcal{F}_+(T), \delta)$ is a compact metric space.*

(2) *A sequence* $(F_n)_n$ *converges to* F *in* $(\mathcal{F}_+(T), \delta)$ *if and only if* $\mathrm{Li}\,(F_n) = F = \mathrm{Ls}\,(F_n)$.

(3) *Let* $\mathcal{S}$ *be a family of subsets of the space* $\mathcal{F}_+(T)$*, whose elements* $\mathcal{S}(F, G)$ *are defined by* $\mathcal{S}(F, G) = \{S \in \mathcal{F}_+(T) \mid S \subset T - F \text{ and } S \cap G \neq \emptyset\}$ *where* F *is closed and* G *open in the space* (T, ρ)*. Then,* $\mathcal{S}$ *is a subbase for the topology on the space* $\mathcal{F}_+(T)$ *induced by the Hausdorff metric* δ*.*

Proof (1) $(\mathcal{F}_+(T), \delta)$ is a metric space by Mathematical Proposition 3.2. Since a compact metric space (T, ρ) is separable, the compactness of the space $\mathcal{F}_+(T)$ follows from Mathematical Proposition 3.1 if we prove the property (2) above.

(2) If $(F_n)_n$ converges to F in $(\mathcal{F}_+(T), \delta)$, then it follows from Lemma 3.2 that we have $F \subset \mathrm{Li}\,(F_n)$ and $\mathrm{Ls}\,(F_n) \subset F$ simultaneously. Therefore, one has $\mathrm{Li}\,(F_n) = F = \mathrm{Ls}\,(F_n)$.

Conversely, let F be a closed limit of a sequence $(F_n)_n$ in $\mathcal{F}_+(T)$. The set $F = \mathrm{Ls}\,(F_n) = \mathrm{Li}\,(F_n)$ is non-empty as T is compact. Thus, it follows from Lemma 3.2 that $\delta(F_n, F) \to 0$.

(3) To verify that $\mathcal{S}$ is indeed a subbase for the topology on $\mathcal{F}_+(T)$, we first show that each member $\mathcal{S}(F, G)$ of $\mathcal{S}$ is open, and secondly, for each point $S \in \mathcal{F}_+(T)$ and a neighborhood $\mathcal{U}$ of S in $\mathcal{F}_+(T)$, there are a finite number of elements in $\mathcal{S}$ whose intersection is contained in $\mathcal{U}$.

Fix a closed set F and an open set G. Since $\mathcal{S}(F, \emptyset) = \emptyset$ and $\mathcal{S}(\emptyset, T) = \mathcal{F}_+(T)$, we show $\mathcal{S}(F, G)$ is open for $F \neq \emptyset$ and $G \neq \emptyset$. Let $S \in \mathcal{S}(F, G)$. Since $S \subset T - F$ and $S \cap G \neq \emptyset$, there are $x \in S \cap G$ and $\varepsilon > 0$ such that $U_\varepsilon(x) \subset G$ and $N_\varepsilon(S) \subset T - F$. Let $\mathcal{V}_\varepsilon(S)$ be the ε-neighborhood of S in $(\mathcal{F}_+(T), \delta)$. We want to show $\mathcal{V}_\varepsilon(S) \subset \mathcal{S}(F, G)$. If $E \in \mathcal{V}_\varepsilon(S)$, then $\delta(S, E) < \varepsilon$ so that $S \subset N_\varepsilon(E)$. This implies $E \cap U_\varepsilon(x) \neq \emptyset$, which in turn implies $E \cap G \neq \emptyset$. One also has $E \subset N_\varepsilon(S) \subset T - F$. It follows that $E \in \mathcal{S}(F, G)$. Hence, $\mathcal{V}_\varepsilon(S) \subset \mathcal{S}(F, G)$. Thus, $\mathcal{S}(F, G)$ is open.

It remains to show that given $S \in \mathcal{F}_+(T)$ and $\varepsilon > 0$, there are a finite number of elements $\mathcal{S}(F_j, G_j)$, $j = 1, \ldots, k$, in $\mathcal{S}$ such that $S \in \cap_{j=1}^k \mathcal{S}(F_j, G_j) \subset \mathcal{V}_\varepsilon(S)$. Take an open cover $\{U_{\varepsilon/2}(x)\}_{x \in E}$ of S in (T, ρ). Since S is compact, there is a finite subcover $\{U_{\varepsilon/2}(x_1), \ldots, U_{\varepsilon/2}(x_k)\}$. Set $F_j = T - N_\varepsilon(S), G_j = U_{\varepsilon/2}(x_j)$ for $j = 1, \ldots, k$. We have $S \in \cap_{j=1}^k \mathcal{S}(F_j, G_j)$. Let $E \in \cap_{j=1}^k \mathcal{S}(F_j, G_j)$; then, $E \subset N_\varepsilon(S)$, and for each $j = 1, \ldots, k$, one has $E \cap U_{\varepsilon/2}(x_j) \neq \emptyset$. Thus, $S \subset \cup_j U_{\varepsilon/2}(x_j) \subset N_\varepsilon(E)$. Hence, $E \in \mathcal{V}_\varepsilon(S)$. Therefore, we proved $S \in \cap_{j=1}^k \mathcal{S}(F, G_j) \subset \mathcal{V}_\varepsilon(S)$. □

In (2) of the above proposition the concept of the closed convergence of a sequence of subsets in a metric space is equivalently connected with that of the convergence of a corresponding sequence in the Hausdorff metric space of non-empty subsets. One may note the interplay between the topological concepts in the space (T, ρ) and those in the space $(\mathcal{F}_+(T), \delta)$ in the above proof (3).

Closed Convergence Topology

For the purpose of introducing a topological structure in the space of economic agents' characteristics we are interested in an extension of Mathematical Proposition

3.3 to spaces of *all closed sets of a metric space which may not be compact.* In this respect the following observations may be helpful:

- Since the Haudorff distances are only defined between non-empty subsets, it seems better to introduce a topology on a space of all closed sets of a metric space directly by defining a subbase as in (3) of the proposition.
- Subbasic open sets $S(F,G)$ are defined using closed sets F and open sets G. In view of the proof of (3), closed sets F need be replaced by compact sets K in the way in which a neighborhood base of a one-point compactification is defined in a non-compact space.

Definition 3.5 (Closed ConvergenceTopology) Given a locally compact topological space $(T,\mathcal{T})$, let $\mathcal{S}(K,G) = \{S \in \mathcal{F}(T) \mid S \subset T - K \text{ and } S \cap G \neq \emptyset\}$ and define a subfamily $\mathcal{S}$ of subsets of $\mathcal{F}(T)$ by

$$\mathcal{S} = \{\mathcal{S}(K,G) \subset \mathcal{F}(T) \mid K \text{ is compact, and } G \neq \emptyset \text{ is open in } T\}.$$

The *closed convergence topology* $\mathcal{T}_c$ on $\mathcal{F}(T)$ is the unique topology generated by a subbase $\mathcal{S}$.

Remark 3.1 If the topological space $(T,\mathcal{T})$ is locally compact, then topology $\mathcal{T}_c$ of closed convergence is Hausdorff, i.e., T_2. Indeed, let $S_1, S_2 \in \mathcal{F}(T)$ and $S_1 \neq S_2$. Without loss of generality, we can pick a point $x \in S_1 - S_2$. There is a compact neighborhood K_x of x such that $K_x \subset T - S_2$. Then, $\mathcal{S}(\emptyset, \text{int}\, K_x)$ is a neighborhood of S_1, and $\mathcal{S}(K_x, T)$ $(= \{S \in \mathcal{F}(T) \mid S \subset T - K_x \text{ and } S \cap T \neq \emptyset\} = \{S \mid S \subset T - K_x\})$, a neighborhood of S_2. These neighborhoods of S_1 and S_2 have an empty intersection.

Mathematical Theorem 3.1 [Closed Convergence Topology] *Let (T,ρ) be a locally compact separable metric space.*

(1) *$(\mathcal{F}(T), \mathcal{T}_c)$ is a compact metrizable space.*
(2) *A sequence $(F_n)_n$ converges to F in $(\mathcal{F}(T), \mathcal{T}_c)$ if and only if* $\text{Li}\,(F_n) = F = \text{Ls}\,(F_n)$.

Proof Let $T_{\{\infty\}}$ be the one-point compactification of the locally compact metric space (T,ρ), and set $z_\infty \in T_{\{\infty\}} - T$. Since (T,ρ) is separable, the compact space $T_{\{\infty\}}$ is metrizable. Let ρ^∞ denote a metric on $T_{\{\infty\}}$. Apply Mathematical Proposition 3.3 to the compact metric space $(T_{\{\infty\}}, \rho^\infty)$ and obtain a compact metric space $(\mathcal{F}_+(T_{\{\infty\}}), \delta^\infty)$ with the Hausdorff metric δ^∞.

Define a subset $\mathcal{F}(T)^\infty$ of $\mathcal{F}_+(T_{\{\infty\}})$ by $\mathcal{F}(T)^\infty = \{F_\infty \in \mathcal{F}_+(T_{\{\infty\}}) \mid z_\infty \in F_\infty\}$. Write $F = F_\infty - \{z_\infty\}$ for $F_\infty \in \mathcal{F}(T)^\infty$. If $F_\infty \in \mathcal{F}(T)^\infty, T^\infty - F_\infty = T - F$ is open in $(T_{\{\infty\}}, \rho^\infty)$ and does not contain z_∞. Thus, $T - F$ is open in (T,ρ) or F is closed in (T,ρ). On the other hand, for F closed in (T,ρ), the set $F_\infty = F \cup \{z_\infty\}$ is closed in $(T_{\{\infty\}}, \rho^\infty)$. It follows that the mapping $h : F \mapsto F \cup \{z_\infty\}$ is a bijection between $\mathcal{F}(T)$ and $\mathcal{F}(T)^\infty$. Since $\mathcal{F}(T)^\infty$ is a subspace of $\mathcal{F}_+(T_{\{\infty\}})$, we define a topology $\mathcal{T}_{\delta^\infty}$ on $\mathcal{F}(T)$ as the one induced by the Haudorff metric δ^∞ on $\mathcal{F}_+(T_{\{\infty\}})$, that is, $\mathcal{U} \in \mathcal{T}_{\delta^\infty}$ whenever there is an open set $\mathcal{V}$ in $\left(\mathcal{F}_+(T_{\{\infty\}}), \delta^\infty\right)$ such that $\mathcal{U} = h^{-1}(\mathcal{V})$.

We show $\mathcal{F}(T)^\infty$ is closed in $\mathcal{F}_+(T_{\{\infty\}})$. Let $F_n \in \mathcal{F}(T)^\infty$ for each n and $\delta^\infty(F_n, F_0) \to_n 0$. Then, by Mathematical Proposition 3.3, one has $\mathrm{Li}\,(F_n) = F_0 = \mathrm{Ls}\,(F_n)$, and hence, F_0 is closed in $T_{\{\infty\}}$, and $z_\infty \in F_0$ as $z_\infty \in F_n$ for all n so that $F_0 \in \mathcal{F}(T)^\infty$. Hence, $\mathcal{F}(T)^\infty$ is compact. It follows that $(\mathcal{F}(T), \mathcal{T}_{\delta^\infty})$ is a compact metrizable space.

To show $\mathcal{T}_c \subset \mathcal{T}_{\delta^\infty}$, let $\mathcal{S}(K,G) \in \mathcal{T}_c$ for K compact and G open in T. Let $\mathcal{V}(K,G)$ in $(\mathcal{F}_+(T_{\{\infty\}}), \delta^\infty)$ be defined by

$$\begin{aligned}\mathcal{V}(K,G) &= \{S \cup \{z_\infty\} \in \mathcal{F}_+(T_{\{\infty\}}) \mid S \in \mathcal{F}(T), S \cap K = \emptyset, \text{ and } S \cap G \neq \emptyset\} \\ &= \{S \cup \{z_\infty\} \in \mathcal{F}(T)^\infty \mid S \in \mathcal{S}(K,G)\}.\end{aligned}$$

Then, it follows from Mathematical Proposition 3.3 that $\mathcal{V}(K,G)$ is a subbasic open set in $(\mathcal{F}_+(T_{\{\infty\}}), \delta^\infty)$ since the complement of a compact set K in T is open in $T_{\{\infty\}}$ so that K is closed in $T_{\{\infty\}}$ and since an open set G in T is also open in $T_{\{\infty\}}$. Moreover, as one has $\mathcal{S}(K,G) = h^{-1}(\mathcal{V}(K,G))$, one obtains $\mathcal{S}(K,G) \in \mathcal{T}_{\delta^\infty}$.

Now, as $(\mathcal{F}(T), \mathcal{T}_c)$ is Hausdorff, cf. Remark 3.1, and $(\mathcal{F}(T), \mathcal{T}_{\delta^\infty})$ is compact, $\mathcal{T}_c \subset \mathcal{T}_{\delta^\infty}$ implies by Mathematical Property 1.9 (2), p. 32, that $\mathcal{T}_c = \mathcal{T}_{\delta^\infty}$ on $\mathcal{F}(T)$. Then, define a metric τ on $\mathcal{F}(T)$ by $\tau(F_1, F_2) = \delta^\infty(F_1 \cup \{\infty\}, F_2 \cup \{\infty\})$ for $F_1, F_2 \in \mathcal{F}(T)$. By the way the topology $\mathcal{T}_{\delta^\infty}$ is defined, $\mathcal{T}_c$ is the metric topology induced by τ.

If $(F_n)_n \to_n F$ in $(\mathcal{F}(T), \mathcal{T}_c)$, then the sequence $(F_n \cup \{z_\infty\})_n \to_n F \cup \{z_\infty\}$ in $(\mathcal{F}(T)^\infty, \delta^\infty)$. It thus follows from Mathematical Proposition 3.3 $\mathrm{Li}\,(F_n \cup \{z_\infty\}) = F \cup \{z_\infty\} = \mathrm{Ls}\,(F_n \cup \{z_\infty\})$. Therefore, one must have $\mathrm{Li}\,(F_n) = F = \mathrm{Ls}\,(F_n)$.

Conversely, if $\mathrm{Li}\,(F_n) = F = \mathrm{Ls}\,(F_n)$, then $\mathrm{Li}\,(F_n \cup \{z_\infty\}) = F \cup \{z_\infty\} = \mathrm{Ls}\,(F_n \cup \{z_\infty\})$. By Mathematical Proposition 3.3, $(F_n \cup \{z_\infty\})_n \to_n F \cup \{z_\infty\}$ in $(\mathcal{F}(T)^\infty, \delta^\infty)$. Therefore, $(F_n)_n$ converges to F in $(\mathcal{F}(T), \mathcal{T}_c)$. □

Remark 3.2 Two different metrics inducing the same topology on the space T may lead to different topologies on $\mathcal{F}_+(T)$. Thus, the topology on $\mathcal{F}_+(T)$ defined by the Hausdorff distance is not determined by the topology of the metric space T. (See [55, p.17] for an example.) However, they induce the same topology if the two metrics induce the same uniformity.

3.2 Topological Structure on the Universal Preference Set $\mathcal{P}$

We are now ready to introduce a topological structure on the universal preference set $\mathcal{P}$. A topological structure to be introduced on $\mathcal{P}$ will turn out to be a very reasonable one from an economic point of view.

Let us endow the family $\mathcal{F}(\mathbb{G}\times\mathbb{G})$ of all closed subsets in the product space $\mathbb{G}\times\mathbb{G}$ of the commodity spaces with the closed convergence topology $\mathcal{T}_c$. It is a compact metrizable space by Mathematical Theorem 3.1. Recall that for each $(X, \succ) \in \mathcal{P}$, we defined $\precsim = \{(x,y) \in X \times X \mid (x,y) \notin \succ\}$. Since $\precsim \in \mathcal{F}(\mathbb{G}\times\mathbb{G})$, one can now define a mapping $\pi : \mathcal{P} \to \mathcal{F}(\mathbb{G}\times\mathbb{G})$ by $(X, \succ) \mapsto \precsim$. One easily checks that π is an injection. Indeed, let $(X_1, \succ_1) \neq (X_2, \succ_2)$ in $\mathcal{P}$ and assume we had $\precsim_1 = \precsim_2$. If

$X_1 = X_2$, then it follows from $X_1 \times X_1 - \succ_1 = X_2 \times X_2 - \succ_2$ that we have $\succ_1 = \succ_2$ in contradiction to $(X_1, \succ_1) \neq (X_2, \succ_2)$. If $X_1 \neq X_2$, assume without loss of generality $X_1 - X_2 \neq \emptyset$. It follows from $X_1 \times X_1 - \succ_1 = X_2 \times X_2 - \succ_2$ that for any $z \in X_1 - X_2$ we have $(z, z) \in \succ_1$ in contradiction to irreflexivity. Thus, we must have $\precsim_1 \neq \precsim_2$ if $(X_1, \succ_1) \neq (X_2, \succ_2)$.

Define a topology $\mathscr{T}$ on $\mathcal{P}$ by $\mathscr{T} = \{\pi^{-1}(G) \mid G \in \mathscr{T}_c\}$. (That is, open sets in $\mathcal{P}$ are the inverse images of open sets in $\mathcal{F}(\mathbb{G} \times \mathbb{G})$ with respect to the mapping π.) $\mathscr{T}$ in fact defines a topology on $\mathcal{P}$. It is the weakest or the coarsest topology on $\mathcal{P}$ making the map π continuous. The topology $\mathscr{T}$ on $\mathcal{P}$ is the topology induced by the closed convergence topology on $\mathcal{F}(\mathbb{G} \times \mathbb{G})$ and will be called the *closed convergence topology* on $\mathcal{P}$; or more precisely, it is the closed convergence topology on the subspace $\{\precsim \subset \mathbb{G} \times \mathbb{G} \mid (X, \succ) \in \mathcal{P}\} \subset \mathcal{F}(\mathbb{G} \times \mathbb{G})$. The basic properties of the closed convergence topology $\mathscr{T}$ introduced on the universal preference set $\mathcal{P}$ are described in the following theorem:

Theorem 3.1 (Closed Convergence Topology $\mathscr{T}$ on the Universal Preference Space $\mathcal{P}$)

(1) *The universal preference space $\mathcal{P}$ endowed with the closed convergence topology $\mathscr{T}$ is a compact metrizable space (and hence, a Polish space).*
(2) *A sequence $(X_n, \succ_n)_n$ of preferences converges to $(X, \succ)$ in $(\mathcal{P}, \mathscr{T})$ if and only if* $\mathrm{Li}\,(\precsim_n) = \precsim = \mathrm{Ls}\,(\precsim_n)$.
(3) *The closed convergence topology $\mathscr{T}$ is the weakest (or the coarsest) Hausdorff topology on $\mathcal{P}$ having the property that the set $\{((X, \succ), x, y) \in \mathcal{P} \times \mathbb{G} \times \mathbb{G} \mid x, y \in X, x \not\succ y\}$ is closed.*

Proof (1) The space $\mathcal{F}(\mathbb{G} \times \mathbb{G})$ endowed with $\mathscr{T}_c$ is compact and metrizable by Mathematical Theorem 3.1. To prove that the topological space $(\mathcal{P}, \mathscr{T})$ is compact and metrizable, we need only to show that $\pi(\mathcal{P})$ is closed in $\mathcal{F}(\mathbb{G} \times \mathbb{G})$. Thus, consider a sequence $(\precsim_n)_n$ in $\pi(\mathcal{P})$ such that $\precsim_n \to_n \precsim$ and $\precsim_n = \pi(X_n, \succ_n)$, $(X_n, \succ_n) \in \mathcal{P}$ for each n. Put $X = \{x \in \mathbb{G} \mid (x, x) \in \precsim\}$ and $\succ = X \times X - \precsim$. We need to show that $(X, \succ) \in \mathcal{P}$ and $\pi(X, \succ) = \precsim$.

Since $\mathrm{Li}\,(\precsim_n) = \precsim = \mathrm{Ls}\,(\precsim_n)$ by Mathematical Theorem 3.1, it follows from the irreflexivity of each $(X_n, \succ_n)$ that $\mathrm{Li}\,(X_n) = X = \mathrm{Ls}\,(X_n)$. Moreover, if we put $K = \{x = (x_j)_{1,\ldots,\ell} \in \mathbb{R}^\ell \mid x_j \leqq \overline{\beta} \text{ for all } j\}$, $X_n \cap K \neq \emptyset$ for all n implies $X \cap K \neq \emptyset$. Similarly, $\underline{\underline{\beta}} \leqq X_n$ (i.e., $\underline{\beta} \leqq x$ for any $x \in X_n$) for all n implies $\underline{\beta} \leqq X$. (See p. 75 for $\underline{\beta}, \overline{\beta}$.) X is a closed set as $\precsim \in \mathcal{F}(\mathbb{G} \times \mathbb{G})$. Therefore, $X \in \mathscr{X}$. (This argument also shows that $\mathscr{X}$ with the topology of closed convergence on $\mathscr{X}$ is compact.)

Since $\precsim$ is closed in $\mathbb{G} \times \mathbb{G}$, the preference relation $(X, \succ)$ is continuous, i.e., open in $X \times X$. We shall show that $\succ$ is asymmetric. Assume we had $x \succ y$ and $y \succ x$. There is a sequence $(x_n, y_n)_n$ such that $(x_n, y_n) \in X_n \times X_n$ and $(x_n, y_n) \to_n (x, y)$. If $x_n \not\succ y_n$ infinitely often so that $(x_n, y_n)_n \in \precsim_n$ infinitely often, it would follow that $(x, y) \in \mathrm{Ls}\,(\precsim_n) = \precsim$ which leads to a contradiction $x \not\succ y$. Thus, we must have $x_n \succ y_n$ for all n large enough. Since $\succ_n$ is asymmetric, $y_n \not\succ_n x_n$ so that

$(y_n, x_n) \in \precsim_n$ for all n large enough. It follows that we have $(y, x) \in \text{Li}(\precsim_n) = \precsim$, i.e., $y \not\succ x$ in contradiction to $y \succ x$.

If $(X, \succ)$ satisfies the property (FD), then it follows from Proposition 2.1 (1) that $(X, \succ) \in \mathcal{P}$ as $\mathcal{P} = \mathcal{P}_{FD}$. So, let $p \in \mathbb{R}^\ell$ and a non-empty finite subset $Q \subset X$ be given. Assume the set $B(X, \pi_p(Q), p)$ is compact. As $X_n \to_n X$, there is a sequence $(Q_n)_n$ of non-empty finite subset of X_n such that $Q_n \to_n Q$. Letting $w_n = \pi_p(Q_n), w = \pi_p(Q)$, we have $w_n \to_n w$ as $Q_n \to_n Q$. It follows from the compactness of $B(X, w, p)$ that the sets $B(X_n, w_n, p)$ are compact for n sufficiently large. Now, $(X_n, \succ_n) \in \mathcal{P}_{FD}$ for all n. Thus, for n large enough there is $x_n^* \in B(X_n, w_n, p)$ satisfying $z \not\succ x_n^*$ for any $z \in Q_n$. Since $X_n \to_n X$ and $w_n \to_n w$, one can assume without loss of generality that $x_n^* \to_n x^* \in B(X, w, p)$. We want to show $z \not\succ x^*$ for any $z \in Q$. Suppose we had $z \succ x^*$ for some $z \in Q$. There would be a sequence $(z_n)_n$ such that $z_n \in Q_n$ for each n and $z_n \to_n z$. It then follows from the continuity of $(X, \succ)$ and $\precsim_n \to_n \precsim$ that one would have $z_n \succ_n x_n^*$ for n large enough contradicting the way x_n^* was chosen. $(X, \succ)$ thus satisfies the property (FD).

We proved $(X, \succ) \in \mathcal{P}$. Since $\pi(X, \succ) = \precsim$, we obtain $\precsim \in \pi(\mathcal{P})$. Hence, the set $\pi(\mathcal{P})$ is closed.

(2) Let $(X_n, \succ_n) \to_n (X, \succ)$ in $\mathcal{T}$. Then, given an open neighborhood $\pi^{-1}(G)$, $G \in \mathcal{T}_c$, of $(X, \succ)$, there is an integer n_0 such that for $n \geqq n_0$, $(X_n, \succ_n) \in \pi^{-1}(G)$. This is equivalent to saying that given an open neighborhood $G \in \mathcal{T}_c$ of $\pi(X, \succ)$, there is n_0 such that for $n \geqq n_0$, $\pi(X_n, \succ_n) \in G$. It means that one has $\pi(X_n, \succ_n) = \precsim_n \to_n \precsim = \pi(X, \succ)$ in $\mathcal{T}_c$. Therefore, it follows from (2) in Mathematical Theorem 3.1 that $(X_n, \succ_n) \to_n (X, \succ)$ if and only if $\text{Li}(\precsim_n) = \precsim = \text{Ls}(\precsim_n)$.

(3) Since $(\mathcal{P}, \mathcal{T})$ is a compact space, every Hausdorff coarser topology on $\mathcal{P}$ coincides with $\mathcal{T}$. Thus we only need to show that the set

$$\{((X, \succ), x, y) \in \mathcal{P} \times \mathbb{G} \times \mathbb{G} \mid x, y \in X \text{ and } x \not\succ y\}$$

is closed. Let $((X_n, \succ_n), x_n, y_n) \to_n ((X, \succ), x, y)$ with $x_n, y_n \in X_n$ and $x_n \not\succ_n y_n$. Then, as $(x_n, y_n) \in \precsim_n$ for every n, one must have $x, y \in \text{Ls}(X_n) = X$, $(x, y) \in \text{Ls}(\precsim_n) = \precsim$, that is, $x \not\succ y$. □

Remark 3.3 (Topology on $\mathcal{P}$ induced by the closed convergence topology on the space $\mathcal{X} \times \mathcal{F}(\mathbb{G} \times \mathbb{G})$) By the way topology $\mathcal{T}$ is defined on the universal preference set $\mathcal{P}$, the meaning of the convergence $(X_n, \succ_n)_n \to_n (X, \succ)$ is that one has $\precsim_n \to_n \precsim$ in the topology $\mathcal{T}_c$ of closed convergence on the space $\mathcal{F}(\mathbb{G} \times \mathbb{G})$. As is noted in (1) of the proof of Theorem 3.1, $\precsim_n \to_n \precsim$ in $\mathcal{T}_c$ implies by the irreflexivity of preferences $X_n \to_n X$ in the closed convergence topology on $\mathcal{X}$. Thus, the closed convergence topology $\mathcal{T}$ on $\mathcal{P}$ is equivalent to the one induced by the product of closed convergence topologies on the space $\mathcal{X}$ and $\mathcal{F}(\mathbb{G} \times \mathbb{G})$.

Corollary 3.1 *If a sequence $(X_n, \succ_n)_n$ of preferences converges to $(X, \succ)$ in $(\mathcal{P}, \mathcal{T})$, then* $\text{Li}(X_n) = X = \text{Ls}(X_n)$.

Corollary 3.2 (Similar Preferences)

Let $(X, \succ) \in \mathcal{P}, x, y \in X$*, and* $x \succ y$*. Then, there are neighborhoods* $\mathcal{U}, V_x, V_y$ *of* $(X, \succ)$ *in* $\mathcal{P}$*,* x *and* y *in* $\mathbb{G}$*, respectively, such that*

$$(\forall (X', \succ') \in \mathcal{U})(\forall x' \in V_x \cap X')(\forall y' \in V_y \cap X')\, x' \succ' y'.$$

Proof Since the set

$$\mathcal{P} \times \mathbb{G} \times \mathbb{G} - \{((X', \succ'), x', y') \mid x', y' \in X \text{ and } x' \not\succ y'\}$$

is open by Theorem 3.1, and $((X, \succ), x, y)$ belongs to the set, the conclusion follows.
□

In a casual expression Corollary 3.2 says that "agents having similar tastes show similar preferences over similar commodity bundles." In this sense the closed convergence topology introduced on the universal preference set $\mathcal{P}$ has a very reasonable property from an intuitive point of view (cf. the paragraph on p. 149 immediately following Theorem 4.2).

3.3 Measurable Structure on the Universal Preference Set $\mathcal{P}$

In general a standard way to introduce a measurable structure on a topological space $(T, \mathcal{T})$ is to use as a σ-algebra on T the σ-algebra generated by the topology $\mathcal{T}$, that is, the Borel σ-algebra $\mathcal{B}(T)$. Since it is the smallest σ-algebra containing the topology $\mathcal{T}$, it is natural to use $\mathcal{B}(T)$ as its measurable structure when one wishes to treat mappings beyond continuous ones. Accordingly, a natural measurable structure is introduced on the universal preference set $\mathcal{P}$ by using the topological structure on $\mathcal{P}$. It is the σ-algebra generated by the closed convergence topology in the space $(\mathcal{P}, \mathscr{T})$, that is, the Borel σ-algebra $\mathcal{B}(\mathcal{P})$.

Now, as a corollary to Theorem 3.1 we obtain the Borel measurability of pairs consisting of preference relations and pairs of consumption vectors with one preferred to another. A precise statement is given below.

Proposition 3.1 (Borel Property of Pairs of Preferences and Consumption Vectors)

The set

$$\{(X, \succ, x, y) \in \mathcal{P} \times \mathbb{G} \times \mathbb{G} \mid x, y \in X, x \succ y\}$$

is a Borel subset in $\mathcal{P} \times \mathbb{G} \times \mathbb{G}$*.*

Proof The set in question is equal to the intersection of the following two sets, S_1 and S_2:

$$S_1 = \{(X, \succ, x, y) \in \mathcal{P} \times \mathbb{G} \times \mathbb{G} \mid x \in X \text{ and } y \in X\},$$
$$S_2 = \mathcal{P} \times \mathbb{G} \times \mathbb{G} - F \text{ where } F = \{(X, \succ, x, y) \in \mathcal{P} \times \mathbb{G} \times \mathbb{G} \mid x, y \in X, x \not\succ y\}.$$

S_1 is closed. F is closed by (3) of Theorem 3.1 and hence S_2 is open. Thus, the set is a Borel set as the intersection of a closed set and an open set. □

3.4 Properties of Subsets of $\mathcal{P}$

Having introduced topological as well as measurable structures on the universal preference set $\mathcal{P}$, let us clarify some of the properties of subsets of $\mathcal{P}$ in relation to its topological and measurable structures that will become useful later in this book.

Proposition 3.2 (Properties of $\mathcal{P}_t$ and $\mathcal{P}_{ac}$)

(1) $\mathcal{P}_t$ *is compact in* $(\mathcal{P}, \mathcal{T})$.
(2) $\mathcal{P}_{ac}$ *is compact in* $(\mathcal{P}, \mathcal{T})$.

Proof (1) Let $\pi : \mathcal{P} \to (\mathcal{F}(\mathbb{G}\times\mathbb{G}), \mathcal{T}_c)$ be a mapping defined by $(X, \succ) \mapsto \precsim$ as in Section 3.2. Since $(\mathcal{F}(\mathbb{G}\times\mathbb{G}), \mathcal{T}_c)$ is a compact metrizable space by Mathematical Theorem 3.1, we shall prove that $\pi(\mathcal{P}_t)$ is closed in $(\mathcal{F}(\mathbb{G}\times\mathbb{G}), \mathcal{T}_c)$. Consider a sequence $(\precsim_n)_n$ in $\pi(\mathcal{P}_t)$ such that $\precsim_n \to_n \precsim$ and $\precsim_n = \pi(X_n, \succ_n)$ with $(X_n, \succ_n) \in \mathcal{P}_t$. Put $X = \{x \in \mathbb{G} \mid (x, x) \in \precsim\}$ and $\succ = X \times X - \precsim$. Then, we know $\precsim = \pi(X, \succ)$, and $(X, \succ) \in \mathcal{P}$ by Theorem 3.1. Thus, it remains to show that $(X, \succ)$ is transitive. Let $x \succ y$ and $y \succ z$. Assume $x \not\succ z$, that is, $(x, z) \in \precsim$. Since $\mathrm{Li}\,(\precsim_n) = \precsim$, there is a sequence $(x_n, z_n) \to_n (x, z)$ with $(x_n, z_n) \in \precsim_n$ for each n. As we have $\mathrm{Li}\,(X_n) = X$, there is a sequence $(y_n)_n \to_n y$ such that $y_n \in X_n$ for each n. It follows from Corollary 3.2 that for n large enough we have $x_n \succ_n y_n$ and $y_n \succ_n z_n$. Hence, by the transitivity of $\succ_n$ for each n, we obtain $(x_n, z_n) \notin \precsim_n$ for n large enough, a contradiction.

(2) Similarly, let us show that $\pi(\mathcal{P}_{ac})$ is closed in $\mathcal{F}(\mathbb{G}\times\mathbb{G})$. Let $(\precsim_n)_n$ be a sequence in $\pi(\mathcal{P}_{ac})$ satisfying $\precsim_n \to_n \precsim$, $\precsim_n = \pi(X_n, \succ_n)$ and $(X_n, \succ_n) \in \mathcal{P}_{ac}$ for each n. Define a preference relation $(X, \succ)$ as in (1). Since $\pi(X, \succ) = \precsim$, $\precsim \in \pi(\mathcal{P}_{ac})$ if $(X, \succ) \in \mathcal{P}_{ac}$. Let $\{x_1, \ldots, x_k\}$ be a finite subset of X such that $x_1 \succ x_2, \ldots, x_{k-1} \succ x_k$. We want to show $x_k \not\succ x_1$. Since $X = \mathrm{Li}\,(X_n)$, for each $m = 1, \ldots, k$, there is a sequence $(x_{m,n})_n$ such that $x_{m,n} \in X_n$ and $(x_{m,n})_n \to_n x_m$. By Corollary 3.2, one has $x_{m,n} \succ_n x_{m+1,n}$ for all n large enough and for each $m = 1, \ldots, k-1$. It follows from the acyclicity of $(X_n, \succ_n)$ for each n that we have $x_{k,n} \not\succ_n x_{1,n}$ for all n large enough. Therefore, one obtains $x_k \not\succ x_1$ by Theorem 3.1. □

Proposition 3.3 (Properties of $\mathcal{P}_{lns}$ and $\mathcal{P}_{co}$)

(1) $\mathcal{P}_{lns}$ *is a* G_δ *set.*
(2) $\mathcal{P}_{co}$ *is a* G_δ *set.*

Proof For any $(X, \succ) \in \mathcal{P}$ and subsets $S, T \subset X$, let $S \succ T$ to mean $x \succ y$ for any $x \in S$ and $y \in T$.

(1) For any pair of positive integers m, n define a subset $\mathcal{P}_{mn}$ of $\mathcal{P}$ by

$$\mathcal{P}_{mn} = \{(X, \succ) \in \mathcal{P} \mid (\exists x \in X)[\|x\| \leqq m, (\forall y \in U_{1/n}(x) \cap X)\, y \not\succ x]\}.$$

Then, one can check that we have $\mathcal{P} - \mathcal{P}_{lns} = \cup_{m=1}^{\infty} \cup_{n=1}^{\infty} \mathcal{P}_{mn}$. Thus, the set $\mathcal{P}_{lns}$ is G_δ if $\mathcal{P}_{mn}$ is shown to be closed for each m, n.

Let $(X_k, \succ_k)_k \to_k (X, \succ)$ be a convergent sequence in $\mathcal{P}$ such that $(X_k, \succ_k) \in \mathcal{P}_{mn}$ for all k for any fixed m, n. Then, there is a sequence $(x_k)_k$ satisfying, for each k, $\|x_k\| \leqq m$ and $y \not\succ_k x_k$ for all $y \in U_{1/n}(x_k) \cap X_k$. Extract a convergent subsequence still denoted by $(x_k)_k$ satisfying all of these conditions. Let $(x_k)_k \to_k x$. As $\mathrm{Ls}\,(X_k) = X$, $x \in X$ and $\|x\| \leqq m$. Suppose we had $y \succ x$ for some $y \in U_{1/n}(x) \cap X$. Since $\mathrm{Li}\,(X_k) = X$, there is a convergent sequence $(y_k)_k \to y$ with $y_k \in X_k$ for every k. As $\|y_k - x_k\| \leqq \|y_k - y\| + \|y - x\|$, one has $\|y_k - x_k\| < 1/n$ for all k sufficiently large. It follows from Corollary 3.2 and $y \succ x$ that $y_k \succ x_k$ for all k large enough, which is contradictory to $(X_k, \succ_k) \in \mathcal{P}_{mn}$ for all k. Therefore, it must be $(X, \succ) \in \mathcal{P}_{mn}$, showing the closedness of $\mathcal{P}_{mn}$.

(2) For any quadruplet of positive integers m, n, r, s, define a subset $\mathcal{P}_{mnrs}$ of $\mathcal{P}$ by

$$\begin{aligned}\mathcal{P}_{mnrs} = \{(X, \succ) \in \mathcal{P} \mid (\exists x, y, z \in X)[&\|x\| \leqq m, \|y\| \leqq m, \|z\| \leqq m, \|x - y\| \geqq 1/n, \\ &\|x - z\| \geqq 1/n, 1/r \leqq \alpha \leqq 1 - (1/r), \alpha y + (1 - \alpha) z \not\succ x, \\ &U_{1/s}(y) \cap X \succ U_{1/s}(x) \cap X, U_{1/s}(z) \cap X \succ U_{1/s}(x) \cap X]\}.\end{aligned}$$

Then, one can check $\mathcal{P} - \mathcal{P}_{co} = \cup_{m=1}^{\infty} \cup_{n=1}^{\infty} \cup_{r=1}^{\infty} \cup_{s=1}^{\infty} \mathcal{P}_{mnrs}$. Hence, the set $\mathcal{P}_{co}$ is G_δ if $\mathcal{P}_{mnrs}$ is closed for each m, n, r, s.

Let $(X_k, \succ_k)_k$ be a convergent sequence in $\mathcal{P}_{mnrs}$ with its limit $(X, \succ) \in \mathcal{P}$. Then, there is a sequence consisting of quadruplets of positive integers, $(x_k, y_k, z_k, \alpha_k)_k$, satisfying, for each k,

$$\begin{aligned}&\|x_k\|, \|y_k\|, \|y_k\| \leqq m, \|x_k - y_k\|, \|x_k - z_k\| \geqq 1/n, \\ &1/r \leqq \alpha_k \leqq 1 - (1/r), \alpha_k y_k + (1 - \alpha_k) z_k \not\succ x_k, \\ &U_{1/s}(y_k) \cap X_k \succ U_{1/s}(x_k) \cap X_k, U_{1/s}(z_k) \cap X_k \succ U_{1/s}(x_k) \cap X_k.\end{aligned}$$

Extract a convergent subsequence still denoted by $(x_k, y_k, z_k, \alpha_k)_k$ satisfying all of these conditions. Let $(x_k, y_k, z_k, \alpha_k)_k \to_k (x, y, z, \alpha)$. Then,

$$\|x\|, \|y\|, \|z\| \leqq m, \|x - y\|, \|x - z\| \geqq 1/n, 1/r \leqq \alpha \leqq 1 - (1/r), \alpha y + (1 - \alpha) z \not\succ x.$$

If we show $U_{1/s}(y) \cap X \succ U_{1/s}(x) \cap X, U_{1/s}(z) \cap X \succ U_{1/s}(x) \cap X$, then $(X, \succ) \in \mathcal{P}_{mnrs}$ so that $\mathcal{P}_{mnrs}$ is closed. So, suppose we fail to have $U_{1/s}(y) \cap X \succ U_{1/s}(x) \cap X$. There would exist $x' \in U_{1/s}(x) \cap X$, $y' \in U_{1/s}(y) \cap X$ such that $y' \not\succ x'$. As $(\precsim_k)_k$ closed converges to $\precsim$, there is a convergent sequence $(y'_k, x'_k)_k \to_k (y, x)$ with $(y'_k, x'_k) \in \precsim_k$ for each k. Since $\|x_k - x'_k\| \leqq \|x_k - x\| + \|x - x'_k\|$, $\|y_k - y'_k\| \leqq \|y_k - y\| + \|y - y'_k\|$, one would have $\|x_k - x'_k\|, \|y_k - y'_k\| < 1/s$ for all k sufficiently large, contradictory to $U_{1/s}(y) \cap X \succ U_{1/s}(x) \cap X$ for all k. Therefore, $U_{1/s}(y) \cap X \succ U_{1/s}(x) \cap X$. By replacing in this argument y, y_k, y', y'_k by z, z_k, z', z'_k, $U_{1/s}(z) \cap X \succ U_{1/s}(x) \cap X$ is confirmed. □

Exercise 3.1

1. Confirm $\mathcal{P} - \mathcal{P}_{lns} = \cup_{m=1}^{\infty} \cup_{n=1}^{\infty} \mathcal{P}_{mn}$ as in the above proof.
2. Confirm $\mathcal{P} - \mathcal{P}_{co} = \cup_{m=1}^{\infty} \cup_{n=1}^{\infty} \cup_{r=1}^{\infty} \cup_{s=1}^{\infty} \mathcal{P}_{mnrs}$ as in the above proof.

3.5 Further Topics on Measurable Structures

This section presents some further mathematical properties concerning measurable structures including measurable selections for correspondences and measurable relations that will play an important role in the measure theoretic approach to general equilibrium analysis with a large population. Let us begin with a topic on the existence of a measurable selection.

Measurable Selections

Measurable Everywhere Selections

Definition 3.6 (Measurable Selections) Let φ be a correspondence from a measure space $(M, \mathcal{M}, \mu)$ to a measurable space $(S, \mathcal{S})$. A measurable mapping $f : (M, \mathcal{M}) \to (S, \mathcal{S})$ is a *measurable selection* of φ if $f(x) \in \varphi(x)$, μ-almost everywhere. In the case $f(x) \in \varphi(x)$ for every $x \in M$, f is said to be a *measurable everywhere selection.*

Mathematical Proposition 3.4 [Measurable Everywhere Selections] *Let φ be a closed-valued correspondence from a measurable space $(M, \mathcal{M})$ to a Polish space T such that its inferior inverse image is measurable for every closed subset F of T: that is, $\varphi_w^{-1}(F) = \{x \in M \mid \varphi(x) \cap F \neq \emptyset\} \in \mathcal{M}$. Then, there exists a measurable everywhere selection of the correspondence φ.*

Proof For a countably infinite dense subset $\{s_j \in T \mid j \in \mathbb{N}_+\}$ of T, define a family $\{U(j,n) \subset T \mid U(j,n) = \mathrm{cl}\, U_{1/n}(s_j)$ for each $j, n \in \mathcal{N}_+\}$ of closed $(1/n)$-neighborhoods of s_j. Let $(\varphi_n)_n$ be a sequence of correspondences defined inductively as follows: set $\varphi_1 = \varphi$; then, having defined φ_n, define φ_{n+1} by $\varphi_{n+1}(x) = \varphi_n(x) \cap U(j(n,x), n)$ where $j(n,x) = \min\{j \in \mathbb{N}_+ \mid \varphi_n(x) \cap U(j,n) \neq \emptyset\}$. For every $x \in M$, the diameter of $\varphi_n(x)$ tends to zero as $n \to \infty$. Thus, it follows from the completeness of the metric space T that $\cap_{n \in \mathbb{N}_+} \varphi_n(x)$ consists of exactly one point. Define a mapping $f : M \to T$ by $f(x) = \cap_{n \in \mathbb{N}_+} \varphi_n(x)$ for each $x \in M$. Then, $f(x) \in \varphi(x)$ for all $x \in M$. Hence, f is an everywhere selection of the correspondence φ.

It remains to show that f is measurable. Let us confirm by an induction argument that for every n and for any closed set F in T, we have $\{x \in M \mid \varphi_n(x) \cap F \neq \emptyset\} \in \mathcal{M}$. Indeed, it is true for $\varphi_1 = \varphi$. Assume it is true for φ_n. Then,

$$\{x \in M \mid \varphi_{n+1}(x) \cap F \neq \emptyset\} = \{x \in M \mid \varphi_n(x) \cap U(j(n,x),n) \cap F \neq \emptyset\}$$
$$= \bigcup_{j \in \mathbb{N}_+} \left[\{x \in M \mid \varphi_n(x) \cap U(j,n) \cap F \neq \emptyset\} \cap \{x \in M \mid j(n,x) = j\} \right].$$

The first term belongs to $\mathscr{M}$ by the induction hypothesis, and the second term

$$\{x \in M \mid j(n,x) = j\} = \bigcap_{k=1}^{j-1} \left[\{x \in M \mid \varphi_n(x) \cap U(k,n) = \emptyset\} \right.$$
$$\left. \bigcap \{x \in M \mid \varphi_n(x) \cap U(j,n) \neq \emptyset\} \right]$$

also belongs to $\mathscr{M}$ by the induction hypothesis.

Since T is a complete metric space, we can verify that for every closed set F we have $f^{-1}(F) = \bigcap_{n \in \mathbb{N}_+} \{x \in M \mid \varphi_n(x) \cap F \neq \emptyset\}$, and hence $f^{-1}(F) \in \mathscr{M}$. Therefore, f is measurable (cf. Mathematical Fact 1.28 (1), p. 49). □

Remark 3.4 Let φ be a correspondence from a complete measure space $(M, \mathscr{M}, \mu)$ to a Polish space T satisfying $\varphi \in \mathscr{M} \otimes \mathcal{B}(T)$. Then, by the Projection Theorem (cf. Mathematical Fact 1.29, p. 53), the condition of the above proposition that for every closed subset F of T, $\varphi_w^{-1}(F) = \{x \in M \mid \varphi(x) \cap F \neq \emptyset\} \in \mathscr{M}$ is met.

Mathematical Proposition 3.5 [Spanning by Measurable Maps] *Let φ be a correspondence from a measurable space $(M, \mathscr{M})$ to a Polish space T satisfying the assumptions of Mathematical Proposition 3.4. Then, there is a countable family $\{f_n \mid n \in \mathbb{N}_+\}$ of measurable maps from $(M, \mathscr{M})$ to T satisfying $\varphi(x) = \mathrm{cl}\,\{f_n(x) \mid n \in \mathbb{N}_+\}$ for every $x \in M$.*

Proof Let $\{G_n \mid n \in \mathbb{N}_+\}$ be a countable base of open sets in the Polish space T. Define a correspondence φ_n from M to T for each $n \in \mathbb{N}_+$ by

$$\varphi_n(x) = \begin{cases} \varphi(x) \cap \mathrm{cl}\, G_n & \text{if } \varphi(x) \cap \mathrm{cl}\, G_n \neq \emptyset \\ \varphi(x) & \text{if } \varphi(x) \cap \mathrm{cl}\, G_n = \emptyset. \end{cases}$$

It follows from Mathematical Proposition 3.4 that there is a measurable everywhere selection f_n for φ_n for each $n \in \mathbb{N}_+$. These f_n satisfy the required property. □

Capacitability and Measurable Selections

Let φ be a correspondence from a measurable space $(M, \mathscr{M})$ to a Polish space T. If $\varphi \subset M \times T$ is a $(\mathscr{M} \times \mathcal{B}(T))$-Souslin set, then a measurable selection of φ exists. We wish to present this Measurable Selection Theorem along the line of a proof by Hildenbrand [55, pp. 54–58] using Choquet's Capacitability Theorem.

Definition 3.7 (Capacity, Capacitability) Let $\mathscr{M}$ be a family of subsets of a set M containing the empty set $\emptyset$ and closed under finite unions and intersections. An extended real-valued set function τ defined on $\mathbb{P}(M)$ is called a (Choquet) $\mathscr{M}$-*capacity* or simply, a *capacity* on M if it satisfies the conditions below:

(a) [Monotonicity] $\tau(S_1) \leqq \tau(S_2)$ for $S_1 \subset S_2, S_1, S_2 \in \mathbb{P}(M)$.
(b) [Continuity from below in $\mathbb{P}(M)$] For every increasing sequence $(S_n)_n$ of sets in $\mathbb{P}(M)$, $\tau(\cup_n S_n) = \sup\{\tau(S_n) \mid n \in \mathbb{N}_+\}$ $(= \lim_n \tau(S_n))$.
(c) [Continuity from above in $\mathscr{M}$] For every decreasing sequence $(S_n)_n$ of sets in $\mathscr{M}$, $\tau(\cap_n S_n) = \inf\{\tau(S_n) \mid n \in \mathbb{N}_+\}$ $(= \lim_n \tau(S_n))$.

Given an $\mathscr{M}$-capacity τ on M, a subset S of M is said to be *capacitable* (relative to τ) if $\tau(S) = \sup\{\tau(C) \mid C \subset S, C \in \mathscr{M}_\delta\}$.

Mathematical Theorem 3.2 [Choquet Capacitability Theorem] *Let τ be an $\mathscr{M}$-capacity on M. Every $\mathscr{M}$-Souslin set is capacitable relative to τ.*

Proof The proof is done in two steps. In the first step the following statement is proved:

[Step 1] *Every $S \in \mathscr{M}_{\sigma\delta}$ is capacitable relative to τ.*

If $\tau(S) = -\infty$, then $\tau(C) = -\infty$ for all $C \subset S$. It follows that $\sup\{\tau(C) \mid C \subset S, C \in \mathscr{M}_\delta\} = -\infty = \tau(S)$. Hence, S is capacitable. Now, let $\tau(S) > -\infty$. If the following property is proved, then S is shown to be capacitable relative to τ:

(*) *For any real number t such that $\tau(S) > t$, there is $C \in \mathscr{M}_\delta$ such that $\tau(C) \geqq t, C \subset S$.*

Indeed, suppose that it stands true but S is not capacitable. Then, there is a real number t such that $\tau(S) > t > \sup\{\tau(C) \mid C \subset S, C \in \mathscr{M}_\delta\}$. By applying the property (*) to this t, one gets a contradiction.

As we have $S \in \mathscr{M}_{\sigma\delta}$, there are a decreasing sequence $(S_n)_n$ of sets in $\mathscr{M}_\sigma$ and an increasing sequence $(S_{nm})_m$ of sets in $\mathscr{M}$ such that $S = \cap_{n\in\mathbb{N}_+} S_n$ and $S_n = \cup_{m\in\mathbb{N}_+} S_{nm}$ for each $n \in \mathbb{N}_+$. To prepare for confirming the statement (*), let us first show the following property:

There is a sequence $(C_n)_n$ of sets in $\mathscr{M}$ satisfying $C_n \subset S_n$ and $\tau(T_n) > t$ with $T_n = S \cap C_1 \cap \cdots \cap C_n$ for each n.

Since $S \subset S_1$, $\tau(S \cap S_1) = \sup\{\tau(S \cap S_{1m}) \mid m \in \mathbb{N}_+\}$ by the continuity of τ from below. Take m sufficiently large so that $\tau(S \cap S_{1m}) > t$. Set $C_1 = S_{1m}$, and define a sequence $(C_n)_n$ inductively. Assuming $C_1, C_2, \ldots, C_{n-1}$ have been defined, let us show how C_n is defined. By hypothesis, $T_{n-1} \subset S$ and $\tau(T_{n-1}) > t$. It follows from $T_{n-1} \subset S \subset S_n$ and the τ's continuity from below that we have $\tau(T_{n-1}) = \tau(T_{n-1} \cap S_n) = \sup\{\tau(T_{n-1} \cap S_{nm}) \mid m \in \mathbb{N}_+\}$. Select m large enough so that $\tau(T_{n-1} \cap S_{nm}) > t$ and set $C_n = S_{nm}$. Then, we have $\tau(T_n) = \tau(T_{n-1} \cap S_{nm}) > t$. Thus, the existence of the sequence $(C_n)_n$ is shown.

To confirm the assertion (*), take a sequence $(C_n)_n$ of sets in the previous paragraph, and define a sequence $(C'_n)_n$ and a set C by putting $C'_n = C_1 \cap \cdots \cap C_n$

and $C = \cap_n C_n = \cap_n C'_n$. $(C'_n)_n$ is a decreasing sequence in $\mathcal{M}$. Since $T_n = S \cap C_1 \cap \cdots \cap C_n = S \cap C'_n$ for each n, we have $\tau(C'_n) \geqq \tau(T_n) > t$. Thus, one obtains from the τ's continuity from above in $\mathcal{M}$ that $\tau(C) \geqq t$. One also has $C \subset S$ as $C_n \subset S_n$ for all n. Therefore, the set C satisfies the condition of (*). This completes the proof of Step 1. In the next step we prove the following conclusion of the theorem:

[Step 2] *If S is an $\mathcal{M}$-Souslin set, then it is capacitable relative to τ.*

Given an $\mathcal{M}$-Souslin set S, by the property of Souslin sets (cf. Mathematical Property 1.25, p. 51) there is a countably compact family $\mathcal{K}$ of subsets of a set F with $\emptyset \in \mathcal{K}$ such that $S = \text{pr}_M(T)$ for some set $T \in (\mathcal{M} \times \mathcal{K})_{\sigma\delta}$. Let us introduce a capacity η on $M \times F$ by way of the $\mathcal{M}$-capacity τ on M induced by the projection map $\text{pr}_M : M \times F \to M$. Let $\mathcal{G}$ be the class generated by closing the class of rectangles $\mathcal{M} \times \mathcal{K}$ under finite unions and intersections. For any subset Q of $M \times F$, define $\eta(Q) = \tau(\text{pr}_M(Q))$. Then, *$\eta$ is a $\mathcal{G}$-capacity on $M \times F$*. In fact, the monotonicity of η follows from that of τ since, for any $Q_1, Q_2 \in \mathcal{G}$ with $Q_1 \subset Q_2$, one has $\text{pr}_M(Q_1) \subset \text{pr}_M(Q_2)$. Next, we show its continuity from below in $\mathbb{P}(M \times F)$. For an increasing sequence $(Q_n)_n$ in $\mathbb{P}(M \times F)$, one has

$$\begin{aligned}\eta(\cup_n Q_n) &= \tau(pr_M(\cup_n Q_n)) = \tau(\cup_n pr_M(Q_n)) \\ &= \sup\{\tau(pr_M(Q_n)) \mid n \in \mathbb{N}_+\} = \sup\{\eta(Q_n) \mid n \in \mathbb{N}_+\}\end{aligned}$$

which shows η's continuity from below.

Let $(Q_n)_n$ be a decreasing sequence in $\mathcal{G}$. If we can confirm $pr_M(\cap_n Q_n) = \cap_n pr_M(Q_n)$, then η's continuity from above in $\mathcal{G}$, i.e., $\eta(\cap_n Q_n) = \inf\{\eta(Q_n) | n \in \mathbb{N}_+\}$, follows from τ's continuity from above in $\mathcal{M}$. Obviously $pr_M(\cap_n Q_n) \subset \cap_n pr_M(Q_n)$. So, let $x \in \cap_n pr_M(Q_n)$. Noting that $\mathcal{K}$ is countably compact, so is the class $\{pr_M^{-1}(x) \cap Q \mid Q \in \mathcal{G}\}$. Since the class $\{pr_M^{-1}(x) \cap Q_n \mid n \in \mathbb{N}_+\}$ has the finite intersection property, its intersection $\cap_n \left(pr_M^{-1}(x) \cap Q_n\right)$ is non-empty. Choosing z from its intersection, we have $pr_M(z) = x$ which shows $x \in \cap_n pr_M(Q_n)$. This completes the proof that η is a $\mathcal{G}$-capacity on $M \times F$.

Let us now proceed to complete the proof of the theorem. Since η is a $\mathcal{G}$-capacity, by applying the result of Step 1 we know the set $T \in (\mathcal{M} \times \mathcal{K})_{\sigma\delta}$ is capacitable relative to η. Hence, for any given $\varepsilon > 0$, there is a set $Q \in \mathcal{G}_\delta, Q \subset T$ such that $\eta(Q) \geqq \eta(T) - \varepsilon$. Let $(Q_n)_n$ be a decreasing sequence of sets in $\mathcal{G}$ such that $Q = \cap_n Q_n$. Setting $C = \text{pr}_M(Q)$ and $C_n = \text{pr}_M(Q_n)$, we obtain $C = \cap_n C_n$ as $\cap_n pr_M(Q_n) = pr_M(\cap_n Q_n)$. Thus, $C \in \mathcal{M}_\delta$. Since $S = \text{pr}_M(T)$, one also has $C \subset S$ and $\tau(C) \geqq \tau(S) - \varepsilon$. Therefore, $\tau(S) = \sup\{\tau(C) \mid C \subset S,\ C \in \mathcal{M}_\delta\}$ which proves that S is capacitable. □

Mathematical Theorem 3.3 [Measurable Selection Theorem] *Let φ be a correspondence from a finite measure space $(M, \mathcal{M}, \mu)$ to a G_δ subset T of $\mathbb{R}^m$ such that $\varphi \subset M \times T$ is $(\mathcal{M} \times \mathcal{B}(T))$-Souslin set. Then, there is a measurable selection $f : M \to T$ of the correspondence φ.*

Proof Given a correspondence φ satisfying the conditions of the theorem, we first confirm the following statement:

(*) *There exists a compact-valued correspondence* ψ *from* M *to* T *such that* $\psi(x) \subset \varphi(x)$, *a.e.* $x \in M$.

Let $G = M \times T$ and $\mathcal{K}$ be the set of all compact subsets of T. Let $\mathcal{G}$ be the class generated by closing the class of rectangles $\mathcal{M} \times \mathcal{K}$ under finite unions and countable intersections. Then, we have $\mathcal{G} \subset \mathcal{M} \otimes \mathcal{B}(T) \subset \mathcal{S}(\mathcal{M} \times \mathcal{B}(T)) = \mathcal{S}(\mathcal{M} \times \mathcal{K})$.

Now, define an extended real-valued set function τ on G by $Q \mapsto \tau(Q) = \mu^*(\text{pr}_M(Q))$ where μ^* is the outer measure induced by the measure μ on $\mathcal{M}$ and $\text{pr}_M : G \to M$ the projection map. We shall show that τ is a $\mathcal{G}$-capacity. The monotonicity of τ follows from that of the outer measure μ^*. To show the continuity from below, let $(Q_n)_n$ be an increasing sequence of subsets of G. Then, it follows from the continuity from below of μ^* that

$$\begin{aligned}\sup\{\tau(Q_n) \mid n \in \mathbb{N}_+\} &= \sup\{\mu^*(\text{pr}_M(Q_n)) \mid n \in \mathbb{N}_+\} \\ &= \mu^*(\cup_n \text{pr}_M(Q_n)) = \mu^*(\text{pr}_M(\cup_n Q_n)) \\ &= \tau(\cup_n Q_n),\end{aligned}$$

which shows that τ is continuous from below. We show next its continuity from above in $\mathcal{G}$. Let $(Q_n)_n$ be a decreasing sequence in $\mathcal{G}$. Just as in the proof of Mathematical Theorem 3.2, one has $\cap_n pr_M(Q_n) = pr_M(\cap_n Q_n)$. It follows from the Projection Theorem (cf. Mathematical Fact 1.29, p. 53) that $\text{pr}_M(S) \in \mathcal{M}_\mu$ for every $S \in \mathcal{G}$. Let $\bar{\mu}$ be the completion of μ. Then, by the properties related to the completion $\bar{\mu}$ (cf. Mathematical Fact 1.26, p. 48, and Mathematical Fact 1.25, p. 47), one obtains

$$\begin{aligned}\inf\{\tau(Q_n) \mid n \in \mathbb{N}_+\} &= \inf\{\mu^*(\text{pr}_M(Q_n)) \mid n \in \mathbb{N}_+\} = \inf\{\bar{\mu}(\text{pr}_M(Q_n)) \mid n \in \mathbb{N}_+\} \\ &= \bar{\mu}\left(\cap_n \text{pr}_M(Q_n)\right) = \bar{\mu}\left(\text{pr}_M(\cap_n Q_n)\right) = \mu^*\left(\text{pr}_M(\cap_n Q_n)\right) \\ &= \tau(\cap_n Q_n),\end{aligned}$$

showing the continuity from above in $\mathcal{G}$. Hence, τ is a $\mathcal{G}$-capacity.

In order to confirm the validity of the statement (*), define a relation $\varphi_n \subset \varphi$ inductively for each $n \in \mathbb{N}_+$ by applying the Choquet Capacitability Theorem repeatedly. Let $(\varepsilon_n)_n$ be a sequence of real numbers satisfying $0 < \varepsilon_n < \tau(\varphi)$ and $\varepsilon_n \to_n 0$. Since $\varphi \in \mathcal{S}(\mathcal{M} \times \mathcal{B}(T))$, φ is capacitable. Therefore, there is a set $\varphi_1 \in \mathcal{G}, \varphi_1 \subset \varphi$, such that $\tau(\varphi_1) = \bar{\mu}(\text{pr}_M(\varphi_1)) \geqq \tau(\varphi) - \varepsilon_1$. This defines φ_1. Assuming $\varphi_1, \ldots, \varphi_{n-1}$ have been defined, let us proceed to define φ_n. Set $G_{n-1} = [(M - \text{pr}_M(\cup_{j=1}^{n-1} \varphi_j)) \times T] \cap \varphi$. Then, $G_{n-1} \in \mathcal{S}(\mathcal{M} \times \mathcal{B}(T))$. Hence, there is a set $\varphi_n \in \mathcal{G}$ such that $\varphi_n \subset G_{n-1}$ and $\tau(\varphi_n) \geqq \tau(G_{n-1}) - \varepsilon_n$. As $\tau(G_{n-1}) = \bar{\mu}(M - \text{pr}_M(\cup_{j=1}^{n-1} \varphi_j)) = \bar{\mu}(M) - \sum_{j=1}^{n-1} \bar{\mu}(\text{pr}_M(\varphi_j)) = \tau(\varphi) - \sum_{j=1}^{n-1} \tau(\varphi_j)$, one obtains $\tau(\cup_{j=1}^{n} \varphi_j) \geq \tau(\varphi) - \varepsilon_n$. It thus follows $\tau(\cup_n \varphi_n) = \tau(\varphi)$, which in turn implies $\bar{\mu}(\text{pr}_M(\cup_n \varphi_n)) = \mu(M)$. Consequently, one has $M - \text{pr}_M(\cup_n \varphi_n) \subset N$ for some μ-negligible set $N \subset M$. Let $K \subset T$ be a non-empty compact set, and define a correspondence ψ from M to T by $\psi = (\cup_n \varphi_n) \cup (N \times K)$. The measurability of ψ follows from $\varphi_n \in \mathcal{G} \subset \mathcal{M} \otimes \mathcal{B}(T)$ for each n. By the way ψ is defined, $\psi(x) \subset \varphi(x)$ for any $x \in M - N$. Since $\{\text{pr}_M(\varphi_n) \mid n \in \mathbb{N}_+\}$ is a disjoint family of sets, for each $x \in M - N$ there is a unique n for which $x \in \text{pr}_M(\varphi_n)$. Thus, $\psi(x) = \varphi_n(x) \in \mathcal{K}$ as

$\varphi_n \in \mathcal{G}$. For $x \in N$, one has either $\psi(x) = K$ or $\psi(x) = K \cup \varphi_n(x)$. It means $\psi(x)$ is everywhere compact-valued. This completes the proof of the statement (*).

Finally, to complete the proof of the theorem, let ψ be the correspondence as in (*) for the given correspondence φ. For every $x \in M$, $\psi(x) \subset T \subset \mathbb{R}^m$ is compact, and hence, closed. Thus, the measurable correspondence ψ satisfies the conditions of Mathematical Proposition 3.4. Therefore, there exists a measurable selection f of the correspondence ψ such that $f(x) \in \varphi(x)$, a.e. $x \in M$. □

Remark 3.5 The assertion of Mathematical Theorem 3.3 can be proved under more general conditions. Let φ be a correspondence from a measure space M to a Polish space T. (Here, one can take T to be a G_δ set of a Polish space S because, then, T itself becomes a Polish space. This means that in the above theorem one can replace $\mathbb{R}^m$ by a Polish space S.) If the space T is a countable union of compact sets, then the above proof goes through as it stands since we have $\mathcal{S}(\mathcal{M} \times \mathcal{K}) = \mathcal{S}(\mathcal{M} \times \mathcal{B}(T))$ (cf. Sion and Bressler [114, 5.2.3, p. 215]).

In the case of a general Polish space T, by Mathematical Property 1.13, p. 34, the space T is homeomorphic to a G_δ-subset F of the Hilbert cube $\mathbb{I}^\infty$. Let $\mathcal{K}_\infty$ be the class of all compact subsets of $\mathbb{I}^\infty$. By Tychonov's Theorem, $\mathbb{I}^\infty$ is a compact metric space. Identify T with F. Then, if a correspondence $\varphi : M \to T$ is a $(\mathcal{M} \times \mathcal{B}(T))$-Souslin set, $\varphi : M \to T \subset \mathbb{I}^\infty$ becomes a $(\mathcal{M} \times \mathcal{K}_\infty)$-Souslin set. Thus, the proof of Mathematical Theorem 3.3 also applies to this case.

Measurable Relations

Generally speaking, most textbooks on measure theory or probability theory seem to restrict their coverage of the topic of measurability to measurable functions and mappings, and do not present materials concerning measurable relations. Yet, in applied mathematics including economic theory and game theory, correspondences along with functions come to play a basic role in its analysis. In addition to the measurable selection theorem presented so far, we shall present some mathematical facts and propositions on measurable relations that will be used later in the theoretical analysis presented in this book.

Definition 3.8 (Measurable Relations) Let $(M_i, \mathcal{M}_i), i = 1, 2$, be measurable spaces and φ a relation from M_1 to M_2. The relation $\varphi \subset M_1 \times M_2$ is called *measurable* if $\varphi \in \mathcal{M}_1 \otimes \mathcal{M}_2$. Then, it is a *measurable relation* from M_1 to M_2. In particular, it is called a *measurable correspondence* if φ is a correspondence.

Mathematical Proposition 3.6 [Sup-Measurability] *Let φ be a measurable correspondence from a complete measure space $(M, \mathcal{M}, \mu)$ to a Polish space T, and f a measurable function from T to $\mathbb{R}$. Then, the function* $\sup f \circ \varphi$ *from M to $\mathbb{R}$, $x \mapsto \sup\{f(t) \mid t \in \varphi(x)\}$, is measurable, and the relation φ_f from M to T defined by $x \mapsto \{t \in \varphi(x) \mid f(t) = \sup f \circ \varphi(x)\}$ is measurable.*

Proof If, for every $\alpha \in \mathbb{R}$, $M_\alpha = \big\{x \in M \mid \sup\{f(t) \mid t \in \varphi(x)\} > \alpha\big\} \in \mathcal{M}$, then the function $\sup f \circ \varphi$ is measurable. Since $M_\alpha = \mathrm{pr}_M(\{(x,t) \in \varphi \mid f(t) > \alpha\})$,

$\{(x,t) \in \varphi \mid f(t) > \alpha\} \in \mathcal{M} \otimes \mathcal{B}(T)$ by the measurability of φ and f. Hence, it follows from the Projection Theorem that one has $M_\alpha \in \mathcal{M}$. Furthermore, the function $(x,t) \mapsto f(t) - \sup f \circ \varphi(x)$ is $\mathcal{M} \otimes \mathcal{B}(T)$-measurable (cf. (3) and (4) of Mathematical Fact 1.28, p. 49). Hence, one obtains

$$V = \{(x,t) \in M \times T \mid f(t) = \sup f \circ \varphi(x)\} \in \mathcal{M} \otimes \mathcal{B}(T).$$

Therefore, we must have $\varphi_f = \varphi \cap V \in \mathcal{M} \otimes \mathcal{B}(T)$. □

Mathematical Proposition 3.7 [Measurability of Convex Hull of Closed-Valued Correspondences] *Let φ be a measurable correspondence from a complete measure space $(M, \mathcal{M}, \mu)$ to $\mathbb{R}^m$ such that $\varphi(x)$ is closed for every $x \in M$. Then, the induced convex hull correspondence* $\operatorname{co}\varphi : M \to \mathbb{R}^m$, *defined by $x \mapsto \operatorname{co}\varphi(x)$, is measurable.*

Proof Put $\varphi_k(x) = \{u \in \varphi(x) \mid \|x\| \leqq k\}$ for each $k \in \mathbb{N}_+$. Then, $\operatorname{co}\varphi(x) = \cup_{k\in\mathbb{N}_+} \operatorname{co}\varphi_k(x)$. For each $q \in \mathbb{R}^m$, define a correspondence H_q from M to $\mathbb{R}^m$ by $x \mapsto H_q(x) = \{z \in \mathbb{R}^m \mid q \cdot z \leqq \sup q \cdot \varphi_k(x)\}$ where $\sup q \cdot \varphi_k(x) = \sup\{q \cdot y \mid y \in \varphi_k(x)\}$. By Mathematical Proposition 3.6, the map $x \mapsto \sup q \cdot \varphi_k(x)$ is measurable. If follows that the correspondence H_q is measurable. Let W be a countable dense subset of $\mathbb{R}^m$. Then, we have $\operatorname{co}\varphi_k(x) = \cap_{q\in W} H_q(x)$ (cf. Mathematical Property 1.18, p. 42). This implies that the correspondence $\operatorname{co}\varphi_k$ is measurable for each k. Consequently, the correspondence $\operatorname{co}\varphi$ is measurable. □

Mathematical Proposition 3.8 [Inferior Measurability] *Let φ be a correspondence from a measure space $(M, \mathcal{M}, \mu)$ to a Polish space (T, ρ).*

(1) *If $\varphi \in \mathcal{M} \otimes \mathcal{B}(T)$, then $\varphi_w^{-1}(S) = \{x \in M \mid \varphi(x) \cap S \neq \emptyset\} \in \mathcal{M}_\mu$ for any $S \in \mathcal{B}(T)$.*
(2) *Suppose $\varphi_w^{-1}(G) = \{x \in M \mid \varphi(x) \cap G \neq \emptyset\} \in \mathcal{M}$ for every open set $G \subset T$, then the correspondence* $\operatorname{cl}\varphi$ *from M to T, defined by $x \mapsto \operatorname{cl}\varphi(x)$, is measurable.*

Proof (1) For any $S \in \mathcal{B}(T)$, we have

$$\varphi_w^{-1}(S) = \{x \in M \mid \varphi(x) \cap S \neq \emptyset\} = \operatorname{pr}_M(\varphi \cap (M \times S)) \in \mathcal{M}_\mu$$

by the Projection Theorem.

(2) Take a countable dense subset W of T. One can check that

$$\operatorname{cl}\varphi = \cap_{n\in\mathbb{N}_+} \cup_{t\in W} \left(\{x \in M \mid \operatorname{dist}(t, \varphi(x)) < 1/n\} \times \{s \in T \mid \rho(s,t) < 1/n\}\right).$$

As $\{s \in T \mid \rho(s,t) < 1/n\} \in \mathcal{B}(T)$, $\operatorname{cl}\varphi \in \mathcal{M} \otimes \mathcal{B}(T)$ follows from

$$\begin{aligned}\{x \in M \mid \operatorname{dist}(t, \varphi(x)) < 1/n\} \\ = \left\{x \in M \mid \varphi(x) \cap \{s \in T \mid \rho(s,t) < 1/n\} \neq \emptyset\right\} \in \mathcal{M}.\end{aligned}$$

□

Mathematical Proposition 3.9 [Composite Measurability] *Let φ be a measurable correspondence from a measure space $(M, \mathcal{M}, \mu)$ to $\mathbb{R}^m$.*

(1) *If $f : M \to \mathbb{R}^m$ is a measurable function, the correspondence $\varphi + f$ defined by $x \mapsto \varphi(x) + f(x)$ is measurable.*
(2) *If g is a measurable mapping from a measurable space $(S, \mathcal{S})$ to M, the composite correspondence $\varphi \circ g$ from M to $\mathbb{R}^m$, given by $x \mapsto \varphi(g(x))$, is measurable.*

Proof (1) Define a mapping $h : M \times \mathbb{R}^m \to M \times \mathbb{R}^m$ by $(x, y) \mapsto (x, y - f(x))$. h is measurable (cf. Mathematical Fact 1.28 (3) and (4), p. 49). Therefore, $\varphi + f = h^{-1}(\varphi) \in \mathcal{M} \otimes \mathcal{B}(\mathbb{R}^m)$.
(2) Define a measurable mapping $h : S \times \mathbb{R}^m \to M \times \mathbb{R}^m$ by $(x, y) \mapsto (g(x), y)$. Then, $\varphi \circ g = h^{-1}(\varphi) \in \mathcal{S} \otimes \mathcal{B}(\mathbb{R}^m)$. □

Mathematical Proposition 3.10 [Composite Souslin] *Let φ be a measurable correspondence from a measurable space $(M, \mathcal{M})$ to a Polish space S and g be a measurable mapping from S to a topological space T. Then, the correspondence $g \circ \varphi : x \to g(\varphi(x))$ is a $(\mathcal{M} \times \mathcal{B}(T))$-Souslin set.*

Proof Since both the correspondence φ and the mapping g are measurable, the set $G = \{(x, y, z) \in M \times S \times T \mid y \in \varphi(x), z = g(y)\}$ belongs to $\mathcal{M} \otimes \mathcal{B}(S) \otimes \mathcal{B}(T)$. Let $\mathrm{pr}_{M \times T} : M \times S \times T \to M \times T$ be the projection map. Then, we have $\mathrm{pr}_{M \times T}(G) = \{(x, z) \in M \times T \mid z = g(y) \text{ for some } y \in \varphi(x)\} = g \circ \varphi \subset M \times T$. It then follows from Mathematical Property 1.28, p. 52, that $g \circ \varphi$ is a $(\mathcal{M} \times \mathcal{B}(T))$-Souslin set. □

Mathematical Proposition 3.11 [Souslin Convex Hull Correspondences] *Let φ be a measurable correspondence from a measure space $(M, \mathcal{M}, \mu)$ to $\mathbb{R}^m$. Then, the convex hull correspondence $\operatorname{co} \varphi : x \to \operatorname{co} \varphi(x)$ is a $(\mathcal{M} \times \mathcal{B}(\mathbb{R}^m))$-Souslin set.*

Proof Define a subset S of $\mathbb{R}^{m+1}$ by $S = \left\{(t_1, \ldots, t_{m+1}) \in \mathbb{R}_+^{m+1} \mid \sum_{i=1}^{m+1} t_i = 1\right\}$ and a correspondence $\psi : M \to \mathbb{R}^{m(m+1)} \times S$ by $x \mapsto \varphi(x) \times \cdots \times \varphi(x) \times S$. Since the correspondence φ is measurable, ψ is measurable. Let $g : \mathbb{R}^{m(m+1)} \times S \to \mathbb{R}^m$ be the mapping given by $(z_1, \ldots, z_{m+1}, (t_1, \ldots, t_{m+1}))) \mapsto \sum_{j=1}^{m+1} t_i z_i \in \mathbb{R}^m$. g is continuous. Since we have $\operatorname{co} \varphi = g \circ \psi$ by Carathéodory's Theorem, p. 41, Mathematical Proposition 3.10 implies that $\operatorname{co} \varphi$ is a $(\mathcal{M} \times \mathcal{B}(\mathbb{R}^m))$-Souslin set. □

3.6 Purely Non-Atomic Vector Measures and Liapounoff Convexity

A well-known theorem of Liapounoff [74] on the convexity of the range of a purely non-atomic vector measure is one of the key elements in the measure theoretic approach to economic theory. In this section a characterization of purely non-atomic bounded m-dimensional vector measures is presented in relation to the convexity of the range of a purely non-atomic vector measure.

Purely Non-Atomic Vector Measures

Non-Atomic and Atomless Measures

Let μ be a (signed) measure on a measurable space $(M, \mathscr{M})$. An *atom* of the measure μ is a measurable set S which satisfies $\mu(S) \neq 0$ and either $\mu(C) = \mu(S)$ or $\mu(C) = 0$ for every measurable subset $C \subset S$. The measure μ as well as the space $(M, \mathscr{M}, \mu)$ is called *non-atomic* if μ has no atoms. There is an alternative way to express the non-atomicity of a (signed) measure. μ is called *atomless* if, for any measurable set S, there is a measurable subset C of S such that $\mu(C) = t\mu(S)$ for some $0 < t < 1$. When the measure μ is non-atomic, μ and the space $(M, \mathscr{M}, \mu)$ are often called *atomless*.

Here are some of the well-known facts about atoms and an atomless measure space.

Mathematical Fact 3.1 [Atoms and Atomlessness] *Let μ be a (signed) measure on a measurable space $(M, \mathscr{M})$.*

(1) *μ is non-atomic if and only if it is atomless.*
(2) *If S_1 and S_2 are atoms, then either $\mu(S_1 \cap S_2) = 0$ or $\mu(S_1 \Delta S_2) = 0$.*
(3) *A Polish space always has an atomless measure if it is uncountable.*
(4) *If μ is a finite positive measure, then there are at most countably many disjoint atoms. Thus, a positive measure space can be decomposed into a countable union of atoms and a non-atomic set.*
(5) *If M is a separable metric space, μ is atomless if and only if $\mu(x) = 0$ for all $x \in M$.*

Characterizing Properties of Purely Non-Atomic Vector Measures

Now, let $\mu = (\mu_1, \ldots, \mu_m)$ be an $\mathbb{R}^m$-valued vector measure on a measurable space $(M, \mathscr{M})$. $\mu = (\mu_1, \ldots, \mu_m)$ is called *purely non-atomic* if none of its coordinates $\mu_i, i = 1, \ldots, m$, has any atoms. Four properties of a vector measure $\mu = (\mu_1, \ldots, \mu_m)$ on $(M, \mathscr{M})$ are given below, each of which is shown to be equivalent to the purely non-atomic property of μ.

The first one is well-known and very often used as the defining property of a non-atomic numerical measure.

(1) [Atomlessness] A vector measure $\mu = (\mu_1, \ldots, \mu_m)$ is *atomless* if for any measurable set S, there is a measurable subset C of S such that $\mu_i(C) = t\mu_i(S), i = 1, \ldots, m$, for some $0 < t < 1$.

For a measurable set $S \in \mathscr{M}$, let $\mathscr{M}_S$ denote the subfamily of $\mathscr{M}$ given by $\mathscr{M}_S = \{C \in \mathscr{M} | \; C \subset S\}$. Let I denote the closed unit interval $[0, 1]$ of the real line. The second property is the following:

(2) [Chain Convexity] Given a measurable set S, there is a maximal chain C_S in $\mathscr{M}_S$ linearly ordered by inclusion with $\emptyset, S \in C_S$ such that for each $t \in I$ there is a measurable set $C_t \in C_S$ satisfying $\mu_i(C_t) = t\mu_i(S), i = 1, \ldots, m$.

The third property is the following:

(3) [Functional Convexity] Given a measurable set S, there is a real-valued measurable function $\varphi_S : S \longrightarrow I$ satisfying, for each $t \in I$,

$$\mu_i(\{x \in S | \varphi_S(x) < t\}) = t\mu_i(S), \; i = 1, \dots, m.$$

The last property reflects the Liapounoff Theorem on the range of a vector measure.

(4) [Liapounoff Convexity] The range of a vector measure $\mu = (\mu_1, \dots, \mu_m)$,

$$\mu(\mathscr{M}) = \{v \in \mathbb{R}^m | v = \mu(C) \text{ for some } C \in \mathscr{M}\},$$

is a convex set in $\mathbb{R}^m$, i.e., for any $v, w \in \mu(\mathscr{M})$ and for each $t \in I$, $(1-t)v + tw \in \mu(\mathscr{M}) \subset \mathbb{R}^m$.

Each of these four properties will be shown to characterize bounded purely non-atomic vector measures.

Mathematical Proposition 3.12 [Characterization of Purely Non-Atomic Vector Measures] *Let $\mu = (\mu_1, \dots, \mu_m)$ be a bounded $\mathbb{R}^m$-valued vector measure on a measurable space $(M, \mathscr{M})$. Then, μ is purely non-atomic if and only if it satisfies one of the four properties: Atomlessness, Chain Convexity, Functional Convexity, or Liapounoff Convexity.*

Characterization of Purely Non-Atomic Vector Measures

Let $\mu = (\mu_1, \dots, \mu_m)$ be a bounded $\mathbb{R}^m$-valued vector measure on a measurable space $(M, \mathscr{M})$. Since the atomlessness of μ immediately implies that it is purely non-atomic, a proof of Mathematical Proposition 3.12 is given in the following four steps, each explained below:

1. The atomlessness is equivalent to the chain convexity.
2. The chain convexity implies the functional convexity.
3. A purely non-atomic μ satisfies the functional convexity.
4. The functional convexity implies the Liapounoff convexity which, in turn, implies the atomlessness.

1. *Equivalence between Atomlessness and Chain Convexity*

The chain convexity of $\mu = (\mu_1, \dots, \mu_m)$ immediately implies that μ is atomless. So, assuming that μ is atomless, we show that it satisfies the chain convexity.

Given a measurable set S, let a subset $\mathscr{M}_S(\mu)$ of $\mathscr{M}_S \times I$ be defined by

$$\mathscr{M}_S(\mu) = \{(C, t) \in \mathscr{M}_S \times I | \; \mu_i(C) = t\mu_i(S), i = 1, \dots, m\}.$$

$\mathscr{M}_S(\mu) \neq \emptyset$ as $(\emptyset, 0), (S, 1) \in \mathscr{M}_S(\mu)$. Define a subfamily $\mathscr{N}_S$ of $\mathscr{M}_S$ by

$$\mathscr{N}_S = \{C \in \mathscr{M}_S \mid (C,t) \in \mathscr{M}_S(\mu) \text{ for some } t \in I\},$$

and partially order $\mathscr{N}_S$ by inclusion. By the Hausdorff Maximal Principle (cf. p. 18 and Mathematical Fact 1.4, p. 19) there is a maximal chain $\mathscr{C}_S$ in $\mathscr{N}_S$ containing $\{\emptyset, S\}$. Write $I_S = \{t \in I \mid (C,t) \in \mathscr{M}_S(\mu) \text{ for some } C \in \mathscr{C}_S\}$. Obviously, $0, 1 \in I_S \cap I$. If we show, in fact, that $I_S = I$, then, μ satisfies the chain convexity. Assume to the contrary that there is $\underline{t} \in I - I_S$. Trivially, $\underline{t} \neq 0$ or 1. Let

$$t^- = \sup\{t \in I \mid t \in I_S \cap [0, \underline{t})\} \leqq \underline{t},$$
$$t^+ = \inf\{t \in I \mid t \in I_S \cap (\underline{t}, 1]\} \geqq \underline{t}.$$

There are an increasing sequence $\{C_n^-\}_n$ and a decreasing sequence $\{C_n^+\}_n$ of sets in $\mathscr{C}_S$ such that $t_n^- \to_n t^-$ and $t_n^+ \to_n t^+$ for $t_n^- = \mu_i(C_n^-)/\mu_i(S), t_n^+ = \mu_i(C_n^+)/\mu_i(S), i = 1, \ldots, m$. Setting $C^- = \lim_n C_n^-$ and $C^+ = \lim_n C_n^+$, we have $C^-, C^+ \in \mathscr{M}_S$ and, for $i = 1, \ldots, m$,

$$\mu_i(C^-) = \mu_i(\lim_n C_n^-) = \lim_n t_n^- \mu_i(S) = t^- \mu_i(S),$$
$$\mu_i(C^+) = \mu_i(\lim_n C_n^+) = \lim_n t_n^+ \mu_i(S) = t^+ \mu_i(S),$$
$$t^- \leqq \underline{t} \leqq t^+.$$

If $t^- = \underline{t}$ or $t^+ = \underline{t}$, we would have $\mu_i(C^-) = \underline{t}\mu_i(S)$ or $\mu_i(C^+) = \underline{t}\mu_i(S), i = 1, \ldots, m$, so that $\underline{t} \in I_S$. Hence, it must be that $t^- < \underline{t} < t^+$. It then follows from the atomlessness of μ that there is a measurable subset $C_\Delta \subset C^+ - C^-, C_\Delta \in \mathscr{M}_S$ such that $\mu_i(C_\Delta) = t_\Delta \mu_i(C^+ - C^-), i = 1, \ldots, m$, for some $0 < t_\Delta < 1$. Letting $\overline{C} = C^- \cup C_\Delta$, $\overline{C} \in \mathscr{M}_S$, and $\bar{t} = t^- + t_\Delta(t^+ - t^-)$, we have $\overline{C} \subset S, t^- < \bar{t} < t^+$, and

$$\begin{aligned}\mu_i(\overline{C}) &= \mu_i(C^-) + \mu_i(C_\Delta)\\ &= [t^- + t_\Delta(t^+ - t^-)]\mu_i(S)\\ &= \bar{t}\mu_i(S), i = 1, \ldots, m.\end{aligned}$$

It follows that $(\overline{C}, \bar{t}) \in \mathscr{M}_S(\mu)$ and $\overline{C} \in \mathscr{N}_S$. Thus, a chain $\mathscr{C}_S \cup \{C^- \cup C_\Delta\}$ properly extends the maximal chain $\mathscr{C}_S$ in $\mathscr{N}_S$, leading to a contradiction. Therefore, I_S must be equal to I, showing the chain convexity of μ.

2. *Chain Convexity Implying Functional Convexity*

Assuming the chain convexity of μ, we show that it satisfies the functional convexity. Let $S \in \mathscr{M}$ be given. Then, there is a maximal chain $\mathscr{C}_S$ in $\mathscr{M}_S$ linearly ordered by inclusion with $\emptyset, S \in \mathscr{C}_S$ such that for each $t \in I$ there is a measurable set $C_t \in \mathscr{C}_S$ satisfying $\mu_i(C_t) = t\mu_i(S), i = 1, \ldots, m$, and $C_{t'} \subset C_{t''}$ for $t' \leqq t''$.

Now, define a measurable function $\varphi_S : S \to I$ and measurable sets S_t^φ by setting for each $x \in S$ and $t \in I$

$$\varphi_S(x) = \inf\{t' \in I|\ x \in C_{t'}, C_{t'} \in \mathscr{C}_S\}, \text{ and}$$
$$S_t^{\varphi} = \{x \in S|\ \varphi_S(x) < t\}.$$

For $t \in I$, if $t = 0$, $\mu_i(S_t^{\varphi}) = t\mu_i(S), i = 1, \ldots, m$, holds trivially. Thus, let $t > 0$ be given, and $\{t_k\}_{k=1,2,\ldots}$ be an increasing sequence $t_k \rightarrow_k t$ in I. Then, $S_t^{\varphi} = \cup_k C_{t_k}$. In fact, if $x \in S_t^{\varphi}$, there is $t_{k'}$ such that $\varphi_S(x) < t_{k'} < t$. As $x \in C_{t_{k'}}$ for $t_{k'} > \varphi_S(x)$, $x \in \cup_k C_{t_k}$. Conversely, if $x \in \cup_k C_{t_k}$, $x \in C_{t_{k_0}}$ for some $t_{k_0} < t$, and thus, $\varphi_S(x) < t$, showing that $x \in S_t^{\varphi}$.

Since $\{C_{t_k}\}_k$ is increasing and $t_k \rightarrow_k t$, for each $i = 1, \ldots, m$,

$$\begin{aligned}\mu_i(S_t^{\varphi}) &= \mu_i(\cup_k C_{t_k}) = \mu_i(\lim_k C_{t_k}) \\ &= \lim_k \mu_i(C_{t_k}) = \lim_k t_k \mu_i(S) = t\mu_i(S).\end{aligned}$$

This proves the functional convexity of μ.

3. *Functional Convexity of Purely Non-Atomic Measure*

Assuming that the measure $\mu = (\mu_1, \ldots, \mu_m)$ is purely non-atomic, we shall show that it satisfies the functional convexity by introducing its associated vector measure $\widetilde{\mu}$ and proving its functional convexity by the induction argument on m.

Define first an associated vector measure $\widetilde{\mu} = (\widetilde{\mu}_1, \ldots, \widetilde{\mu}_m)$ by setting $\widetilde{\mu}_i = \mu_i + \cdots + \mu_m, i = 1, \ldots, m$. Then, it is immediate that each coordinate measure of $\widetilde{\mu}$, $\widetilde{\mu}_i$, is absolutely continuous with respect to its predecessor, i.e., $\widetilde{\mu}_i \ll \widetilde{\mu}_{i-1}, i = 2, \ldots, m$, and moreover, μ satisfies the functional convexity if and only if so does $\widetilde{\mu}$.

Indeed, if μ satisfies the functional convexity, given $S \in \mathscr{M}$, there is a measurable $\varphi_S : S \rightarrow I$ such that $\mu_i(S_t^{\varphi}) = t\mu_i(S), i = 1, \ldots, m$, for each $t \in I$, where $S_t^{\varphi} = \{x \in S|\ \varphi_S(x) < t\}$. Thus, we have

$$\begin{aligned}\widetilde{\mu}_i(S_t^{\varphi}) &= (\mu_i + \cdots + \mu_m)(S_t^{\varphi}) \\ &= t(\mu_i + \cdots + \mu_m)(S) \\ &= t\widetilde{\mu}_i(S)\,, \quad i = 1, \ldots, m,\end{aligned}$$

showing that $\widetilde{\mu}$ satisfies the functional convexity.

Conversely, if $\widetilde{\mu}$ satisfies the functional convexity, there is a measurable $\widetilde{\varphi}_S : S \rightarrow I$ such that $\widetilde{\mu}_i(S_t^{\widetilde{\varphi}}) = t\widetilde{\mu}_i(S), i = 1, \ldots, m$, for each $t \in I$, where $S_t^{\widetilde{\varphi}}$ is defined similarly to S_t^{φ}. It follows immediately that $\mu_i(S_t^{\widetilde{\varphi}}) = t\mu_i(S), i = 1, \ldots, m$, for each $t \in I$. Hence μ satisfies the functional convexity.

We now proceed with the induction argument. For $m = 1$, μ, non-atomic, is equivalent to μ, atomless. It in turn is equivalent to the chain convexity which implies the functional convexity of μ. Thus, $\widetilde{\mu}$ satisfies the functional convexity when $m = 1$. Suppose that the $(m-1)$-dimensional vector measure $\check{\mu} = (\widetilde{\mu}_1, \ldots, \widetilde{\mu}_{m-1})$ satisfies the functional convexity. If we show that $\widetilde{\mu} = (\widetilde{\mu}_1, \ldots, \widetilde{\mu}_m)$ is atomless, then, by the arguments in step 1, $\widetilde{\mu}$ satisfies the chain convexity which in turn implies the functional convexity of $\widetilde{\mu}$ by the arguments in step 2.

Let us note first that for any measurable set C and a measurable function $\varphi_C : C \to I$ such that $\tilde{\mu}_i(\{x|\ \varphi_C(x) < k\}) = k\tilde{\mu}_i(C), i = 1, \ldots, m-1$, for each $k \in I$, $\tilde{\mu}_m(\{x|\ \varphi_C(x) < k\})$ is a continuous function of $k \in I$. In fact, since $\tilde{\mu}_m$ is absolutely continuous with respect to $\tilde{\mu}_{m-1}$, for any given $\varepsilon > 0$, we can choose a $\delta > 0$ so that $\tilde{\mu}_m(D) < \varepsilon$ for every measurable set D for which $\tilde{\mu}_{m-1}(D) < \delta$. If $0 \leqq k_1 < k_2 \leqq 1$ and $(k_2 - k_1)\tilde{\mu}_{m-1}(C) < \delta$, then

$$\tilde{\mu}_{m-1}(\{x|\ k_1 \leqq \varphi_C(x) < k_2\}) = (k_2 - k_1)\tilde{\mu}_{m-1}(C) < \delta$$

and thus

$$\tilde{\mu}_m(\{x|\ \varphi_C(x) < k_2\}) - \tilde{\mu}_m(\{x|\ \varphi_C(x) < k_1\}) < \varepsilon,$$

showing that $\tilde{\mu}_m(\{x|\ \varphi_C(x) < k\})$ is continuous in k.

Given a measurable set S, it contains measurable subsets S_t such that $\tilde{\mu}_i(S_t) = t\tilde{\mu}_i(S), i = 1, \ldots, m-1$, for each $t \in I$, as $\check{\mu}$ satisfies the functional convexity. So, in particular, take $t = 1/2$, and write $C_0 = S_{1/2}$ so that $\tilde{\mu}_i(C_0) = (1/2)\tilde{\mu}_i(S), i = 1, \ldots, m-1$. If $\tilde{\mu}_m(C_0) = (1/2)\tilde{\mu}_m(S)$, $\tilde{\mu}$ is atomless. Hence, without loss of generality, assume that $\tilde{\mu}_m(C_0) < (1/2)\tilde{\mu}_m(S)$, and set $D_0 = S - C_0$. Again, by the functional convexity of $\check{\mu}$, there are measurable functions $\check{\varphi}_{C_0} : C_0 \to I$ and $\check{\varphi}_{D_0} : D_0 \to I$ such that $\tilde{\mu}_i(\{x|\ \check{\varphi}_{C_0}(x) < k\}) = k\tilde{\mu}_i(C_0)$, and $\tilde{\mu}_i(\{x|\ \check{\varphi}_{D_0}(x) < k\}) = k\tilde{\mu}_i(D_0), i = 1, \ldots, m-1$, for each $k \in I$. Define measurable subsets $C(k) \subset S$ which depend upon $k \in I$ by setting

$$C(k) = \{x|\ \check{\varphi}_{C_0}(x) < 1 - k\} \cup \{x|\ \check{\varphi}_{D_0}(x) < k\}.$$

Then, $\tilde{\mu}_i(C(0)) = \tilde{\mu}_i(C_0) = \tilde{\mu}_i(D_0) = \tilde{\mu}_i(C(1)), i = 1, \ldots, m-1$. Since $\tilde{\mu}_m(C(k))$ is a continuous function of k and $\tilde{\mu}_m(C(0)) = \tilde{\mu}_m(C_0) < (1/2)\tilde{\mu}_m(S) < \tilde{\mu}_m(D_0) = \tilde{\mu}_m(C(1))$, there is some $0 < k_0 < 1$ satisfying $\tilde{\mu}_m(C(k_0)) = (1/2)\tilde{\mu}_m(S)$. Moreover, for $i = 1, \ldots, m-1$,

$$\begin{aligned}\tilde{\mu}_i(C(k_0)) &= (1 - k_0)\tilde{\mu}_i(C_0) + k_0\tilde{\mu}_i(D_0)\\ &= \tilde{\mu}_i(C_0) + k_0(\tilde{\mu}_i(D_0) - \tilde{\mu}_i(C_0))\\ &= \tilde{\mu}_i(C_0) = \frac{1}{2}\tilde{\mu}_i(S).\end{aligned}$$

Therefore, we obtain $\tilde{\mu}_i(C(k_0)) = (1/2)\tilde{\mu}_i(S)$ for all $i = 1, \ldots, m$, and hence, $\tilde{\mu}$ is atomless which implies its functional convexity. Consequently, the inductive proof of the functional convexity of $\tilde{\mu}$ is complete.

4. *Liapounoff Convexity*

We first show that the functional convexity of μ implies the Liapounoff convexity. Let $v, w \in \mu(\mathscr{M}), v \neq w$; then, for some sets $C, D \in \mathscr{M}$, $v = \mu(C)$ and $w = \mu(D)$. By the functional convexity of μ, there are measurable functions $\varphi_{C-D} : C - D \to I$ and $\varphi_{D-C} : D - C \to I$ such that $\mu_i\{x \in C - D|\ \varphi_{C-D}(x) < 1 - t\} = (1 - t)\mu_i(C - D)$ and $\mu_i\{x \in D - C|\ \varphi_{D-C}(x) < t\} = t\mu_i(D - C), i = 1, \ldots, m$, for any $t \in I$. Let

$$C(t) = \{x \in C - D | \varphi_{C-D}(x) < 1 - t\} \cup \{x \in D - C | \varphi_{D-C}(x) < t\} \cup (C \cap D).$$

Then, $C(t)$ is measurable, and we have

$$\begin{aligned}\mu_i(C(t)) &= (1-t)\big(\mu_i(C - D) + \mu_i(C \cap D)\big) + t\big(\mu_i(D - C) + \mu_i(C \cap D)\big) \\ &= (1-t)\mu_i(C) + t\mu_i(D), i = 1, \ldots, m.\end{aligned}$$

It follows that

$$(1-t)v + tw = \mu(C(t)) \in \mu(\mathscr{M}),$$

which shows the Liapounoff convexity.

Finally, we check that the Liapounoff convexity immediately leads to the atomlessness. If μ satisfies the Liapounoff convexity, for $S \in \mathscr{M}$, let $v = \mu(\emptyset), w = \mu(S) \in \mu(\mathscr{M})$. Then, for each $0 < t < 1$, we have $(1-t)v + tw = t\mu(S) \in \mu(\mathscr{M})$ so that for some $C \in \mathscr{M}$ $\mu_i(C) = t\mu_i(S), i = 1, \ldots, m$. Therefore, μ is atomless.

□

3.7 Integral of Correspondences

For the presentation of analysis in subsequent chapters, the *aggregation of individual economic agents' behaviors* comes to play a very important part of our analysis. The mathematical concept of integrals of correspondences gives expression to the aggregation of individual behaviors. Thus, we like to spend some time on the topic of integrals of correspondences before we proceed with our economic analysis.

Integrable Selections and Integral of Correspondences

If f is a mapping from a measure space $(M, \mathscr{M}, \mu)$ to $\mathbb{R}^m$, f can be expressed coordinatewise such that $f = (f_1, \ldots, f_m), f_i : M \to \mathbb{R}, i = 1, \ldots, m$. Then, f is said to be *integrable* if each f_i is. The *integral* of f is defined by

$$\int f d\mu = \left(\int f_1 d\mu, \ldots, \int f_m d\mu\right).$$

$\int f d\mu$ is also written as $\int_M f d\mu$.

Now, let φ be a correspondence from M to $T \subset \mathbb{R}^m$. If a measurable selection $f : M \to T$ of φ is integrable, f is called an *integrable selection*. The set of all integrable selections of φ will be denoted by $\mathcal{L}_\varphi$. The set $\left\{\int f d\mu \in \mathbb{R}^m \,\middle|\, f \in \mathcal{L}_\varphi\right\}$ is called the *integral* of the correspondence φ, and is denoted by $\int \varphi d\mu$ or $\int_M \varphi d\mu$. If φ is a relation instead of a correspondence, then its *integral* is defined as that of its restriction to its domain dom φ, i.e., $\int \varphi d\mu = \int_{\text{dom}\,\varphi} \varphi d\mu$.

There are three properties of integrals of the correspondences each of which is either basic or useful in pursuing the mathematical analysis in economic theory, in particular, in general equilibrium analysis.

(1) $\int \varphi d\mu \neq \emptyset$. This is a most essential property in that it gives meaning to the integral of a correspondence.
(2) The integral $\int \varphi d\mu \subset \mathbb{R}^m$ of a correspondence becomes a convex set under an appropriate condition in general equilibrium analysis.
(3) Given a topological space S and a class of correspondences $\{\varphi_p |\ p \in S\}$, if each $p \mapsto \varphi_p(x), x \in M$, possesses a continuity property (in particular, upper hemi-continuity) μ-almost everywhere, its continuity property is preserved by the correspondence $p \mapsto \int \varphi_p d\mu$ of S into $\mathbb{R}^m$. It is essentially related to a continuity issue of the aggregate demands of an economy and the determinateness of Walrasian equilibrium of an economy.

The three properties above are ensured respectively by the Measurable Selection Theorem, "Liapounoff Convexity Theorem," and Multidimensional Fatou's Lemma. Unquestionably, these theorems provide, so to speak, three "golden tools" in the measure theoretic approach to general equilibrium analysis. Since the Measurable Selection Theorem and the characterizing properties of purely non-atomic vector measures in relation to the Liapounoff Convexity Theorem are presented in previous sections (cf. Section 3.5 and 3.6), in the remaining part of the present chapter various properties of the integral of correspondences including the Multidimensional Fatou's Lemma are presented.

Definition 3.9 (Integrable Boundedness) A relation φ from a measure space $(M, \mathscr{M}, \mu)$ to $\mathbb{R}^m$ is *integrably bounded above* if there is an integrable function g from M to $\mathbb{R}^m$ such that, for every $x \in M$, one has $z \leqq g(x)$ for all $z \in \varphi(x)$. If in this condition one has $z \geqq -g(x)$ in place of $z \leqq g(x)$, φ is *integrably bounded below*. φ is *integrably bounded* if it is both integrably bounded above and below.

The most important implication of Mathematical Theorem 3.3 (Measurable Selection Theorem) is the following property which provides the basis for the integral of correspondences.

Mathematical Proposition 3.13 [Integral of Correspondences] *Let φ be an integrably bounded correspondence from a measure space $(M, \mathscr{M}, \mu)$ to a G_δ set $T \subset \mathbb{R}^m$ such that $\varphi \subset M \times T$ is a $(\mathscr{M} \times \mathscr{B}(T))$ Souslin set. Then, $\int \varphi d\mu \neq \emptyset$.*

Mathematical Proposition 3.14 [Richter's Proposition] *Let φ be a correspondence from an atomless finite measure space $(M, \mathscr{M}, \mu)$ to a subset $T \subset \mathbb{R}^m$. Then, the integral of φ, $\int \varphi d\mu$, is a convex subset of $\mathbb{R}^m$.*

Proof If $\int \varphi d\mu$ is a singleton or empty, there is nothing to prove. So, let $x, y \in \int \varphi d\mu$ and $x \neq y$. We want to show $[x, y] \subset \int \varphi d\mu$. There are integrable selections $f, g : M \to T$ of φ such that $x = \int f d\mu,\ y = \int g d\mu$. Define an $\mathbb{R}^{2m}$-valued set function ν on $\mathscr{M}$ by $\nu(S) = \left(\int_S f d\mu, \int_S g d\mu\right)$ for each $S \in \mathscr{M}$. Since the measure space $(M, \mathcal{M}, \mu)$ is atomless, ν is a bounded purely non-atomic vector measure. It follows from Mathematical Proposition 3.12 that ν satisfies the Liapounoff convexity so that the range of ν, i.e. $\left\{\nu(S) = \left(\int_S f d\mu, \int_S g d\mu\right) \in \mathbb{R}^{2m} \mid S \in \mathcal{M}\right\}$ is a convex set in $\mathbb{R}^{2m}$. Hence, for any $0 \leqq t \leqq 1$, there is a set $S_t \in \mathcal{M}$ such that $(tx, ty) = \left(\int_{S_t} f d\mu, \int_{S_t} g d\mu\right)$, as both $(0, 0)$ and (x, y) belong to the range of ν.

To show that there is an integrable selection $h : M \to T$ of φ for which $tx + (1-t)y = \int h d\mu \in \int \varphi d\mu$, let $h(z) = f(z)$ if $z \in S_t$ and $h(z) = g(z)$ if $z \in M - S_t$. Then, h is an integrable selection of φ, and $\int h d\mu = \int_{S_t} f d\mu + \int_{M-S_t} g d\mu = tx + (1-t)y \in \int \varphi d\mu$. This proves $[x, y] \subset \int \varphi d\mu$. □

Mathematical Proposition 3.15 [Vind's Proposition] *Let φ be a correspondence from an atomless measurable space $(M, \mathcal{M}, \mu)$ to $T \subset \mathbb{R}^m$, and, for each $f \in \mathcal{L}_\varphi$, ν_f a set function $\nu_f : S \mapsto \int_S f d\mu \in \mathbb{R}^m$ for $S \in \mathcal{M}$. Then, the set $F_\varphi = \left\{\nu_f(S) \in \mathbb{R}^m \mid f \in \mathcal{L}_\varphi, S \in \mathcal{M}, \text{ and } \mu(S) > 0\right\}$ is a convex subset of $\mathbb{R}^m$.*

Proof Let $x, y \in F_\varphi$. We want to show $[x, y] \subset F_\varphi$. There are $f, g \in \mathcal{L}_\varphi$, and sets $X, Y \in \mathcal{M}$ with $\mu(X), \mu(Y) > 0$ such that $x = \nu_f(X)$ and $y = \nu_g(Y)$. Since ν_f and ν_g are bounded purely non-atomic vector measures, they satisfy the Liapounoff convexity (cf. Mathematical Proposition 3.12) so that the sets $F_{X-Y} = \left\{\nu_f(S) \in \mathbb{R}^m \mid S \subset X - Y, S \in \mathcal{M}\right\}$, $F_{Y-X} = \left\{\nu_g(S) \in \mathbb{R}^m \mid S \subset Y - X, S \in \mathcal{M}\right\}$, and $F_{X\cap Y} = \left\{(\nu_f(S), \nu_g(S)) \in \mathbb{R}^{2m} \mid S \subset X \cap Y, S \in \mathcal{M}\right\}$ are convex. Thus, for any $0 \leqq t \leqq 1$ there are sets X_t, Y_t and W_t in $\mathcal{M}$ such that $\nu_f(X_t) = t\nu_f(X - Y)$, $\nu_g(Y_t) = t\nu_g(Y - X)$, $\nu_f(W_t) = t\nu_f(X \cap Y)$, $\nu_g(W_t) = t\nu_f(X \cap Y)$, $X_t \subset X - Y$, $Y_t \subset Y - X$, and $W_t \subset X \cap Y$.

Now, define an integrable selection $h : M \to T$ of φ by

$$h(z) = \begin{cases} f(z) & \text{if } z \in X_t \cup W_t, \\ g(z) & \text{if } z \in M - (X_t \cup W_t), \end{cases}$$

and set $W = X_t \cup Y_t \cup (X \cap Y)$. Then, one has

$$\begin{aligned} tx + (1-t)y &= t\nu_f(X) + (1-t)\nu_g(Y) \\ &= t\nu_f((X - Y) \cup (X \cap Y)) + (1-t)\nu_g((Y - X) \cup (X \cap Y)) \\ &= \nu_f(X_t \cup W_t) + \nu_g(Y_t \cup ((X \cap Y) - W_t)) \\ &= \nu_h(W) \in F_\varphi \end{aligned}$$

This proves $[x, y] \subset F_\varphi$. □

Mathematical Proposition 3.16 [Hildenbrand's Lemma] *Let φ be a measurable correspondence from a measure space $(M, \mathcal{M}, \mu)$ to $\mathbb{R}^m$ such that $\int \varphi d\mu \neq \emptyset$. Then, for every vector $p \in \mathbb{R}^m$, we have*

$$\sup\left\{p \cdot z \mid z \in \int \varphi d\mu\right\} = \int \sup\left\{p \cdot x \mid x \in \varphi(u)\right\} d\mu(u).$$

Proof The left-hand side of the above equality is not greater than its right-hand side. Thus, we shall show the opposite. By Mathematical Proposition 3.6, the mapping $u \mapsto \sup p \cdot \varphi(u)$ is $\mathscr{M}_\mu$-measurable. Since $\int \varphi d\mu \neq \emptyset$, there is an integrable selection of φ, and the right-hand side is well defined (it may be $+\infty$). For each u, set $g(u) = \sup p \cdot \varphi(u)$, and let $t < \int g d\mu$. If we show, for some $f \in \mathcal{L}_\varphi$, we have $t < p \cdot \int f$, then the left-hand side $\geqq$ the right-hand side. For this purpose, let $h \in \mathcal{L}_\varphi$ and define a measurable truncated correspondence φ_n for each $n \subset \mathbb{N}_+$ by $\varphi_n(u) = \{x \in \varphi(u) \mid \|x - h(u)\| \leqq n\}$. By Mathematical Proposition 3.6, the function $g_n : u \mapsto \sup p \cdot \varphi_n(u)$ is $\mathscr{M}_\mu$-measurable. The integrability of h implies that of g_n for each n. At every $u \in M$ the sequence $(g_n(u))_n$ is monotone increasing and converges to $g(u)$. Thus, it follows from the Monotone Convergence Theorem (cf. Mathematical Property 1.32 (1), p. 56) that $\lim_n \int g_n d\mu = \int g d\mu$. (If g is not finite on a set of positive measure, then on that set $g_n(u) \to \infty$; hence, $\int g_n d\mu \to \infty$. This means that the equality still holds.) Hence, for n sufficiently large, $\int g_n d\mu > t$. Fix n for such a g_n. There is an integrable function $k : M \to \mathbb{R}$ such that $\int k d\mu > t$ and $g_n(u) > k(u)$ for each $u \in M$. Define a correspondence $\psi(u) = \{x \in \varphi_n(u) | \ p \cdot x > k(u)\}$. $\psi(u) \neq \emptyset$, and ψ is measurable. Thus by Mathematical Theorem 3.3 there is a measurable selection f of ψ. By the way ψ is defined, it is integrable and $f \in \mathcal{L}_\varphi$. Since $p \cdot f(u) > k(u)$, $u \in M$, we obtain $p \cdot \int f d\mu > \int k d\mu$. Therefore, $p \cdot \int f d\mu > t$, which was to be shown. □

Mathematical Theorem 3.4 [Aumann's Theorem] *Let φ be a measurable correspondence from a measure space $(M, \mathscr{M}, \mu)$ to $\mathbb{R}^m$, which is integrably bounded from below. Then,* co $\int \varphi d\mu = \int$ co $\varphi d\mu$. *In particular, if the measure space is atomless,* $\int \varphi d\mu = \int$ co $\varphi d\mu$.

Proof We prove the theorem by an induction argument on the dimension of $\mathbb{R}^m$. If $m = 0$, the conclusion is trivial. Assuming that the conclusion holds for $m - 1$, to show it is also true for m, let us first confirm:

$$\int \text{co}\, \varphi d\mu = \emptyset \text{ if and only if co} \int \varphi = \emptyset. \tag{3.1}$$

If $\int \text{co}\, \varphi d\mu = \emptyset$, then $\int \varphi d\mu = \emptyset$. Thus, co $\int \varphi d\mu = \emptyset$. Conversely, let co $\int \varphi d\mu = \emptyset$; then, if $\int \text{co}\, \varphi d\mu \neq \emptyset$, there would exist an integrable selection f of co φ. Fix a vector $q \in \mathbb{R}^m_{++}$ and define a relation $\psi : M \to \mathbb{R}^m$ by $x \mapsto \psi(x) = \{z \in \varphi(x) \mid q \cdot z \leqq q \cdot f(x)\}$. Since $f(x) \in \text{co}\, \varphi(x)$ a.e. $x \in M$, $\psi(x) \neq \emptyset$ a.e. $x \in M$. Thus, without loss of generality one can assume ψ to be a correspondence from M to $\mathbb{R}^m$. The measurability of φ and f imply that of the correspondence ψ. Therefore, by Mathematical Theorem 3.3, there is a measurable selection h of ψ. h is integrable due to the fact that $q \in \mathbb{R}^m_{++}$, f is integrable, and φ is integrably bounded below. Hence, one would obtain $\int \varphi d\mu \neq \emptyset$, contradictory to co $\int \varphi d\mu = \emptyset$

Since (3.1) holds, from now on we assume $\int \varphi d\mu \neq \emptyset$. Let us show:

$$\text{co} \int \varphi d\mu \text{ and } \int \text{co}\, \varphi d\mu \text{ have the identical closure.} \tag{3.2}$$

Note that the following series of inequalities and an equality holds true for each $q \in \mathbb{R}^m$:

$$\sup q \cdot \int \varphi d\mu \leqq \sup q \cdot \int \operatorname{co} \varphi d\mu \leqq \int \sup q \cdot \varphi d\mu = \sup q \cdot \int \varphi d\mu.$$

Indeed, the first inequality is trivial, and the second follows from $q \cdot \operatorname{co} \varphi(x) \leqq \sup q \cdot \varphi(x)$, whereas the last equality comes from Mathematical Proposition 3.16. Thus, for every $q \in \mathbb{R}^m$, we have

$$\sup q \cdot \operatorname{co} \int \varphi d\mu = \sup q \cdot \int \varphi d\mu = \sup q \cdot \int \operatorname{co} \varphi d\mu. \tag{3.3}$$

Since the both of the sets in (3.2) are convex, (3.3) implies (3.2) by a property of convex sets in $\mathbb{R}^m$ (cf. Mathematical Property 1.18, p. 42).

Now, for any $q \in \mathbb{R}^m$ and $S \subset \mathbb{R}^m$, let us define a subset $S_q \subset S$ by $S_q = \{z \in S \mid q \cdot z = \sup q \cdot S\}$. If it is shown that

$$\left(\int \operatorname{co} \varphi d\mu\right)_q = \left(\operatorname{co} \int \varphi d\mu\right)_q$$

is true for each $q \in \mathbb{R}^m - \{0\}$, then the conclusion of the theorem will follow from Mathematical Proposition 3.14 and the characterizing property of convex sets (Mathematical Property 1.18).

Given any correspondence ψ from $(M, \mathcal{M}, \mu)$ to $\mathbb{R}^m$ for which $\psi \subset M \times \mathbb{R}^m$ is a $(\mathcal{M} \times \mathcal{B}(\mathbb{R}^m))$-Souslin set and $\int \psi d\mu \neq \emptyset$, let a relation $\psi_q : M \to \mathbb{R}^m$ be defined by $\psi_q : x \mapsto (\psi(x))_q$ for each $q \in \mathbb{R}^m - \{0\}$. One can check without difficulty the equivalence of the following conditions (1) and (2) using the Measurable Selection Theorem and Hildenbrand's Lemma (cf. Mathematical Proposition 3.3 and 3.16).

(1) $\left(\int \psi d\mu\right)_q \neq \emptyset$;

(2) $\psi_q(x) \neq \emptyset$, a.e. $x \in M$, and $\int \psi_q d\mu = \left(\int \psi d\mu\right)_q$.

One can also verify $(\operatorname{co} S)_q = \operatorname{co} S_q$ for any $S(\neq \emptyset) \subset \mathbb{R}^m$. The correspondence φ is measurable, and the correspondence $\operatorname{co} \varphi$ is $(\mathcal{M} \times \mathcal{B}(\mathbb{R}^m))$-Souslin by Mathematical Proposition 3.11. By applying these equivalent conditions to the correspondences φ and $\operatorname{co} \varphi$, one obtains

$$\left(\operatorname{co} \int \varphi d\mu\right)_q = \operatorname{co} \left(\int \varphi d\mu\right)_q = \operatorname{co} \int \varphi_q d\mu,$$

and similarly

$$\left(\int \operatorname{co} \varphi d\mu\right)_q = \int \left(\operatorname{co} \varphi\right)_q d\mu = \int \operatorname{co} \varphi_q d\mu.$$

Hence, to complete the proof it remains to show

$$\text{co} \int \varphi_q d\mu = \int \text{co}\, \varphi_q d\mu. \tag{3.4}$$

It follows from Mathematical Proposition 3.6 that we have

$$\varphi_q \in \mathcal{M}_\mu \otimes \mathcal{B}(\mathbb{R}^m). \tag{3.5}$$

Since $\int \text{co}\, \varphi_q d\mu = \emptyset$ if and only if $\int \varphi_q d\mu = \emptyset$ by (3.1), we prove (3.4) in the case $\int \varphi_q d\mu \neq \emptyset$.

Consider a hyperplane $H(q, 0)$ in $\mathbb{R}^m$. There is at least one unit vector $(0, \ldots, 0, \overset{i}{1}, 0, \ldots, 0)$ that is not contained in $H(q, 0)$. Fix one of them, and denote it by $\mathbf{e}_i$. Letting L be the one-dimensional linear subspace generated by $\mathbf{e}_i$, pr_L is the (oblique) projection from $\mathbb{R}^m$ to $H(q, 0)$ parallel to L. (That is, $\text{pr}_L(z) = z_L$, where $\{z_L\} = H(q, 0) \cap (z + L)$ for $z \in \mathbb{R}^m$.) By (3.5) there is a measurable selection $g : M \to \mathbb{R}^m$ of φ_q (cf. Mathematical Theorem 3.3). Define a mapping $h : M \to \mathbb{R}^m$ by $x \mapsto h(x) = g(x) - pr_L(g(x))$; then, h is $\mathcal{M}_\mu$-measurable. Since $\varphi_q(x) \subset H(q, q \cdot g(x))$, a.e. $x \in M$, we have $\varphi_q(x) = pr_L(\varphi_q(x)) + h(x)$, a.e. $x \in M$. One thus obtains that

$$\text{co} \int \varphi_q d\mu = \text{co} \int (pr_L \circ \varphi_q + h) d\mu = \text{co} \int pr_L \circ \varphi_q d\mu + \int h d\mu,$$

and

$$\int \text{co}\, \varphi_q d\mu = \int \text{co}\, (pr_L \circ \varphi_q + h) d\mu = \int \text{co}\, pr_L \cdot \varphi_q d\mu + \int h d\mu.$$

Therefore, showing (3.4) reduces to showing

$$\text{co} \int pr_L \circ \varphi_q d\mu = \int \text{co}\, \text{pr}_L \circ \varphi_q d\mu, \tag{3.6}$$

which is a consequence of the induction hypothesis. In fact, the $(m - 1)$ vectors, $pr_L(e_j)$, $(j = 1, \ldots, i-1, i+1, \ldots, m)$, form a basis in the linear subspace $H(q, 0)$ where $\mathbf{e}_j$ is the j-th unit vector in $\mathbb{R}^m$. With respect to this basis, the correspondence $\text{pr}_L \circ \varphi_q$ becomes a correspondence $\hat{\varphi}_q$ of M into $\mathbb{R}^{m-1}$. $\text{pr}_L \circ \varphi_q \in \mathcal{M} \otimes \mathcal{B}(\mathbb{R}^m)$ implies $\hat{\varphi}_q \in \mathcal{M} \otimes \mathcal{B}(\mathbb{R}^{m-1})$. Since φ is integrably bounded from below, so is $\hat{\varphi}_q$. Therefore, by the induction hypothesis, we obtain

$$\text{co} \int \hat{\varphi}_q d\mu = \int \text{co}\, \hat{\varphi}_q d\mu. \tag{3.7}$$

Letting $p : (z^1, \ldots, z^{m-1}) \mapsto \sum_j z^j pr_L(e_j)$ be a linear injection from $\mathbb{R}^{m-1}$ to $\mathbb{R}^m$, we have

$$pr_L \circ \varphi_q(x) = p \circ \hat{\varphi}_q(x). \tag{3.8}$$

Hence, it follows from (3.7) that

$$\mathrm{co} \int p \circ \hat{\varphi}_q d\mu = p\left(\mathrm{co} \int \hat{\varphi}_q d\mu\right) = p\left(\int \mathrm{co}\, \hat{\varphi}_q d\mu\right) = \int \mathrm{co}\,(p \circ \hat{\varphi}_q) d\mu. \quad (3.9)$$

Therefore, (3.6) follows from (3.8) and (3.9). This completes the proof of the theorem. □

Mathematical Theorem 3.5 [Change of Variable for Integration of Correspondences] *Let g be a measurable mapping from a finite measure space $(M_1, \mathscr{M}_1, \mu)$ to a measurable space $(M_2, \mathscr{M}_2)$, and φ a measurable correspondence from M_2 to $\mathbb{R}^m$ such that $\varphi(y)$ is closed and convex for every $y \in M_2$ and contains no straight lines. Then,*

$$\int_{M_2} \varphi d(\mu \circ g^{-1}) = \int_{M_1} \varphi \circ g d\mu.$$

Proof Let $z \in \int_{M_2} \varphi d(\mu \circ g^{-1})$; then, there is an integrable selection h of φ such that $z = \int_{M_2} h d(\mu \circ g^{-1}) = \int_{M_1} h \circ g d\mu$. But since $h \circ g$ is a measurable selection of $\varphi \circ g$, it follows that $z \in \int_{M_1} \varphi \circ g d\mu$. Thus, $\int_{M_2} \varphi d(\mu \circ g^{-1}) \subset \int_{M_1} \varphi \circ g d\mu$.

Conversely, let $z \in \int_{M_1} \varphi \circ g d\mu$. Then, there is an integrable selection $f : M_1 \to \mathbb{R}^m$ of the correspondence $\varphi \circ g$ such that $z = \int_{M_1} f d\mu$. If we show that there is a measurable function $h : M_2 \to \mathbb{R}^m$ satisfying $f = h \circ g$, then $\int_{M_1} \varphi \circ g d\mu \subset \int_{M_2} \varphi d(\mu \circ g^{-1})$ is proved. Such a measurable function h exists if and only if f is $g^{-1}(\mathscr{M}_2)$-measurable (cf. Mathematical Property 1.24, p. 49). Thus, it remains to show that, for any integrable selection f of $\varphi \circ g$, there exists a $g^{-1}(\mathscr{M}_2)$-measurable selection of $\varphi \circ g$ with the same integral.

Let $\mathfrak{Z} = g^{-1}(\mathscr{M}_2) \subset \mathscr{M}_1$, and write the $\mathbb{R}^m$-valued integrable function f coordinatewise as $f = (f_i)_i$ with $f_i : M_1 \to \mathbb{R}, i = 1, \ldots, m$. Given that $\mathfrak{Z}$ is a σ-subalgebra of $\mathscr{M}_1$, by Mathematical Property 1.31, p. 55, there are conditional expectations $\mathbb{E}(f_i|\mathfrak{Z}) : M_1 \to \mathbb{R}$ such that $\int_C \mathbb{E}(f_i|\mathfrak{Z}) d\mu = \int_C f_i d\mu$ for every $C \in \mathfrak{Z}$ and for each i. Setting $\mathbb{E}(f|\mathfrak{Z}) = (\mathbb{E}(f_1|\mathfrak{Z}), \ldots, \mathbb{E}(f_m|\mathfrak{Z}))$, if a $\mathfrak{Z}$-measurable mapping $\mathbb{E}(f|\mathfrak{Z})$ is shown to be a measurable selection of the correspondence $\varphi \circ g$, the proof is completed.

If we put $\psi = \varphi \circ g$, we have $\inf q \cdot \psi(x) \leqq q \cdot f(x)$, a.e. $x \in M_1$, for any $q \in \mathbb{R}^m$ since f is an integrable selection of ψ. As $\psi \in \mathfrak{Z} \otimes \mathscr{B}(\mathbb{R}^m)$ by Mathematical Proposition 3.9, the mapping $x \mapsto \inf q \cdot \psi(x)$ is $\mathfrak{Z}_\mu$-measurable by Mathematical Proposition 3.6. Hence, it is equal to a $\mathfrak{Z}$-measurable function $\mathbb{E}(\inf q \cdot \psi \mid \mathfrak{Z})$ a.e. in M_1. Accordingly, we have for any q

$$\begin{aligned} \inf q \cdot \psi(x) &= \mathbb{E}(\inf q \cdot \psi \mid \mathfrak{Z})(x) \\ &\leqq \mathbb{E}(q \cdot f \mid \mathfrak{Z})(x) = q \cdot \mathbb{E}(f \mid \mathfrak{Z})(x), \\ &\qquad \text{a.e. } x \in M_1. \end{aligned}$$

Letting $Q \subset \mathbb{R}^m$ to be a countable dense subset, one obtains $\inf q \cdot \psi(x) \leqq q \cdot \mathbb{E}(f \mid \mathfrak{Z})(x)$ a.e. in M_1 for all $q \in Q$. Since $\psi(x)$ is closed and convex containing no straight lines, by Mathematical Property 1.18, p. 42, it implies that

$\mathbb{E}(f \mid \mathfrak{Z})(x) \in \psi(x)$ a.e. Therefore, $\mathbb{E}(f \mid \mathfrak{Z}) : M_1 \to \mathbb{R}^m$ is a $\mathfrak{Z}$-measurable selection of ψ. □

Mathematical Theorem 3.6 [Multidimensional Fatou's Lemma] *Let $T \subset \mathbb{R}^m_+$ be closed and $(f_n)_n$ be a sequence of integrable functions from a finite measure space $(M, \mathcal{M}, \mu)$ to T. Assume that* $\lim_n \int f_n d\mu$ *exists; then, there is an integrable function $f : M \to T$ satisfying the following conditions:*

(a) $f(x) \in \mathrm{Ls}\,(f_n(x))$, a.e. $x \in M$;
(b) $\int f d\mu \leqq \lim_n \int f_n d\mu$.

Furthermore, if the sequence $(f_n)_n$ is uniformly integrable, and if the set $\{f_n(x) | n \in \mathbb{N}_+\}$ is bounded for a.e. $x \in M$, *then there exists an integrable selection $f : M \to T$ of the correspondence* $\mathrm{Ls}\,(f_n) : M \to T$ *such that* $\int f d\mu = \lim_{n_k} \int f_{n_k} d\mu$ *holds for some subsequence $(f_{n_k})_{n_k}$ of $(f_n)_n$. If for every function $f : M \to T$ satisfying the above two conditions we have* $\int f d\mu = \lim_{n_k} \int f_{n_k} d\mu$ *for a subsequence $(f_{n_k})_{n_k}$ of $(f_n)_n$, then the sequence $(f_n)_n$ is uniformly integrable.*

Proof Let L^m_p and ba^m denote the m-hold product of $L_p(M, \mathcal{M}, \mu)$, $(p = 1, \infty)$, and $ba(M, \mathcal{M})$ respectively, and call their product topologies the strong topologies. Similarly, the product topologies $\sigma^m(L_1, L_\infty)$ and $\sigma^m(ba, L_\infty)$ where each factor space L_1 and ba are endowed with the weak topology $\sigma(L_1, L_\infty)$ and $\sigma(ba, L_\infty)$, respectively, are called their weak topologies. We consider two cases.

Case I: $(M, \mathcal{M}, \mu)$ is atomless and complete.

Define a set function τ_n on $\mathcal{M}$ by $\tau_n(S) = \int_S f_n d\mu$, $S \in \mathcal{M}$ for each $n \in \mathbb{N}_+$; then, $\tau_n \in ba^m$, $n \in \mathbb{N}_+$. As the sequence $(\tau_n(M))_n$ is bounded, the $\sigma^m(ba, L_\infty)$-closure of the set $\{\tau_n \mid n \in \mathbb{N}_+\} \subset ba^m$ is $\sigma^m(ba, L_\infty)$-compact by the Banach–Alaoglu Theorem, p. 65. Thus, the sequence $(\tau_n)_n$ has a limit point $\tau \in ba^m$ in $\sigma^m(ba, L_\infty)$-topology. It follows that we have $\tau \geqq 0$ and

$$\tau(M) = \lim_n \int f_n d\mu. \tag{3.10}$$

By applying the Yosida–Hewitt Theorem, p. 67, coordinatewise to τ, decompose τ into a countably additive τ_c on $\mathcal{M}$ and a purely finitely additive τ_f on $\mathcal{M}$ so that $\tau = \tau_c + \tau_f$, $\tau_c, \tau_f \geqq 0$, and $\tau_c, \tau_f \in ba^m$. It follows from (3.10) that each coordinate of the measure τ_c is absolutely continuous with respect to μ. So, let $g = (g_1, \ldots, g_m)$ be a $\mathbb{R}^m$-valued function on M with each coordinate being a Radon–Nikodym derivative of the corresponding coordinate of τ_c with respect to μ. Then, it follows from

$$\int g d\mu = \tau_c(M) \leqq \tau(M) = \lim_n \int f_n d\mu \tag{3.11}$$

that we have

$$\int g d\mu \leq \lim_n \int f_n d\mu. \tag{3.12}$$

Again by the Yosida–Hewitt Theorem there is a countable measurable partition $\{S_i \mid i \in \mathbb{N}_+\}$ of M satisfying $\tau_f(S_i) = 0$ for each $i \in \mathbb{N}_+$. It implies that in $L_1^m(S_i, \mathcal{M} \cap S_i, \mu|_{S_i})$ the restriction $g|_{S_i}$ is a $\sigma^m(L_1, L_\infty)$-limit point of the sequence $(f_n|_{S_i})_n$ for every i. Fix S_i, and define sets $F_n = \{f_k|_{S_i} \mid k = n, n+1, \dots\}$ for $n \in \mathbb{N}_+$. In view of Mathematical Property 1.41 (1) and (2), p. 64, there is a sequence of functions $(g_n)_n$ in $\operatorname{co} F_n$ which converges to $g|_{S_i}$ strongly. Hence, there is a subsequence of $(g_n)_n$ that converges a.e. to $g|_{S_i}$. Assume without loss of generality (abbreviated as "w.l.o.g.") $\lim_n g_n(x) = g(x)$, a.e. $x \in S_i$.

We shall show next that

$$g(x) \in \operatorname{co} \operatorname{Ls}(f_n(x)) + \mathbb{R}_+^m, \quad \text{a.e. } x \in M. \tag{3.13}$$

By the way the sequence $(g_n)_n$ has been chosen, $g_n(x) \in \operatorname{co}\{f_n|_{S_i}(x), f_{n+1}|_{S_i}(x), \dots\}$. Hence, by Carathéodory's Theorem, p. 41, we can write $g_n(x)$ as below:

$$g_n(x) = \sum_{i=0}^{m} t_n^i(x) h_n^i(x),$$
$$t_n^0(x), \dots, t_n^m(x) \geqq 0, \ \sum_{i=0}^{m} t_n^i(x) = 1,$$
$$h_n^i(x) \in \{f_n(x), f_{n+1}(x), \dots\}, \ (i = 0, \dots, m).$$

Since vectors $t_n(x) = \left(t_n^0(x), \dots, t_n^m(x)\right)$ lie in a compact set, w.l.o.g. assume that the sequences $t_n(x) \to t(x) = (t^0(x), \dots, t^m(x))$. Then, we have $t^i(x) \geqq 0$ for each i, and $\sum_{i=0}^m t^i(x) = 1$. As $\lim_n g_n(x) = g(x)$, a.e. $x \in S_i$, $t_n^i(x) \geqq 0$ and $h_n^i(x) \geqq 0$ imply that the sequences $(t_n^i(x) h_n^i(x))_n$ are bounded. Therefore, one can assume w.l.o.g. that, for each $i = 0, \dots, m$, $\lim_n t_n^i(x) h_n^i(x)$ exists, and $\lim_n h_n^i(x)$ exists if $t^i(x) > 0$. Thus, for a.e. $x \in S_i$, $g(x)$ can be written as below:

$$\begin{aligned} g(x) &= \lim_n g_n(x) = \lim_n \sum_{i=0}^{m} t_n^i(x) h_n^i(x) \\ &= \sum_{i=0}^{m} \lim_n t_n^i(x) h_n^i(x) \\ &\geqq \sum_{\substack{i=0 \\ t^i(x)>0}}^{m} t^i(x) \lim_n h_n^i(x). \end{aligned} \tag{3.14}$$

By the way $h_n^i(x)$'s are chosen, we have $\lim_n h_n^i(x) \in \operatorname{Ls}(f_n(x))$. Therefore, we obtain

$$g(x) \geqq \sum_{\substack{i=0 \\ t^i(x)>0}}^{m} t^i(x) \lim_n h_n^i(x) \in \operatorname{co} \operatorname{Ls}(f_n(x)).$$

This proves (3.13) as the above arguments are true for any component S_i of the partition $\{S_i \mid i \in \mathbb{N}_+\}$.

Next, define a set $G \subset M \times \mathbb{R}^m$ by

$$G = \{(x, y) \in M \times \mathbb{R}^m | \, g(x) \geqq y \text{ and } y \in \operatorname{co} \operatorname{Ls}(f_n(x))\};$$

then, by Mathematical Proposition 3.7, G is measurable. It follows from (3.13) that $\mu(\mathrm{pr}_M(G)) = \mu(M)$. By applying the Measurable Selection Theorem to G, we obtain a measurable map f^* from M to $\mathbb{R}^m$ such that a.e. in M, $(x, f^*(x)) \in G$. Taking account of (3.12), we have

$$f^*(x) \in \operatorname{co} \operatorname{Ls}(f_n(x)), \quad \text{a. e. } x \in M, \tag{3.15}$$

$$\int f^* d\mu \leqq \lim_n \int f_n d\mu. \tag{3.16}$$

As $(M, \mathscr{M}, \mu)$ is atomless, it follows from Aumann's Theorem that

$$\int \operatorname{Ls}(f_n) d\mu = \int \operatorname{co} \operatorname{Ls}(f_n) d\mu. \tag{3.17}$$

Since the range of each f_n is a closed set $T \subset \mathbb{R}^m_+$, by (3.15) and (3.17) there exists an integrable mapping f from M to T satisfying the conditions

$$f(x) \in \operatorname{Ls}(f_n(x)), \quad \text{a. e. } x \in M, \tag{3.18}$$

$$\int f d\mu \leqq \lim_n \int f_n d\mu. \tag{3.19}$$

Case II: $(M, \mathscr{M}, \mu)$ may not be complete nor atomless.

By Mathematical Fact 3.1 (4), there exists a countable measurable partition of M, $\{M_k \mid k \in \mathbb{N}_+\}$, where M_k is either an atom or atomless. By the assumption of the theorem $\lim_n \int f_n d\mu$ exists and $f_n \geqq 0$. Hence, there exists $z \in \mathbb{R}^m_+$ such that for every $k \in \mathbb{N}_+$ and for every $n \in \mathbb{N}_+$, we have $\int_{M_k} f_n d\mu \leqq z$. Thus, assume w.l.o.g. that $\lim_n \int_{M_k} f_n d\mu$ exists for every k.

Take any k. If M_k is an atom, then the first part of the assertion of the theorem is trivial. If M_k is atomless, apply the arguments of *Case I* to the completion of μ to obtain a $\mathscr{M}_\mu$-measurable function, and modify it on a μ-null set to get an appropriate $\mathscr{M}$-measurable function. Therefore, for each $k \in \mathbb{N}_+$, there exists a function f^k from M_k to T satisfying

$$f^k(x) \in \operatorname{Ls}\,(f_n(x)), \quad \text{a.e. } x \in M_k,$$

$$\int_{M_k} f^k d\mu \leqq \lim_n \int_{M_k} f_n d\mu\,.$$

Now, extend each f^k to be defined on M by setting $f^k(x) = 0$ for $x \in M - M_k$, and define $f = \sum_{k \in \mathbb{N}_+} f^k$. Obviously, we have $\int_{M_k} f d\mu = \int_{M_k} f^k d\mu$ on each M_k. Since we have, a.e. $x \in M$, $f(x) \in \mathrm{Ls}\,(f_n(x))$, it follows from the Monotone Convergence Theorem, p. 56, that $\int f d\mu = \sum_{k \in \mathbb{N}_+} \int f^k d\mu$. Hence, $\int f d\mu \leqq \sum_{k=1}^{\infty} \lim_n \int_{M_k} f_n d\mu$. But the right-hand side of this inequality is equal to

$$\lim_r \left(\sum_{k=1}^{r} \lim_n \int_{M_k} f_n d\mu \right) = \lim_r \left(\lim_n \sum_{k=1}^{r} \int_{M_k} f_n d\mu \right) \leq \lim_n \int_M f_n d\mu .$$

Therefore, one obtains $\int f d\mu \leqq \lim_n \int f_n d\mu$. This completes the proof of the first part of the assertion of the theorem.

Finally, under the additional assumption made in the second part of the assertion of the theorem, we have in (3.12) $\int g d\mu = \lim_n \int f_n d\mu$. And thus in (3.14) we obtain

$$g(x) = \sum_{\substack{i=0 \\ t^i(x)>0}}^{m} t^i(x) \lim_n h_n^i(x)$$

as $\lim_n h_n^i(x)$ exists for each $i = 0, \ldots, m$. Therefore, there is an integrable selection f of $\mathrm{Ls}\,(f_n)$ with $\int f d\mu = \lim_n \int f_n d\mu$.

To prove the last assertion of the theorem, we define a set function τ_n on $\mathcal{M}$ for each n by $\tau_n(S) = \int_S f_n d\mu,\ S \in \mathcal{M}$. If we show that the $\sigma^m(ba, L_\infty)$-closure of the set $\{\tau_1, \tau_2, \ldots\} \subset ba^m$ consists of countably additive set functions only, then by the Generalized Lebesgue's Theorem, p. 62, the last assertion of the theorem is established.

Let τ be a limit point of $\{\tau_n \mid n \in \mathbb{N}_+\}$. Let g, f^*, and f be as in the above proof. Then, by our arguments concerning (3.10), we have

$$0 \leqq \tau_f(M) = \tau(M) - \tau_c(M) = \lim_n \int f_n d\mu - \int g d\mu .$$

It follows from the assumption of the last assertion that

$$\lim_n \int f_n d\mu = \int f d\mu = \int f^* d\mu \leqq \int g d\mu,$$

which implies $\tau_f = 0$. This completes the proof of the theorem. □

Mathematical Proposition 3.17 [Limes Superior under Integration] *Let $(\varphi_n)_n$ be a sequence of correspondences from a measure space $(M, \mathcal{M}, \mu)$ to $\mathbb{R}_+^m$. If there is a sequence $(g_n)_n$ of measurable functions satisfying the following conditions:*

(a) *$\varphi_n(x) \leqq g_n(x)$, a.e. $x \in M$, and*

(b) *the sequence $(g_n)_n$ is uniformly integrable, and the set $\{g_n(x) \mid n \in \mathbb{N}_+\}$ is bounded a.e. $x \in M$;*

then, we have $\mathrm{Ls}\,(\int \varphi_n)\, d\mu \subset \int \mathrm{Ls}\,(\varphi_n)\, d\mu$.

Proof Let $z \in \mathrm{Ls}\,(\int \varphi_n d\mu)$; then, there is a subsequence $(z_{n_k})_{k \in \mathbb{N}_+}$ of a sequence $(z_n)_{n \in \mathbb{N}_+}$ for which $z_n = \int f_n d\mu$, $f_n \in \mathcal{L}_{\varphi_n}$, each n and $(z_{n_k})_k \to_k z$ with $z_{n_k} =$

$\int f_{n_k} d\mu$. It follows from the two conditions of the theorem that the sequence $(f_{n_k})_k$ is uniformly integrable, and the set $\{f_{n_k}(x) \mid k \in \mathbb{N}_+\}$ is bounded a.e. $x \in M$. Thus, by Mathematical Theorem 3.6 there exists an integrable selection f from the correspondence $x \mapsto \mathrm{Ls}\,(f_{n_k}(x))$ such that $z = \int f d\mu = \lim_k \int f_{n_k} d\mu$. Therefore, $z \in \int \mathrm{Ls}\,(\varphi_{n_k}) d\mu \subset \int \mathrm{Ls}\,(\varphi_n) d\mu$. □

By the Generalized Lebesgue's Theorem, p. 62, the second condition in the above mathematical proposition is satisfied if $\lim_n g_n(x)$ exists a.e. in M and $\lim_n \int g_n d\mu = \int \lim_n g_n d\mu$. Of course, if every $\varphi_n, n \in \mathbb{N}_+$, is bounded from above by an identical integrable function g, then both of the conditions of the mathematical proposition are trivially satisfied.

Mathematical Proposition 3.18 [Compact Integrals] *If a correspondence φ from a measure space $(M, \mathscr{M}, \mu)$ to $\mathbb{R}^m$ is closed-valued and integrably bounded, then the integral $\int \varphi d\mu$ is compact.*

Proof As φ is integrably bounded, $\int \varphi d\mu$ is bounded. Let $\varphi_n = \varphi$ for each $n \in \mathbb{N}_+$. Then, $\mathrm{Ls}\,(\varphi_n(x)) = \varphi(x)$ since $\varphi(x)$ is closed. It follows from Mathematical Proposition 3.17 that we have

$$\mathrm{cl}\left(\int \varphi d\mu\right) = \mathrm{Ls}\left(\int \varphi_n d\mu\right) \subset \int \mathrm{Ls}\,(\varphi_n) d\mu = \int \varphi d\mu.$$

Thus $\int \varphi d\mu$ is compact. □

Mathematical Proposition 3.19 [Parametric Closed Graph of Integral] *Let $(M, \mathscr{M}, \mu)$ be a measure space and T a metric space. If φ is a correspondence from $M \times T$ to $\mathbb{R}^m$ satisfying the following properties at $p \in T$:*

(1) $\varphi(\cdot, p) : M \to \mathbb{R}^m$ is integrably bounded on a neighborhood U of p, and
(2) the correspondence $\varphi(x, \cdot) : T \to \mathbb{R}^m$ is closed at $x \in M$ almost everywhere;

then, the relation from T to $\mathbb{R}^m$, $q \mapsto \int \varphi(\cdot, q) d\mu$, is closed at $p \in T$.

Proof Let $(p_n)_n$ be a sequence in T converging to $p \in T$. We show

$$\mathrm{Ls}\left(\int \varphi(\cdot, p_n) d\mu\right) \subset \int \varphi(\cdot, p) d\mu.$$

By the property (2) above, we have $\mathrm{Ls}\,(\varphi(x, p_n)) \subset \varphi(x, p)$, a.e. $x \in M$. It then follows from Mathematical Proposition 3.17 that

$$\mathrm{Ls}\left(\int \varphi(\cdot, p_n) d\mu\right) \subset \int \mathrm{Ls}\,\varphi(\cdot, p_n) d\mu \subset \int \varphi(\cdot, p) d\mu.$$

□

Notes 3

Notes 3.1 [Section 3.1, p. 91]

[Closed convergence, p. 91] Closed convergence was studied by [47, pp. 168–172] and [70, pp. 241–250].
[Hausdorff distance, p. 92] The Hausdorff distance function first appeared in [47, pp. 166–168].
[Closed convergence topology, p. 96] See [123], [38] and [37] in addition to [47] and [70].

Notes 3.2 [Section 3.2, p. 97] The procedure of considering a set of preference relations satisfying certain properties and introducing a metric on the set was first introduced into economic analysis by Kannai's paper, which was later published as [65]. It considered the set of all continuous utility functions $u : \mathbb{R}^\ell_+ \to \mathbb{R}$ that satisfied monotonicity (i.e., $u(x) \geqq u(y)$ for $x, y \in \mathbb{R}^\ell_+$ with $x \geqq y$), and defined a distance function ρ on this set by

$$\rho = \max\left\{ \frac{|\,u(x) - v(x)|}{1 + \sum_{j=1}^{\ell}} x_j^2 \;\middle|\; x \in \mathbb{R}^\ell_+ \right\},$$

where u and v are utility functions. Subsequently, [25] introduced a metric on the set of preference relations using the Hausdorff distance function. The use of the closed convergence topology in economics is attributed to an unpublished paper by J.F. Mertens. Following this, [54] employed the closed convergence topology to prove a limit theorem of the core of an economy. It is no exaggeration to say that [55] established the use of closed convergence topology in economic analysis.

[Closed convergence topology on $\mathcal{P}$, p. 97] The way in which the closed convergence topology is introduced in this section follows that of [42].
[Theorem 3.1, p. 98] This is a version of [55, Theorem 1, p. 96] for the case where the commodity space and the consumption sets are non-convex and the transitivity of preference relations is not assumed.
[Corollary 3.2, p. 100] This is originally due to [25].

Notes 3.3 [Section 3.5, p. 103]

[Measurable selections] Mathematical Proposition 3.4, p. 103, is due to [71]. The method of proof here is originally from [72]. Mathematical Proposition 3.5, p. 103, is a simple consequence of Mathematical Proposition 3.4. It was first used by [19].
[Choquet Capacitability Theorem] Mathematical Theorem 3.2, p. 105, is due to [89] which applied the method of proof by Bourbaki.
[Mathematical Theorem 3.3, p. 106] Measurable Selection Theorem is due to [9]. The proof here using Choquet Capacitability Theorem is due to [55, pp. 54–58]. The textbook [61] also covers this theorem.
[Measurable Relations] The method of proving Mathematical Proposition 3.6, p. 108, and Mathematical Proposition 3.8, p. 109, is due to [24, (4.3) and (4.5)]. For each of the mathematical propositions on measurable relations, we referred to [55, 55–61].

Notes 3.4 [Section 3.6, p. 110]

[Mathematical Fact 3.1, p. 111] See, for example, [97, Theorem 8.1, pp. 53–54] for (3) and [15] for (4).
[Purely non-atomic vector measures and Liapounoff convexity, p. 110] For an elaborate and insightful proof, see [43]. One may also refer to a short proof but with the use of powerful theorems of Banach–Alaoglu and Krein–Milman, see [75]. Another proof without using these theorems is also given by [100].
[Functional convexity, p. 112, p. 115] The proof of the Liapounoff convexity using the functional convexity is due to [43].

Notes 3.5 [Section 3.7, p. 116] Among references concerning the integral of correspondences, basic references related to the measure theoretic approach to general equilibrium analysis are [7], [24] and [5]. The section on the integral of a correspondence in [55, pp. 53–79] brings together the literature on this topic in a form very useful for researchers. Section 3.7 is substantially based on the book [55, D.II].

[Mathematical Proposition 3.14, p. 117] Richter's Proposition is due to [98].
[Mathematical Proposition 3.15, p. 118] Vind's Proposition is from [118, Lemma A].
[Mathematical Proposition 3.16, p. 118] See [51] or [54] for Hildenbrand's Lemma.
[Mathematical Theorem 3.4, p. 119] Aumann's Theorem is from [9].

Chapter 4
Individual Demands

Abstract This chapter addresses the continuity property of demand relations with respect to prices. This property is regarded as an essential element in general equilibrium analysis of market transactions. Potential "discontinuity" of individual demand relations arises from the "discontinuity" of budget set relations. In order to identify the source of the discontinuity of these relations, the points in the consumption sets are first examined. Then, the set of these source points in each consumption set is analyzed. The set is defined as a subset of a consumption set X with respect to a price vector p, and is called the *critical set of the consumption set.* A related concept, that of a *critical set in the wealth space*, is also introduced. We then proceed to clarify the mathematical properties of these critical sets. Based on the concepts of critical sets, we examine the continuity properties of budget set relations and individual demand relations. In the final section of the chapter we bring up the issue of the survival of agents who do not possess sufficient wealth to participate in market transactions to obtain commodities in their consumption sets. This can result in empty demand sets. In order to address the issue of survival and the potential discontinuity of demand relations, the concept of *subdemand* is introduced.

4.1 Critical Sets of Consumption Sets

Concepts of Critical Sets

To begin with, an element that plays a crucial role in affecting the continuity property of budget set relations as well as individual demand relations is introduced by the concepts of a cheaper point and local cheaper points of a consumption vector. Then, critical sets of consumption sets are defined by means of local cheaper points.

A. Yamazaki, *Mathematical Economics of General Equilibrium*, Monographs in Mathematical Economics 8, https://doi.org/10.1007/978-981-96-9497-6_4

Cheaper Points and Local Cheaper Points

Definition 4.1 (Cheaper Points and Local Cheaper Points) Let $X \in \mathscr{X}$ be a consumption set and $p \neq 0 \in \mathbb{R}^\ell$ a price vector. A consumption vector $z \in X$ is a *cheaper point* of a consumption vector $x \in X$ if the value of z is strictly less than the value of x with respect to the price p, that is, if $p \cdot z < p \cdot x$. A consumption vector x in X is said to have *local cheaper points* or *a convergent sequence of cheaper points* if x has a cheaper point in X in any neighborhood of x. That is, for every neighborhood U of x, there exists a consumption vector $z \in U \cap X$ such that $p \cdot z < p \cdot x$.

Definition 4.2 (Critical Sets of Consumption Sets) Given a consumption set $X \in \mathscr{X}$ and a price vector $p \neq 0 \in \mathbb{R}^\ell$, the set of all $x \in X$ such that x does not have a local cheaper point is defined as the *critical set of the consumption set* X at p, and is denoted by $C_p(X)$. A related concept, the *critical set of the consumption set X at p in the wealth space*, is defined as a subset $J_p(X)$ of the real line $\mathbb{R}$ consisting of the wealth levels that represent the values of consumption vectors belonging to the critical set of the consumption set X at p, that is, $J_p(X) = \{w \in \mathbb{R} \mid w = p \cdot x \text{ for some } x \in C_p(X)\}$.

In Section 4.2 examples are provided to illustrate that the critical sets of consumption sets are the sources of discontinuity in budget set relations and individual demand relations.

Identifying the Critical Sets of Consumption Sets

There is a straightforward way to characterize critical sets. Let us define a subset $HN_p(x, \delta)$ of $\mathbb{R}^\ell$ for $p \in \mathbb{R}^\ell, x \in \mathbb{R}^\ell, \delta > 0$, by

$$HN_p(x, \delta) = \{z \in \mathbb{R}^\ell \mid \|x - z\| < \delta \text{ and } p \cdot z < p \cdot x\},$$

and call it the *open half ball (with respect to p) centered at x with radius δ*. (See Fig. 4.1.) Then, the following characterization of the critical sets of consumption sets at a given price vector is obtained immediately.

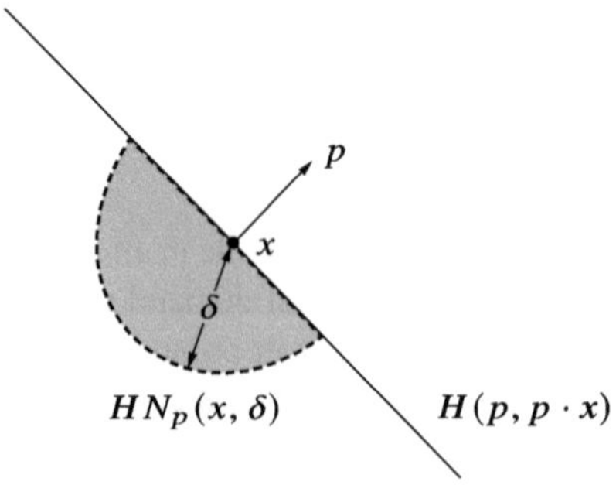

Fig. 4.1 Open half ball

Mathematical Fact 4.1 [Characterization of Critical Sets of Consumption Sets] *Given a consumption set $X \in \mathscr{X}$ and a price vector $p \neq 0 \in \mathbb{R}^\ell$, the critical set of the consumption set X at p is characterized as follows:*

$$C_p(X) = \{x \in X \mid HN_p(x,\delta) \cap X = \emptyset \text{ for some } \delta > 0\}.$$

It is worth noting that within the context of convex consumption sets, the critical set of a consumption set $X \in \mathscr{X}$ coincides with the set of consumption vectors that require the "minimum" expenditure among consumption vectors. Formally, this is expressed as $C_p(X) = \{x \in X | \; p \cdot x = \inf p \cdot X\}$ if $X \in \mathscr{X}$ is a convex set. For future reference, we will now state this fact as the following proposition.

Proposition 4.1 (Critical Sets in Convex Environments)
Let $X \in \mathscr{X}$ be a convex set. Then, $C_p(X) = \{x \in X \mid p \cdot x = \inf p \cdot X\}$.

Proof If $p \cdot x = \inf p \cdot X$, then, for any $\delta > 0$, $HN_p(x,\delta) \cap X = \emptyset$. Thus, $x \in C_p(X)$. Conversely, let $x \in C_p(X)$. Clearly, $p \cdot x \geqq \inf p \cdot X$ as $x \in X$. If $p \cdot x > \inf p \cdot X$, then $p \cdot x > p \cdot z$ for some $z \in X$. By the convexity of X, $[z, x] \subset X$. It follows that there is a sequence $(z_n)_n$ in $[z, x]$ such that $z_n \to_n x$ with $p \cdot z_n < p \cdot x$ for each n. Hence, $x \notin C_p(X)$, a contradiction. Consequently, one must have $p \cdot x = \inf p \cdot X$. □

Examples of Critical Sets of Consumption Sets

Examples [Consumption Vectors without Local Cheaper Points]

In Fig. 4.2 the consumption vector $z \in X$ has local cheaper points at the given price vector p, and thus, $z \notin C_p(X)$. In contrast, for the consumption vector $x \in X$, one has

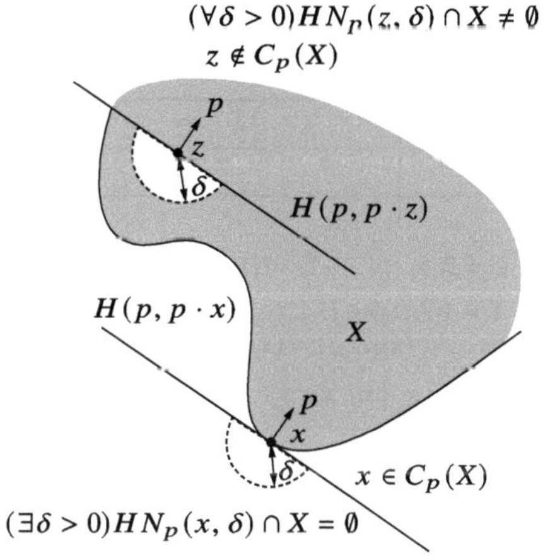

Fig. 4.2 Critical set $C_p(X)$

$HN_p(x, \delta) \cap X = \emptyset$ for any $\delta > 0$. This implies that x does not possess local cheaper points, and thus, $x \in C_p(X)$. The "geometric image" of the consumption vector x belonging to the critical set of the consumption set X at p is that the consumption set X is "tangent" to the hyperplane $H(p, p \cdot x)$ at $x \in X$. This statement is somewhat vague, as X need not be a differentiable manifold.

The following two examples may prove useful in gaining insights into the possible mathematical properties of critical sets.

Example [Critical Sets of Consumption Sets $C_p(X), J_p(X)$]

Let $\mathbb{G} = \mathbb{R} \times \{\dots, -2, -1, 0, 1, 2, \dots\}$ and X be given as in Fig. 4.3. Then, at any $p \in \mathbb{R}^2_{++}$, we have $C_p(X) = \{x_0, x_1, x_2, x_3, x_4, \dots\}$. The characteristics of the sets $C_p(X)$ and $J_p(X)$ in this example are as follows:

(1) The set $C_p(X)$ is countable.
(2) The set $C_p(X)$ does not depend upon $p \in \mathbb{R}^2_{++}$.
(3) The set $J_p(X)$ is countable.

It is evident that the first two properties do not hold in general (see Fig. 4.4). However, it is of significant interest to ascertain whether the third property is true in general.

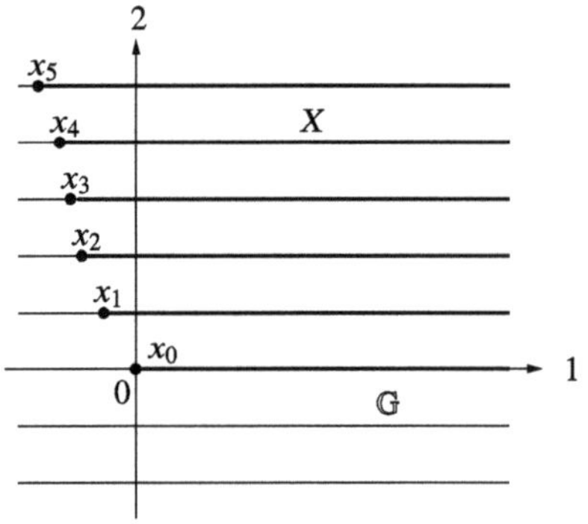

$\mathbb{G} = \mathbb{R} \times \{\dots, -2, -1, 0, 1, 2, \dots\}$
$X = \{(z_1, z_2) \in \mathbb{G} \mid z_1 \geqq x_{j1}, z_2 = j$
for $x_j = (x_{j1}, x_{j2}) = (x_{j1}, j), j = 0, 1, 2, \dots\}$
$C_p(X) = \{x_0, x_1, x_2, \dots\}, p \in \mathbb{R}^2_{++}$

Fig. 4.3

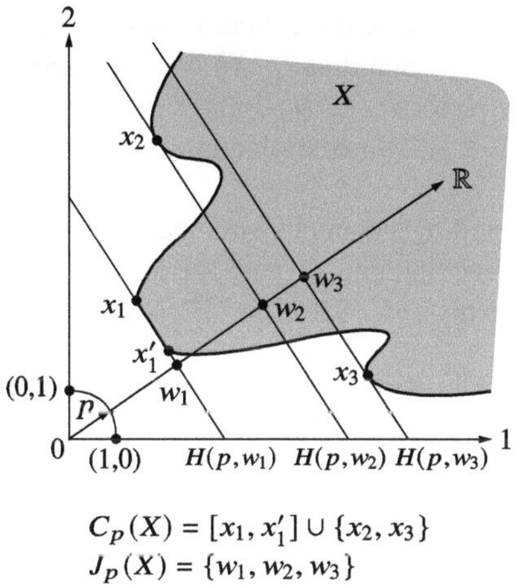

Fig. 4.4

Example [Countability of Critical Sets in the Wealth Space]

Let $\mathbb{G} = \mathbb{R}^2$ and X be given as in Fig. 4.4. For a given price vector $p \in \mathbb{R}^{\ell}_{+}$ as depicted in the figure, the entire points on the line segment $[x_1, x_1']$ belong to $C_p(X)$. Consequently, the set $C_p(X)$ contains a continuum of points, not just a countably many points only. Note, however, that even in this case, the wealth levels in $J_p(X)$, indicated in the figure as points w_1, w_2, w_3 in the real line shown as $\mathbb{R}$, are countable.

Properties of Critical Sets of Consumption Sets

The property (3) in the first of the above two examples concerning the countability of critical sets of consumption sets in the wealth space is demonstrated to be true in general for any $p \neq 0$ and $X \in \mathcal{X}$. This is proven in the following theorem. It plays a pivotal role in the analysis of the continuity properties of the budget set relation and the demand relation in economic environments where the commodity space and/or consumption sets may not be convex.

Countability of Critical Set in the Wealth Space

Theorem 4.1 (Fundamental Theorem on Critical Sets in the Wealth Space)

For any consumption set $X \in \mathcal{X}$ and at any price vector $p(\neq 0) \in \mathbb{R}^{\ell}$, the critical set of the consumption set X at p in the wealth space, $J_p(X)$, is countable.

Proof If this were not the case, there would exist a consumption set $X \in \mathscr{X}$ and a price vector $p(\neq 0) \in \mathbb{R}^\ell$ such that $J_p(X)$ is uncountable. For each $w \in J_p(X)$, select one consumption vector $x_w \in H(p, w) \cap C_p(X)$. The set of such consumption vectors is denoted by $C \subset \mathbb{R}^\ell$. Since each consumption vector in C is distinct, C is uncountable.

Now, $\mathbb{R}^\ell$ can be covered by a countable number of open balls with radius 1. As C is a subset of $\mathbb{R}^\ell$ and is uncountable, at least one of these balls, say N_0, must contain uncountably many elements of C. Set $C_0 = C \cap N_0$.

Claim 1: *There exists an element x in C_0 and a sequence $(x_n)_n$ in C_0 such that $x_n \to_n x$ and $p \cdot x_n < p \cdot x$ for all n* (cf. Fig. 4.5).

If this is not the case, then for ech $x \in C_0$, there exists a positive number δ_x such that $HN_p(x, \delta_x) \cap C_0 = \emptyset$. For each positive integer $n > 0$, define a subset C_n of C_0 as follows:

$$C_n = \{x \in C_0 \mid HN_p(x, 1/n) \cap C_0 = \emptyset\}.$$

Then, $C_0 = \cup_n C_n$. Since C_0 is uncountable, at least one of the C_n's, say C_N, contains an uncountable number of elements of C_0. Therefore, one can pick a sequence $(z_n)_n$ in C_N so that each z_n is a distinct element of C_N. Thus, for n and n' such that $n \neq n'$, one has $p \cdot z_n \neq p \cdot z_{n'}$. Since $C_N \subset C_0 = C \cap N_0$ is a bounded set, there is a convergent subsequence $(z_k)_k$ of $(z_n)_n$. Then, for sufficiently large k and k' such that $k \neq k'$, one has $\|z_k - z_{k'}\| < 1/(2N)$ and $p \cdot z_k \neq p \cdot z_{k'}$, say $p \cdot z_k > p \cdot z_{k'}$. It follows that one has $z_{k'} \in HN_p(z_k, 1/N)$. This contradicts the fact that both z_k and $z_{k'}$ are distinct elements of the set C_N. This completes the proof of Claim 1.

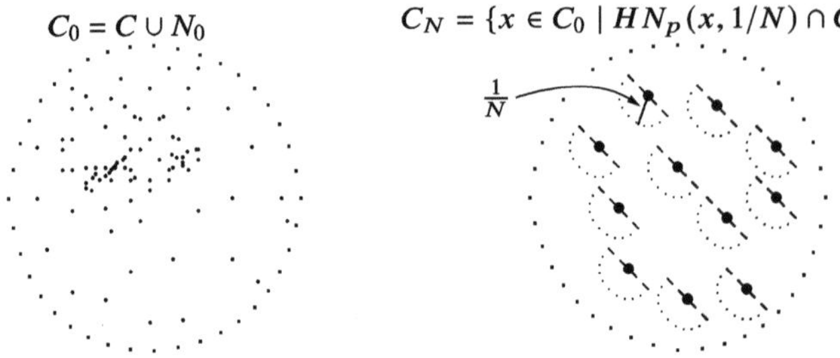

Fig. 4.5 An illustration of the arguments in Claim 1

Let $(x_n)_n$ and x be as in Claim 1. Since x belongs to $C_p(X)$, there is a positive number δ such that $HN_p(x, \delta) \cap X = \emptyset$. But for n sufficiently large, x_n belongs to $HN_p(x, \delta) \cap X$, a contradiction. This proves the assertion of the theorem. □

One can immediately ascertain an essential property of the critical set of a consumption set from the fundamental theorem on critical sets in the wealth space.

Corollary 4.1 (Null Measure of Critical Sets of Consumption Sets)

For any consumption set $X \in \mathscr{X}$ and any price vetor $p(\neq 0) \in \mathbb{R}^\ell$, the critical set $C_p(X)$ of the consumption set X at p is a set of ℓ-dimensional Lebesgue measure zero.

Proof Let $\mathcal{H}_p = \{H(p,w) \subset \mathbb{R}^\ell \mid w \in J_p(X)\}$. As $J_p(X)$ is countable by Theorem 4.1, the set $\cup\mathcal{H}_p$ is a countable union of hyperplanes having $(\ell-1)$- linear dimensions in $\mathbb{R}^\ell$, and thus, it is a set of ℓ-dimensional Lebesgue measure zero. Since the Lebesgue measure is complete, the conclusion follows from the fact that $C_p(X) \subset \cup\mathcal{H}_p$. □

The preceding theorem and corollary illuminate the fundamental mathematical properties of critical sets $J_p(X)$ and $C_p(X)$ for any consumption set $X \in \mathscr{X}$ at any price vector $p(\neq 0) \in \mathbb{R}^\ell$. We now turn to investigating whether these critical sets exhibit a measurable variation as consumption sets vary in $\mathscr{X}$ at a fixed price vector $p(\neq 0) \in \mathbb{R}^\ell$.

Measurability of the Relations $C_p : \mathscr{X} \to \mathbb{G}$ and $J_p : \mathscr{X} \to \mathbb{R}$

Define relations $J_p : \mathscr{X} \to \mathbb{R}$ and $C_p : \mathscr{X} \to \mathbb{G}$ by $X \mapsto J_p(X)$ and $X \mapsto C_p(X)$, respectively. It can be shown that $J_p = \{(X,w) \in \mathscr{X} \times \mathbb{R} \mid w \in J_p(X)\}$ and $C_p = \{(X,x) \in \mathscr{X} \times \mathbb{G} \mid x \in C_p(X)\}$ are measurable relations. To begin, we will demonstrate the measurability of C_p.

Proposition 4.2 (Measurability of C_p)

The relation C_p from $\mathscr{X}$ to $\mathbb{G} \subset \mathbb{R}^\ell$ at a given $p(\neq 0) \in \mathbb{R}^\ell$ is a Borel subset of $\mathscr{X} \times \mathbb{G}$.

Proof Let $p(\neq 0) \in \mathbb{R}^\ell$. Given a vector x in $\mathbb{R}^\ell$, define for each positive integer t and each positive number δ, the *t-truncated closed half ball (with respect to the price vector p) centered at x with radius δ* to be

$$\overline{HN}_p^t(x,\delta) = \left\{z \in \mathbb{R}^\ell \mid \|x-z\| \leqq \delta \text{ and } p\cdot z \leqq p\cdot x - (1/t)\right\}.$$

(See Fig. 4.6.) Note that $\overline{HN}_p^t(x,\delta) \neq \emptyset$ if $t > 1/\delta\|p\|$.

For each positive integer m, let $\mathcal{S}_{m,t} = \{(X,x) \in \mathscr{X}\times\mathbb{G} \mid x \in X$ and $\overline{HN}_p^t(x,1/m) \cap X \neq \emptyset\}$. We show first that $\mathcal{S}_{m,t}$ is closed. Let $(X_n,x_n)_n$ be a sequence in $\mathcal{S}_{m,t}$ which converges to (X_0,x_0). Trivially, $x_0 \in X_0$. To show $\overline{HN}_p^t(x_0,1/m) \cap X_0 \neq \emptyset$, choose a vector $z_n \in \overline{HN}_p^t(x_n,1/m) \cap X_n$ for each $n \in \mathbb{N}_+$ as $\overline{HN}_p^t(x_n,1/m) \cap X_n \neq \emptyset$ for every n. Each z_n satisfies $\|z_n - x_n\| \leqq 1/m$ and $p\cdot z_n \leqq p\cdot x_n - 1/t$. Since the sequence $(x_n)_n$ converges to x_0, $\|z_n - x_n\| \leqq 1/m$ for every n implies that the sequence $(z_n)_n$ is bounded. Thus, there is a convergent subsequence $(z_{n_k})_k$ of the sequence $(z_n)_n$, say $z_{n_k} \to_k z$. For this z, we have $\|z - x_0\| \leqq 1/m$, $p\cdot z \leqq p\cdot x_0 - 1/t$, and $z \in X_0$ as $z_{n_k} \in X_{n_k}$ for all k and $X_{n_k} \to_k X_0$. It follows that $z \in \overline{HN}^t(x_0,1/m) \cap X_0$.

We now claim that we have

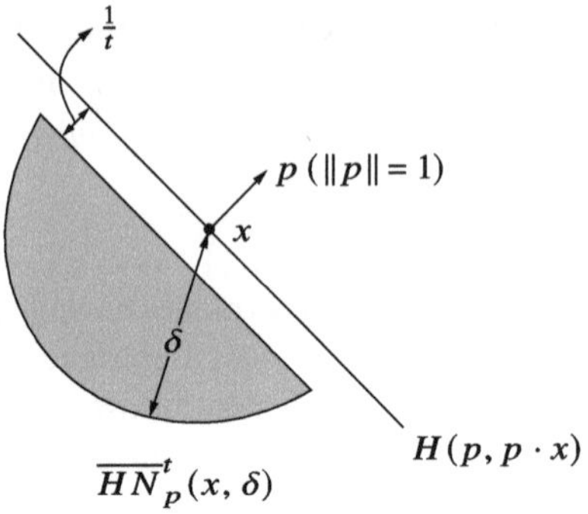

Fig. 4.6 t-truncated closed half ball centered at x with radius δ

$$\mathcal{X} \times \mathbb{G} - C_p = \left(\cap_{m \in \mathbb{N}_+} \cup_{t \in \mathbb{N}_+, t > m/\|p\|} \mathcal{S}_{m,t}\right) \cup \{(X, x) \in \mathcal{X} \times \mathbb{G} \mid x \in \mathbb{G} - X\}.$$

Indeed, let $(X, x) \in \mathcal{X} \times \mathbb{G} - C_p$. Then, either (a) $x \notin X$, or (b) $\overline{HN}_p(x, \delta) \cap X \neq \emptyset$ for any $\delta > 0$, where $\overline{HN}_p(x, \delta)$ is the closure of the set $HN_p(x, \delta)$. In the case of (a), (X, x) belongs to the second term of the right-hand side. In the case of (b), for any $m \in \mathbb{N}_+$, $\overline{HN}^t_p(x, 1/m) \cap X \neq \emptyset$ for t large enough. Thus (X, x) belongs to the first term of the right-hand side. Conversely, let (X, x) belong to the right-hand side. If $x \in \mathbb{G} - X$, $(X, x) \in \mathcal{X} \times \mathbb{G} - C_p$. Now, let (X, x) belong to the first term of the right-hand side; then, it must be $(X, x) \in \mathcal{X} \times \mathbb{G} - C_p$. Otherwise, for some $\delta > 0$ one has $HN_p(x, \delta) \cap X = \emptyset$. Let m be a positive integer greater than $1/\delta$. Then, for some t greater than $m/\|p\|$ we have $\overline{HN}^t_p(x, 1/m) \cap X \neq \emptyset$ by the definition of $\mathcal{S}_{m,t}$. As $\overline{HN}^t_p(x, 1/m) \subset HN_p(x, \delta)$, it follows that $HN_p(x, \delta) \cap X \neq \emptyset$, a contradiction.

It is straightforward to verify that the set $\{(X, x) \in \mathcal{X} \times \mathbb{G} \mid x \in X\}$ is closed. Consequently, it follows from the claim that the set $\mathcal{X} \times \mathbb{G} - C_p$ is the union of a countable intersection of F_σ-sets and an open set. Consequently, C_p is a Borel subset of $\mathcal{X} \times \mathbb{G}$. □

We now turn to the question of the measurability of the relation $J_p : \mathcal{X} \to \mathbb{R}$. As shown in Proposition 4.2, the critical sets of consumption sets as a relation $C_p : \mathcal{X} \to \mathbb{R}$ are measurable. However, the relation J_p may not be measurable if we restrict measurable sets to Borel subsets of $\mathcal{X} \times \mathbb{R}$. Consequently, we extend our considerations to include analytic subsets of $\mathcal{X} \times \mathbb{R}$. The following proposition makes use of the fact that an analytic set in a Polish space is universally measurable (cf. Mathematical Property 1.29 (2)), which may be interpreted as asserting that a class of analytic sets in Polish spaces represents a minimal extension of the class of measurable sets in measure spaces.

Proposition 4.3 (Universal Measurability of J_p)
The relation J_p from $\mathscr{X}$ to $\mathbb{R}$ has the following properties:

(1) *The relation J_p is an analytic subset of $\mathscr{X} \times \mathbb{R}$.*
(2) *The relation J_p belongs to the completion $\mathscr{B}_\mu(\mathscr{X} \times \mathbb{R})$ for any Borel (normalized) measure μ on $\mathscr{B}(\mathscr{X} \times \mathbb{R})$.*

Proof Define a mapping $v_p : \mathscr{X} \times \mathbb{G} \to \mathscr{X} \times \mathbb{R}$ by $(X,x) \mapsto (X, p \cdot x)$. Then, v_p is continuous and satisfies

$$\begin{aligned} v_p(C_p) &= v_p(\{(X,x) \in \mathscr{X} \times \mathbb{G} \mid x \in C_p(X)\}) \\ &= \{(X,w) \in \mathscr{X} \times \mathbb{R} \mid w = p \cdot x \text{ for some } x \in C_p(X)\} \\ &= J_p. \end{aligned}$$

Since C_p is a Borel set by Proposition 4.2, J_p is a continuous image of a Borel set under the mapping v_p. In view of Mathematical Property 1.26 (2), p. 52, and 1.27 (b), p. 52, J_p is an analytic set in $\mathscr{X} \times \mathbb{R}$ by virtue of both $\mathscr{X} \times \mathbb{G}$ and $\mathscr{X} \times \mathbb{R}$ being Polish spaces. Moreover, as an analytic set in a Polish space is universally measurable by Mathematical Property 1.29 (2), p. 53, J_p in particular belongs to $\mathscr{B}_\mu(\mathscr{X} \times \mathbb{R})$ for any given Borel measure μ on $\mathscr{B}(\mathscr{X} \times \mathbb{R})$. □

Remark 4.1 J_p is a continuous image of a Borel set C_p as in the proof of the above proposition. Unless it is a continuous image under an *injection*, a continuous image of a Borel set need not be a Borel set in general.

4.2 Continuity of Individual Demands

We are now prepared to discuss the continuity property of the individual demand relation. When one notes that demand vectors provide a description of optimal consumption choices made by economic agents out of their own budget sets, it is natural that the continuity of the demand relation crucially depends upon that of the budget set relation, given that preference relations are continuous. Accordingly, we first wish to give a close examination of the continuity property as well as the possible discontinuity of the budget set relation.

Discontinuity and Continuity of the Budget Set Relation $B : \mathscr{X} \times \mathbb{R} \times \mathbb{R}^\ell \to \mathbb{G}$

Possible Discontinuities of the Budget Set Relation

In addition to the concept of critical sets of consumption sets, we will introduce further notation and concepts that are applied to the description of the continuity or discontinuity of the budget set relation. A relation φ from a topological space S to another topological space T is said to be *locally uniformly compact-valued* at $x \in S$

if, whenever $\varphi(x) \neq \emptyset$, there exist a neighborhood $U \subset S$ of x and a compact subset $K \subset T$ such that, for any $z \in U$, the set $\varphi(z)$ is compact and $\varphi(z) \subset K$. Although the convexity of the commodity space $\mathbb{G}$ is not assumed in the main mathematical analysis of this book, in the case where the convexity of the commodity space is assumed, $\mathscr{X}_{co}$ denotes the set of all convex consumption sets X in $\mathscr{X}$.

Remark 4.2 The budget set relation $B : \mathscr{X} \times \mathbb{R} \times \mathbb{R}^\ell \twoheadrightarrow \mathbb{G}$ is locally uniformly compact-valued at $(X, w, p) \in \mathscr{X} \times \mathbb{R} \times \mathbb{R}^\ell$ if $p \in \mathbb{R}^\ell_{++}$.

Exercise 4.1 Verify the statement of Remark 4.2.

The following examples illustrate two instances in which the budget set relation fails to be upper hemi-continuous. It is evident that the commodity space's convexity is not a determining factor in its failure of upper hemi-continuity.

Examples [Failure of Upper Hemi-Continuity of Budget Set Relation]

1. (See Fig. 4.7.) Let $\mathbb{G} = \mathbb{R}^2$, and define

$$X_0 = \{x = (x_1, x_2) \in \mathbb{G} \mid x_1 \geqq 10, x_2 \geqq 0, (x_1 - 10)x_2 \geqq 25\}.$$

   Set $X = X_0 \cup \{(10, -5)\}$, and, for each $n \in \mathbb{N}_+$, $X_n = X_0 \cup \{(10 + \frac{1}{n}, -5)\}$. Let $w, w_n = 10, p = (1, 0), p_n = (1, -1/n), n \in \mathbb{N}_+$. Then, $(X_n, w_n, p_n) \to_n (X, w, p)$. Since $B(X, w, p) = \{(10, -5)\}$, let the open ball centered at $(10, -5)$ with radius 1, for example, be an open neighborhood $\mathcal{U}$ of $B(X, w, p)$. Then, as we have $\mathcal{U} \cap X_0 = \emptyset$, $B(X_n, w_n, p_n) \not\subset \mathcal{U}$ for all n. Therefore, the relation B fails to be u.h.c. at (X, w, p).
   This example has the following attributes:

   (a) Although B is compact valued at (X, w, p), it is *not locally uniformly compact-valued* at (X, w, p).
   (b) X and $X_n, n \in \mathbb{N}_+$, are non-convex.

2. (See Fig. 4.8.) Let $\mathbb{G} = \mathbb{R}^2, X, X_n = \mathbb{R}^2_+, w, w_n = 10, p = (1, 0), p_n = (1, -1/n), n \in \mathbb{N}_+$. Then, $(X_n, w_n, p_n) \to_n (X, w, p)$. Define an open neighborhood $\mathcal{U}$ of $B(X, w, p)$ by $\mathcal{U} = \{x = (x_1, x_2) \in \mathbb{G} \mid \text{dist}\,(x, B(X, w, p)) < 1\}$. Then, one has $(12, 2n) \in B(X_n, w_n, p_n)$ but $(12, 2n) \notin \mathcal{U}$ for all n. Hence, B is not u.h.c. at (X, w, p). In this example the consumption sets are convex but the budget sets are not compact.

We will now consider the possibility of failure of lower hemi-continuity of the budget set relation. It should be noted that when wealth levels belong to the critical sets of consumption sets, there is no guarantee that the budget set relation will satisfy the lower hemi-continuity. It is beneficial to comprehend the distinctive characteristics of the examples below, as the establishment of the continuity property

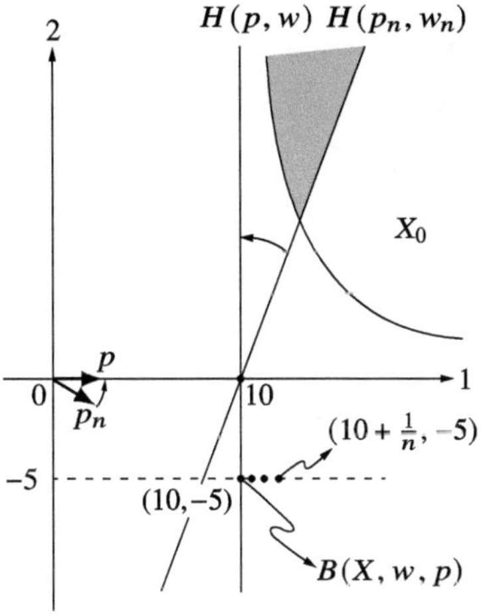

Fig. 4.7 Failure of u.h.c.

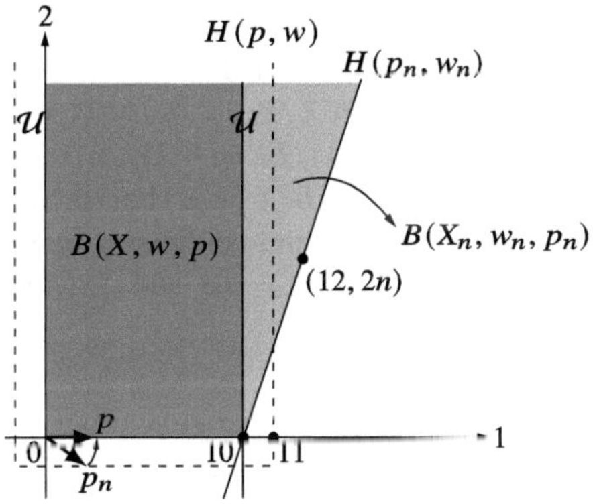

Fig. 4.8 Failure of u.h.c.

of the demand relation $D : \mathcal{P} \times \mathbb{R} \times \mathbb{R}^{\ell} \to \mathbb{G}$ hinges upon the lower hemi-continuity of the budget set relation.

Examples [Failure of Lower Hemi-Continuity of Budget Set Relation]

1. (See Fig. 4.9.) Let $\mathbb{G} = \mathbb{R}^2, X, X_n = \mathbb{R}^2_+, w = 0, w_n = 10^6/n, p = (0, 1), p_n = (1/n, 1 - (1/n)), n \in \mathbb{N}_+$. Then, $(X_n, w_n, p_n) \to_n (X, w, p)$. However, since

$$B(X_n, w_n, p_n) = \{(x_1, x_2) \in \mathbb{R}^2_+ \mid x_1 + (n-1)x_2 \leqq 10^6\},$$

it follows that

$$\text{Li}\,(B(X_n, w_n, p_n)) = \text{Ls}\,(B(X_n, w_n, p_n)) = \{(x_1, x_2) \in \mathbb{R}^2_+ \mid x_1 \in [0, 10^6], x_2 = 0\}.$$

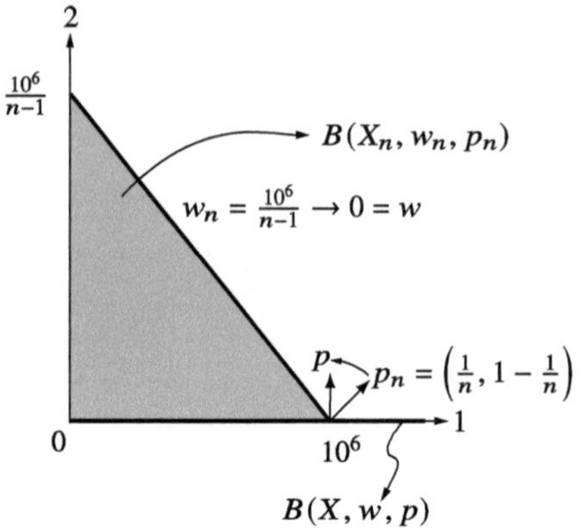

Fig. 4.9 Failure of l.h.c.

Hence, the relation B fails to be lower hemi-continuous at (X, w, p) as one has $B(X, w, p) = \{(x_1, x_2) \in \mathbb{R}^2_+ \mid x_1 \in [0, \infty[, x_2 = 0\}$. Note that in this case we have $C_p(X) = B(X, w, p)$ and $J_p(X) = \{0\}$.

2. (See Fig. 4.10.) Let $\mathbb{G} = \mathbb{R} \times \mathbb{Z}, X, X_n = \mathbb{R}_+ \times \mathbb{N}, w = 5/2, w_n = 5/2 - 5/2n, p = (1/2, 1/2), p_n = ((n-1)/2n, (n+1)/2n))$ for each $n \in \mathbb{N}_+$. Then, $(X_n, w_n, p_n) \to_n (X, w, p)$. $(0, 5) \in B(X, w, p)$ but $(0, 5) \notin \text{Li}\,(B(X_n, w_n, p_n))$. Thus, the relation B is not lower hemi-continuous at (X, w, p). In this case $C_p(X) = \{(x_1, x_2) \in \mathbb{R}_+ \times \mathbb{N} \mid x_1 = 0\}$ and $J_p(X) = \{v \in \mathbb{R}_+ \mid v = (1/2)k \text{ for } k \in \mathbb{N}\}$.

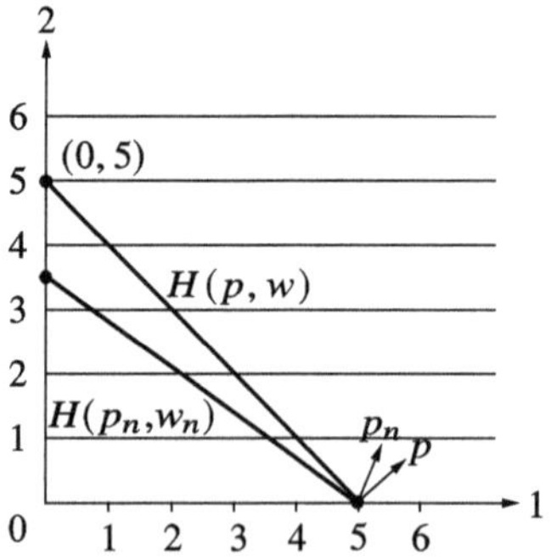

Fig. 4.10 Failure of l.h.c.

3. (See Fig. 4.11.) Let $X, X_n = S, w_n \to_n w, p_n \to_n p, n \in \mathbb{N}_+$, be as in the figure. Since we have $z \in B(X, w, p)$ but $z \notin \text{Li}\,(B(X_n, w_n, p_n))$, the relation B fails to be lower hemi-continuous at (X, w, p).
4. (See Fig. 4.12.) Let $X_n \to_n X, w_n \to_n w, p_n \to_n p, n \in \mathbb{N}_+$, be as in the figure. In this case again, we have $B(X, w, p) \neq \text{Li}\,(B(X_n, w_n, p_n))$, and the relation B fails to be lower hemi-continuous at (X, w, p).

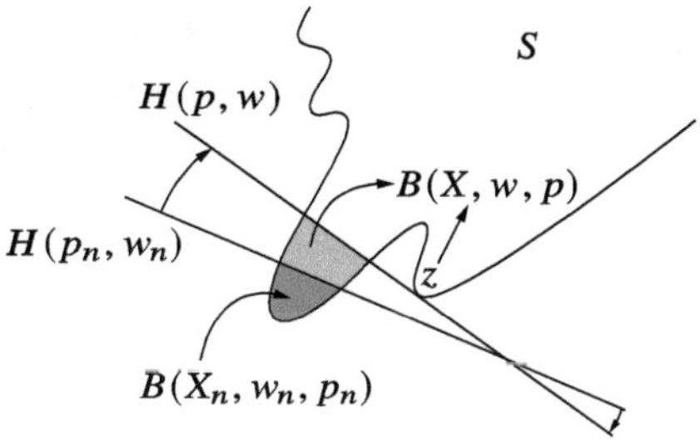

Fig. 4.11 Failure of l.h.c.

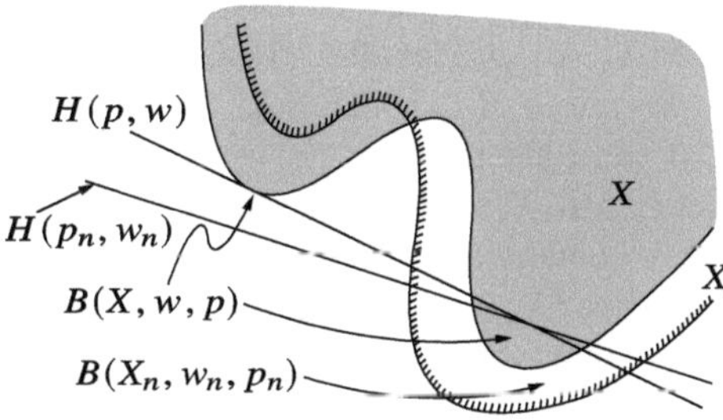

Fig. 4.12 Failure of l.h.c.

Continuity Properties of Budget Set Relation

Having observed potential instances of failure of continuity of the budget set relation, it is necessary to delineate the specific continuity properties of the budget set relation.

Proposition 4.4 (Continuity of Budget Set Relation)

The budget set relation $B : \mathscr{X} \times \mathbb{R} \times \mathbb{R}^{\ell} \to \mathbb{G}$ satisfies the following properties (1) through (3).

(1) *The budget set relation $B : \mathscr{X} \times \mathbb{R} \times \mathbb{R}^{\ell} \to \mathbb{G}$ is lower hemi-continuous off the critical set in the wealth space. That is to say, at every point $(X, w, p) \in \mathscr{X} \times \mathbb{R} \times \mathbb{R}^{\ell}$ with $w \notin J_p(X)$.*

(2) *The budget set relation $B : \mathscr{X} \times \mathbb{R} \times \mathbb{R}^{\ell} \to \mathbb{G}$ is closed, that is, it is a closed set in $\mathscr{X} \times \mathbb{R} \times \mathbb{R}^{\ell} \times \mathbb{G}$.*

(3) *The budget set relation $B : \mathscr{X} \times \mathbb{R} \times \mathbb{R}^{\ell} \to \mathbb{G}$ satisfies the upper hemi-continuity in the following cases:*

(a) *If the budget set relation B is locally uniformly compact-valued at $(X, w, p) \subset \mathscr{X} \times \mathbb{R} \times \mathbb{R}^{\ell}$, then it is upper hemi-continuous at (X, w, p).*

(b) *If the commodity space $\mathbb{G}$ is convex, then the restriction of the budget set relation B to $\mathscr{X}_{co} \times \mathbb{R} \times \mathbb{R}^{\ell}$ is upper hemi-continuous at each point $(X, w, p) \in \mathscr{X}_{co} \times \mathbb{R} \times \mathbb{R}^{\ell}$, where $B(X, w, p) \subset \mathbb{G}$ is compact.*

Proof (1) Let $B(X, w, p) \neq \emptyset, (X_n, w_n, p_n)_n \to_n (X, w, p)$, and $w \notin J_p(X)$. To show that B is l.h.c. at (X, w, p), we prove that $B(X, w, p) \subset \mathrm{Ls}\,(B(X_n, w_n, p_n))$

(cf. Mathematical Proposition 1.2, p. 38). If $x \in B(X, w, p)$ and $p \cdot x < w$, then, since $X = \text{Li}(X_n)$, there is a sequence $(x_n)_n$ such that $x_n \to_n x$ and $x_n \in X_n$ for each n. As $p_n \cdot x_n < w_n$ for large n, $x_n \in B(X_n, w_n, p_n)$ eventually. Thus, $x \in \text{Ls}(B(X_n, w_n, p_n))$.

In general, for any $x \in B(X, w, p)$ with $w \notin J_p(X)$, x has local cheaper points. Thus, there is a sequence $(x_n)_n$ satisfying $x_n \to_n x$ for which $x_n \in X$ and $p \cdot x_n < p \cdot x \leqq w$. It follows that $x \in \text{cl } B_<(X, w, p)$, where $B_<(X, w, p) = \{x \in X \mid p \cdot x < w\}$. This proves $B(X, w, p) = \text{cl } B_<(X, w, p)$ for $w \notin J_p(X)$. Since we showed $B_<(X, w, p) \subset \text{Ls}(B(X_n, w_n, p_n))$, it follows that $B(X, w, p) \subset \text{Ls}(B(X_n, w_n, p_n))$ for $w \notin J_p(X)$.

(2) Let $(X_n, w_n, p_n, x_n)_n$ be a sequence in $\mathscr{X} \times \mathbb{R} \times \mathbb{R}^\ell \times \mathbb{G}$ that converges to (X, w, p, x) with $x_n \in B(X_n, w_n, p_n)$ for all n. Then, $x \in \text{Ls}(X_n) = X$ and $p \cdot x \leqq w$. Thus, $x \in B(X, w, p)$. Therefore, $B \subset \mathscr{X} \times \mathbb{R} \times \mathbb{R}^\ell \times \mathbb{G}$ is a closed set.

(3) Suppose that B were not u.h.c. at $(X, w, p) \in \mathscr{X} \times \mathbb{R} \times \mathbb{R}^\ell$; then, there would exist an open set $U \supset B(X, w, p)$ $(\neq \emptyset)$ such that, for any neighborhood V of $(X, w, p) \in \mathscr{X} \times \mathbb{R} \times \mathbb{R}^\ell$, V contains a point (X', w', p') satisfying $B(X', w', p') \not\subset U$. Thus, one obtains a sequence $(X_n, w_n, p_n)_n$ in $\mathscr{X} \times \mathbb{R} \times \mathbb{R}^\ell$ converging to (X, w, p) and a sequence $(x_n)_n$ in $\mathbb{G}$ satisfying $x_n \in B(X_n, w_n, p_n) - U$ for all n. We consider two cases (a) and (b).

Case (a): *B is locally uniformly compact-valued at (X, w, p).*
There are a neigborhood V of (X, w, p) and a compact subset K of $\mathbb{G}$ such that $B(X'', w'', p'') \subset K$ for all $(X'', w'', p'') \in V$. Thus, the sequence $(x_n)_n$ satisfies $x_n \in K$ for all large n. Hence, one can extract a convergent subsequence $x_{n_k} \to_k z$. It follows from $p_n \cdot x_n \leqq w_n$ and $(X_n, w_n, p_n) \to_n (X, w, p)$ that we have $p \cdot z \leqq w, z \in X$, showing $z \in B(X, w, p)$. Therefore, we must have $x_{n_k} \in U$ for all n_k large enough, contradictory to $x_{n_k} \notin U$ for all n_k.
Case (b): *The commodity space $\mathbb{G}$ is convex, and the budget set relation B is restricted to $\mathscr{X}_{co} \times \mathbb{R} \times \mathbb{R}^\ell$.*
Fix a point $y_0 \in B(X, w, p)$. As $X_n \to_n X$, one can pick a sequence $(y_n)_n$ such that $y_n \in X_n$ for all n and $y_n \to_n y_0$. $[y_n, x_n] \subset X_n$ for each n because $X_n \in \mathscr{X}_{co}$. $B(X, w, p)$ is compact by assumption, and one can assume without loss of generality that U is bounded. Put $t = \text{diam}(U) < \infty$. For each n define z_n as follows: choose $z_n \in [y_n, x_n] - U$ satisfying $\|y_0 - z_n\| \leqq 2t$ if $[y_n, x_n] \cap U \neq \emptyset$, and choose $z_n \in [y_n, x_n] - U$ satisfying $\|y_n - z_n\| \leqq t$ if $[y_n, x_n] \cap U = \emptyset$. The sequence $(z_n)_n$ is bounded as $y_n \to_n y_0$. Extract a convergent subsequence $z_{n_k} \to_k z$. Since $(X_{n_k}, w_{n_k}, p_{n_k})_k \to (X, w, p)$, $p_{n_k} \cdot x_{n_k} \leqq w_{n_k}$, and $p_{n_k} \cdot y_{n_k} \to_k p \cdot y_0$, one obtains $p \cdot z \leqq w, z \in X$, that is, $z \in B(X, w, p)$. Hence, for all k large enough, we have $z_k \in U$ contradictory to $z_{n_k} \notin U$ for all k.

Therefore, in both of the cases (a) and (b), the budget set relation B is u.h.c. at (X, w, p). □

Properties of the Demand Relation $D : \mathcal{P} \times \mathbb{R} \times \mathbb{R}^\ell \to \mathbb{G}$

Having discussed the continuity property of the budget set relation, we will now confirm the measurability of the demand relation before examining the continuity property of the demand relation. Since the discussion focuses on the demand relation D whose domain is a subset of $\mathcal{P} \times \mathbb{R} \times \mathbb{R}^\ell$, it is convenient to identify the budget set relation $B : \mathcal{X} \times \mathbb{R} \times \mathbb{R}^\ell \to \mathbb{G}$ as the relation $B : \mathcal{P} \times \mathbb{R} \times \mathbb{R}^\ell \to \mathbb{G}$. However, instead of writing the budget set at $(X, \succ, w, p)$ as $B(X, \succ, w, p)$, we simply write it as $B(X, w, p)$. It is easy to check that Proposition 4.4 passes with this identification.

Measurability of the Demand Relation

The measurability of the relation $D : \mathcal{P} \times \mathbb{R} \times \mathbb{R}^\ell \to \mathbb{G}$ is confirmed by the following proposition.

Proposition 4.5 (Measurability of Demand Relation)
The demand relation $D : \mathcal{P} \times \mathbb{R} \times \mathbb{R}^\ell \to \mathbb{G}$ is measurable, that is, the set

$$\{(X, \succ, w, p, x) \in \mathcal{P} \times R \times \mathbb{R}^\ell \times \mathbb{G} \mid x \in D(X, \succ, w, p)\}$$

is a Borel subset of $\mathcal{P} \times \mathbb{R} \times \mathbb{R}^\ell \times \mathbb{G}$.

Proof Let $S = \mathcal{P} \times \mathbb{R} \times \mathbb{R}^\ell$ and consider the budget set relation $B : S \to \mathbb{G}$. One can immediately check that its domain, $\operatorname{dom} B$, is closed, and $\operatorname{dom} B \in \mathcal{B}(S)$. It also follows from Proposition 4.4 (2) that $B(s)$ is closed for any $s \in S$, and the inferior inverse image $B_w^{-1}(K) = \{s \in S \mid B(s) \cap K \neq \emptyset\}$ of a compact set is closed for any compact $K \subset \mathbb{G}$. Thus, $B_w^{-1}(K) \in \mathcal{B}(S)$ for every compact K. It follows that $B_w^{-1}(F) = \{s \in S \mid B(s) \cap F \neq \emptyset\} \in \mathcal{B}(S)$ for any closed subset $F \subset \mathbb{G}$. Then, by Mathematical Proposition 3.5, p. 104, there is a countable family $\{f_n \mid n \in \mathbb{N}_+\}$ of measurable maps from $\operatorname{dom} B$ to $\mathbb{G}$ such that $B(s) = \operatorname{cl}\{f_n(s) \mid n \in \mathbb{N}_+\}$ for every $s \in \operatorname{dom} B$. Define a correspondence $D_n : \operatorname{dom} B \to \mathbb{G}$ for each $n \in \mathbb{N}_+$ by $s \mapsto \{x \in B(s) \mid f_n(s) \not\succ x\}$.

Claim: *D_n is a Borel subset of $S \times \mathbb{G}$ for every n.*

To verify the claim, first note that the set $\{(X, \succ, x, y) \in \mathcal{P} \times \mathbb{G} \times \mathbb{G} \mid x, y \in X \text{ and } x \not\succ y\}$ is closed by Theorem 3.1 (3), p. 98. Since the budget set relation B is closed by Proposition 4.4 (2), the set $F = \{(s, x, y) \in (\operatorname{dom} B) \times \mathbb{G} \times \mathbb{G} \mid x, y \in B(s) \subset X_s \text{ and } x \not\succ_s y\}$ is closed in $(\operatorname{dom} B) \times \mathbb{G} \times \mathbb{G}$. (Here, $(X_s, \succ_s)$ is a projection of s on $\mathcal{P}$.) Let us define a mapping $h_n : (\operatorname{dom} B) \times \mathbb{G} \to S \times \mathbb{G} \times \mathbb{G}$ by $(s, x) \mapsto (s, f_n(s), x)$. Then, $D_n = h_n^{-1}(F)$. The measurability of $f_n : \operatorname{dom} B \to \mathbb{G}$ implies that of the mapping h_n. Thus, $D_n \in \mathcal{B}((\operatorname{dom} B) \times \mathbb{G}) \subset \mathcal{B}(S \times \mathbb{G})$. This confirms the above statement.

If we show $D(s) = \bigcap_{n \in \mathbb{N}_+} D_n(s)$ for every $s \in \operatorname{dom} B$, then we have $D = \bigcap_{n \in \mathbb{N}_+} D_n$ and the proof is done. As $f_n(s) \in B(s)$, we must have $D(s) \subset D_n(s)$ for all n. Hence, $D(s) \subset \bigcap_{n \in \mathbb{N}_+} D_n(s)$. On the other hand, let $x \in D_n(s)$ for all $n \in \mathbb{N}_+$. The set $V = \{y \in B(s) \mid y \succ_s x\}$ is open in $B(s)$ by the continuity of the preference

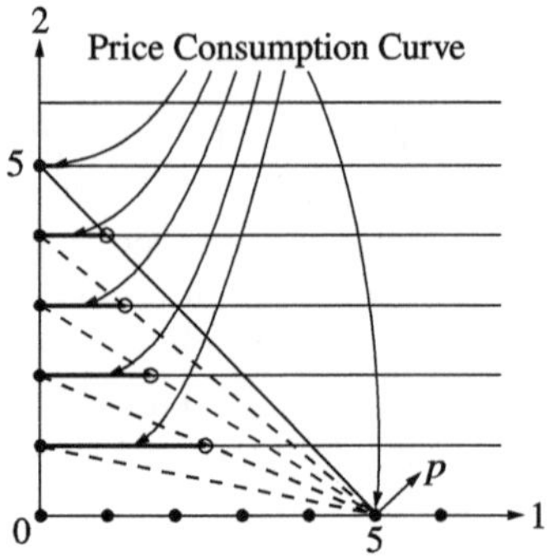

Fig. 4.13 Failure of u.h.c.

relation $\succ_s$. Thus, if $V \neq \emptyset$, there would exist n such that $f_n(s) \in V$, contradictory to $x \in D_n(s)$. Therefore, $V = \emptyset$, meaning that $x \in D(s)$. □

Discontinuity and Continuity of the Demand Relation

Now that the verification of the measurability of the demand relation has been completed, our next step is to discuss its continuity property. First, we review an example of cases in which the demand relation fails to be upper hemi-continuous. Then, conditions are given under which it satisfies some of the continuity properties.

Possible Discontinuities of the Demand Relation

Examples [Failure of Upper Hemi-Continuity of Demand Relation]

1. (See Fig. 4.13.) The commodity space, consumption sets, wealth levels and price vectors are the same as in Fig. 4.10 of the previous example, that is, $\mathbb{G} = \mathbb{R} \times \mathbb{Z}$, $X, X_n = \mathbb{R}_+ \times \mathbb{N}$, $w = 5/2$, $w_n = 5/2 - 5/2n$, $p = (1/2, 1/2)$, $p_n = ((n-1)/2n, (n+1)/2n))$, $n \in \mathbb{N}_+$. Let $\mathbb{T} = \mathbb{R}_+ \times \mathbb{N} \subset \mathbb{G}$, and define the preference relations $\succ, \succ_n$ by
$$\succ, \succ_n = \{((x_1, x_2), (y_1, y_2)) \in \mathbb{T} \times \mathbb{T} \mid x_2 > y_2 \text{ or } [x_2 = y_2 \text{ and } x_1 > y_1]\}.$$
Then $(X, \succ)$, $(X_n, \succ_n) \in \mathcal{P}$. The demand relation D is not u.h.c. at $(X, \succ, w, p)$, as can be seen from the "price consumption curve" in Fig. 4.13.
2. Let X and $\succ$ be as in Fig. 4.14, and put $X_n = X$ and $\succ_n = \succ, n \in \mathbb{N}_+$. Let both $H(p, w)$ and $H(p_n, w_n)$ contain the consumption vector z for all n. If $(p_n, w_n) \to_n (p, w)$ as in the figure, then $D(X_n, \succ_n, w_n, p_n) = \{z\}$ for all n large enough, but $D(\succ, X, w, p) = \{x\}$. Consequently, D fails to be u.h.c. at $(X, \succ, w, p)$.

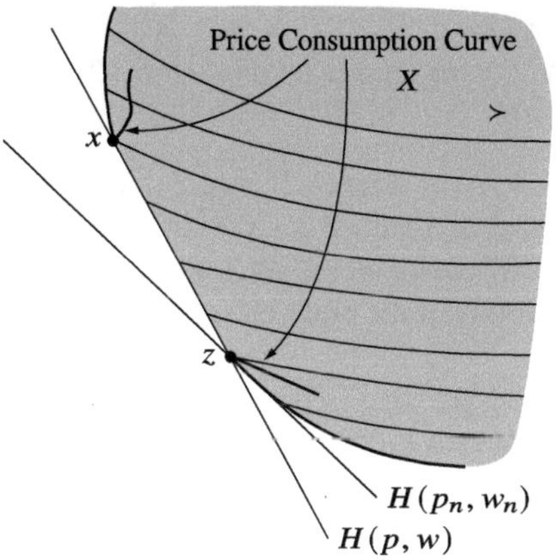

Fig. 4.14 Failure of u.h.c.

Continuity Properties of the Demand Relation

The continuity property of the demand relation initially shown below is its closedness off the critical sets in the wealth space.

Proposition 4.6 (Closedness of Demand Relation)

The demand relation $D : \mathcal{P} \times \mathbb{R} \times \mathbb{R}^\ell \to \mathbb{G}$ *is closed off the critical sets in the wealth space, that is, at every point* $(X, \succ, w, p) \in \operatorname{dom} D$ *with* $w \notin J_p(X)$.

Proof Let $s_n = (X_n, \succ_n, w_n, p_n) \to_n (X, \succ, w, p) = s$, and $x_n \in D(s_n)$ for which $x_n \to_n x$. We want to show $x \in D(s)$. Since the budget set relation is closed by Proposition 4.4 (2), we have $x \in B(s)$.

Let $z \in B(s)$. In the case $p \cdot x < w$, as the budget set relation B is l.h.c. off the critical sets in the wealth space, it follows from Proposition 4.4 (1) that there is a sequence $z_n \to_n z, n \in \mathbb{N}_+$, for which $z_n \in B(s_n)$ for every n, and $p \cdot z_n < w$ eventually, where $z_n \not\succ_n x_n$ for every n owing to $x_n \in D(s_n)$. In the case $p \cdot x = w$, there is a sequence $z_k \to_k z, k \in \mathbb{N}_+$, with $p \cdot z_k < w$ and $z_k \in B(s)$ for every k, since $w \notin J_p(X)$. This implies that for every fixed k there exists a sequence $(z_{k,n})_n$ such that $z_{k,n} \to_n z_k$ and $z_{k,n} \in B(s_n)$ for every k, where $z_{k,n} \not\succ_n x_n$ for every n as $x_n \in D(s_n)$. So, in both of the cases, the continuity of $\succ$ implies $z \not\succ x$. This proves that $x \in D(s)$. □

Note that the above proposition is *not* saying that the demand relation is closed, which means that D is a closed subset of $\mathcal{P} \times \mathbb{R} \times \mathbb{R}^\ell \times \mathbb{G}$ (cf. Mathematical Fact 1.17, p. 37).

Proposition 4.7 (Closed Values/Compact Values of Demand Relation)

The values of the demand relation $D : \mathcal{P} \times \mathbb{R} \times \mathbb{R}^\ell \to \mathbb{G}$ *satisfy the following topological properties:*

(1) *The demand relation* $D : \mathcal{P} \times \mathbb{R} \times \mathbb{R}^\ell \to \mathbb{G}$ *is closed-valued at every* $(X, \succ, w, p) \in \operatorname{dom} D$.

(2) *The demand relation* $D : \mathcal{P} \times \mathbb{R} \times \mathbb{R}^\ell \to \mathbb{G}$ *is compact-valued at* $(X, \succ, w, p) \in \operatorname{dom} D$*, if the budget set relation* $B : \mathcal{P} \times \mathbb{R} \times \mathbb{R}^\ell \to \mathbb{G}$ *is compact-valued at* $(X, \succ, w, p)$.

Proof (1) Let $(X, \succ, w, p) \in \operatorname{dom} D$. If $w \notin J_p(X)$, it follows from Proposition 4.6 that the relation D is closed-valued at $(X, \succ, w, p)$. If not, let $x_n \in D(X, \succ, w, p)$ for each n and $x_n \to_n x$. Then, for all $z \in B(X, w, p)$ and for every n one has $z \not\succ x_n$. It follows from the continuity of the preference relation $(X, \succ)$ that $z \not\succ x$. Therefore, $x \in D(X, \succ, w, p)$.

(2) Since the demand set $D(X, \succ, w, p) \subset B(X, \succ, w, p)$ is closed at every $(X, \succ, w, p) \in \operatorname{dom} D$ by (1), it is compact whenever the budget set $B(X, \succ, w, p)$ is compact. □

The following theorem, which clarifies the conditions under which the demand relation satisfies the upper hemi-continuity at a point, is the main result of this section.

Theorem 4.2 (Upper Hemi-Continuity of Demand Relation)

The demand relation $D : \mathcal{P} \times \mathbb{R} \times \mathbb{R}^\ell \to \mathbb{G}$ *is upper hemi-continuous at* $(X, \succ, w, p) \in \operatorname{dom} D$ *if* $w \notin J_p(X)$ *and the budget set relation* $B : \mathcal{P} \times \mathbb{R} \times \mathbb{R}^\ell \to \mathbb{G}$ *is locally uniformly compact-valued at* $(X, \succ, w, p)$.

Proof Let $B : \mathcal{P} \times \mathbb{R} \times \mathbb{R}^\ell \to \mathbb{G}$ be locally uniformly compact-valued at $(X, \succ, w, p) \in \operatorname{dom} D$ and $w \notin J_p(X)$. If $D(X, \succ, w, p) = B(X, \succ, w, p)$, the upper hemi-continuity of D at $(X, \succ, w, p)$ follows from that of the relation B by Proposition 4.4 (3-a). So let $D(X, \succ, w, p) \neq B(X, \succ, w, p)$. To prove that D is u.h.c. at $(X, \succ, w, p)$, one must show that for a given open set $W_0 \supset D(X, \succ, w, p)$, there is a neighborhood V_0 of $(X, \succ, w, p)$ such that $D(X', \succ', w', p') \subset W_0$ for every $(X', \succ', w', p') \in V_0$. There exists an open subset W of W_0 for which $D(X, \succ, w, p) \subset W$ and $F = (\mathbb{G} - W) \cap B(X, \succ, w, p) \neq \emptyset$ since $D(X, \succ, w, p)$ is a closed set in $\mathbb{G}$ and $D(X, \succ, w, p) \neq B(X, \succ, w, p)$. Since D is closed at $(X, \succ, w, p)$ by Proposition 4.6, there also exist open neighborhoods U_z of z and $V_{s(z)}$ of $(X, \succ, w, p)$ such that $D(V_{s(z)}) \subset \mathbb{G} - U_z$ for every $z \notin D(X, \succ, w, p)$ (cf. Mathematical Fact 1.17 (1-b), p. 37). Now the set $F \subset \mathbb{G} - W$ is compact and a family $\mathcal{U} = \{U_z \mid z \in F\}$ forms an open cover of F. So there exists a finite subcover $\{U_{z_j} \mid j = 1, \ldots, k\}$ of F, so $F \subset \cup_{j=1}^k U_{z_j}$. Set $U = \cup_{j=1}^k U_{z_j}$ and $V = \cap_{j=1}^k V_{z_{s(z_j)}}$. Then, U and V are open neighborhoods of F and $(X, \succ, w, p)$, respectively, with the property that $D(V) \subset \mathbb{G} - U$. Since, by Proposition 4.4 (3-a), the budget set relation B is u.h.c. at $(X, \succ, w, p)$, there exists an open neighborhood V_1 of $(X, \succ, w, p)$ satisfying $B(V_1) \subset W \cup U$. Therefore, for any $(X', \succ', w', p')$ in $V \cap V_1$, we have $D(X', \succ', w', p') \subset (\mathbb{G} - U) \cap (W \cup U) \subset W$. This proves that D is u.h.c. at $(X, \succ, w, p)$. □

An intuitive interpretation from an economic point of view may be in order regarding the result of the above theorem. The upper hemi-continuity of the demand relation D at $(X, \succ, w, p) \in \operatorname{dom} D$ says mathematically that for any open set W in the commodity space $\mathbb{G}$, there exists a neighborhood V of $(X, \succ, w, p) \in \operatorname{dom} D$

such that $D(X', \succ', w', p') \subset W$ for all $(X', \succ', w', p') \in V$. Now it is helpful to think of a neighborhood of a point and an open set containing it in a topological space as an indicator of points that are near or close to the point. From this point of view, for a given open set W, take neighborhoods V_1 of preferences $(X, \succ)$, V_2 of a wealth level w, and V_3 of market prices p such that $D(X', \succ', w', p') \subset W$ for all $(X', \succ', w', p') \in V_1 \times V_2 \times V_3$ for which $w \notin C_p(X)$. Preferences or *tastes* $(X', \succ') \in V_1$ are close or *similar* to $(X, \succ)$, a wealth level $w' \in V_2$ is close or *similar* to w, prices $p' \in V_3$ are close or *similar* to p, and, in the same way, the optimal choices of consumption vectors $z \in D(X', \succ', w', p')$ are close or *similar* to those in $D(X, \succ, w, p)$. In deciphering the optimal choices of individuals, wealth levels, as well as prices and demand vectors are understood to elucidate the situations of individuals and their behavior in a market economy. Accordingly, a colorful way of presenting an economic interpretation of the upper hemi-continuity of the demand relation is to say *individuals facing similar situations behave similarly in markets with respect to their consumption activities, except in rare cases* (cf. Corollary 3.2).

In Theorem 4.2 a basic requirement for the demand relation $D : \mathcal{P} \times \mathbb{R} \times \mathbb{R}^\ell \to \mathbb{G}$ to be u.h.c. at $(\succ, X, w, p) \in \operatorname{dom} D$ is the condition $w \notin J_p(X)$, which depends only on consumption sets under a given price vector. Note that it is *not the necessary condition* for its upper hemi-continuity. Thus, even when $w \in J_p(X)$, preference relations with some very specific properties can avoid situations that could lead to the discontinuity of the demand relation as shown in Fig. 4.13. Of course, such a possibility of avoiding the discontinuity is only realized by a very special class of preference relations. An example of such a class of preference relations is given below.

Examples [A Special Class of Preference Relations Giving Rise to Upper Hemi-Continuity of Demand Relations]

Let $\mathbb{G}$, X, w, p and X_n, w_n, p_n for each $n \in \mathbb{N}_+$ be as in the previous example shown in Fig. 4.13, to compare this example with Fig. 4.13. Let $X = \mathbb{R}_+ \times \mathbb{N} \subset \mathbb{G}$ and define a subset $\mathcal{P}'$ of the universal preference set $\mathcal{P}$ by

$$\begin{aligned}\mathcal{P}' = \{(X', \succ') \in \mathcal{P} \mid X' = X, \text{ and, for } x = (x_1, x_2), y = (y_1, y_2) \in X', \\ x \succ' y \text{ if } [x_1 > 0, y_1 > 0, x_2 > y_2] \text{ or } [x_1 > y_1, x_2 = y_2]\}.\end{aligned}$$

Then, for $(X, \succ), (X', \succ') \in \mathcal{P}'$ we have $J_p(X) = J_p(X')$ for all $p \in \mathbb{R}^\ell_{++}$. Since for any $z \in J_p(X)$ we have $z \not\succ 0$, a consumption vector x in $C_p(X)$ never becomes a demand vector, even if $w \in J_p(X)$ for any $p \in \mathbb{R}^\ell_{++}$. Thus, if one is willing to restrict oneself to the class $\mathcal{P}'$, the discontinuity problem in Fig. 4.13 does not arise. It should also be clear that $\mathcal{P}'$ can be defined analogously in the case of $\mathbb{G} = \mathbb{R}^{\ell-1} \times \mathbb{Z}$.

> **Important [Minimum Income Condition]**

In the analysis of this book, the condition for the demand relation $D : \mathcal{P} \times \mathbb{R} \times \mathbb{R}^\ell \to \mathbb{G}$ to be upper hemi-continuous at $(X, \succ, w, p) \in \operatorname{dom} D$ is that the wealth level w does not belong to the critical set $J_p(X)$. However, it is usually the case in the literature (see, e.g., [88], [87], [50], [95], [23], [22], etc.) that the condition guaranteeing the upper hemi-continuity of the demand relation D at $(X, \succ, w, p)$ is that $w > \min p \cdot X$. Since the *wealth* (or *income*) *level* $\min p \cdot X$ is the *minimum level* that allows individuals with their consumption set X to obtain commodity vectors whose market values strictly exceed the least expensive commodity vectors in the consumption set X, this condition has been called the *minimum income condition*. In a general setting such as this book, the minimum income condition is *misleading* in understanding the true nature of the source of discontinuity in the demand relation. The essence of the requirement should be that wealth levels must not belong to the critical set of the wealth space at the prevailing market price vector p. Whenever consumption sets are convex, the critical set of a consumption set X in wealth space happens to coincide with the set of the wealth level that is equal to the minimum expenditure among consumption vectors, i.e., $J_p(X) = \{w \in \mathbb{R} \mid w = \min p \cdot X\}$ (cf. Proposition 4.1).

4.3 Subdemands

Minimum Income and Survival Issues

The demands of an individual describe its *optimal choices* made within its budget sets with respect to its preference relation. It is evident that the demands of individuals are not well defined unless they have sufficient budgets to be able to buy commodities that they need. This leads to the consideration of the survival problem and the minimum incomes needed by individuals.

Given an individual agent's consumption set $X \in \mathcal{X}$ and a market price vector p, it may be the case that the agent's wealth level w is insufficient to procure any commodity vectors in its consumption set X. In such instances, $B(X, w, p) = \emptyset$, and $D(X, \succ, w, p) = \emptyset$ accordingly. In the real world, this translates to a lack of sufficient food, drink, medicine, various services, and so on to sustain one's life. The standard analytical approach to avoiding such a situation is to impose conditions known as *the minimum income condition* or *the survival condition*. The approach of this book differs in that it does not impose either of these conditions, but instead introduces a new concept of demands, called *subdemands*, which avoid the survival problem and are always non-empty even when the budget sets are empty. I should add a slight modification to this statement, in that in the case of non-empty budget sets the budget sets must be compact.

Minimum Income and Survival Conditions

The *minimum income level* of a consumption set $X \in \mathcal{X}$ at a price vector p is defined as the infimum of expenditure necessary to obtain consumption vectors, that is, $\inf p \cdot X$. A wealth level $w \in \mathbb{R}$ is said to satisfy the *minimum income condition* at a price vector $p \in \mathbb{R}^\ell$ if $w > \inf p \cdot X$. In a market economy, there is no intrinsic mechanism that generates wealth distributions among individuals in a manner that assures the minimum income condition for each individual at a prevailing market price vector.

An individual's initial endowments vector, denoted by $e \in \mathbb{G}$, is said to satisfy the *survival condition* if e is a vector in the individual's consumption set X. In a special case where the survival condition is met, the budget set $B(X, p \cdot e, p)$ is not empty and the demand set $D(X, \succ, p \cdot e, p)$ is not empty whenever the set $B(X, p \cdot e, p)$ is compact. However, it is unlikely that all individuals possess initial endowments that satisfy the survival condition. One of the primary motivations for engaging in market transactions is to obtain a consumption vector within its consumption set. These considerations prompt the introduction of *subdemands*, which assimilate demands and are non-empty even though wealth levels are below the minimum income. This is despite the fact that $B(X, w, p) = \emptyset$. Subdemands also deviate from demands when wealth levels are in the critical set in order to circumvent the discontinuity problems.

Subdemand Relation $D^* \subset \mathcal{P} \times \mathbb{G} \times \mathbb{R}^\ell \times \mathbb{G}$

A supplementary concept of demands termed *subdemands* is introduced below in order to circumvent the continuity issue of demand relations as well as the survival issue of individuals in a general framework of this book.

Definition 4.3 (Subdemands) Let a preference relation $(X, \succ)$, a commodity vector $x \in \mathbb{G}$, and a price vector $p \in \mathbb{R}^\ell$ be given. The *subdemand set* for $(X, \succ)$ at (x, p), denoted by $D^*(X, \succ, x, p)$, is defined by

$$D^*(X,\succ,x,p) = \begin{cases} \{y \in B(X, p \cdot x, p) \mid \succ(y) \cap H_<(p, p \cdot x) = \emptyset\} \\ \quad (= \{y \in B(X, p \cdot x, p) \mid p \cdot z \geqq p \cdot x \text{ for all } z \in X \text{ such that } z \succ y\}, \\ \quad = \{y \in B(X, p \cdot x, p) \mid \succ(y) \subset H_\geqq(p, p \cdot x)\}), \\ \qquad\qquad\qquad\qquad \text{if } p \cdot x \geqq \inf p \cdot X, \\ \{x\}, \qquad\qquad\qquad\quad \text{if } p \cdot x < \inf p \cdot X. \end{cases}$$

In other words, for an economic agent with a commodity vector in $\mathbb{G}$ prior to market transactions, if its market value is at least its minimum income level, then its budget set is non-empty and its subdemand set consists of consumption vectors in the budget set to which no preferred consumption vectors are obtainable with strictly less expenditure. Conversely, if the market value is below the minimum income level, the budget set is empty and the subdemand set comprises the commodity vector prior to market transactions, as there are no other feasible options for obtaining a consumption vector within the budget.

The relation $D^* \subset \mathcal{P} \times \mathbb{G} \times \mathbb{R}^\ell \times \mathbb{G}$ defined by $(X, \succ, x, p) \mapsto D^*(X, \succ, x, p)$, that is, $D^* = \{(X, \succ, x, p, y) \mid y \in D^*(X, \succ, x, p)\}$, is called the *subdemand relation*. A commodity vector in the subdemand set is called a *subdemand vector*.

We start by looking at the basic properties of subdemand sets. The first property is a result of how they are defined.

Proposition 4.8 (Non-Emptiness of Subdemands)

For any preference relation $(X, \succ) \in \mathcal{P}$, the subdemand set $D^*(X, \succ, x, p)$ at $(x, p) \in \mathbb{G} \times \mathbb{R}^\ell$ is non-empty if $B(X, p \cdot x, p)$ is compact when $p \cdot x \geqq \inf p \cdot X$.

Proof If $p \cdot x < \inf p \cdot X$, then $D^*(X, \succ, x, p) = \{x\}$. Otherwise, the budget set $B(X, p \cdot x, p)$ is non-empty and compact, so $D^*(X, \succ, x, p)$ is not empty. □

The next property is the full use of the budget for locally non-satiated preference relations, as in the case of demand sets.

Proposition 4.9 (Full Use of Budget for Subdemands)

Given the subdemand set $D^*(X, \succ, x, p)$ for $(X, \succ) \in \mathcal{P}$ at $(x, p) \in \mathbb{G} \times \mathbb{R}^\ell$ with $p \neq 0$, we have $p \cdot y = p \cdot x$ for any $y \in D^*(X, \succ, x, p)$ if $(X, \succ)$ is a locally non-satiated preference relation in $\mathcal{P}$.

Proof Consider a preference relation $(X, \succ) \in \mathcal{P}_{lns}$ and a pair $(x, p) \in \mathbb{G} \times \mathbb{R}^\ell$ with $p \neq 0$. Let $y \in D^*(X, \succ, x, p)$. If $B(X, p \cdot x, p) = \emptyset$, then $y = x$. Thus, $p \cdot y = p \cdot x$ is trivial. If $B(X, p \cdot x, p) \neq \emptyset$, then we have $p \cdot y \leqq p \cdot x$ since $y \in B(X, p \cdot x, p)$. In the event that $p \cdot y < p \cdot x$, let U be a neighborhood of y such that $p \cdot y' < p \cdot x$ for every $y' \in X \cap U$. By the local non-satiation of the preference relation $(X, \succ)$, there exists a consumption vector $z \in X \cap U$ for which we have $z \succ y$ and $p \cdot z < p \cdot x$. However, given that $y \in D^*(X, \succ, x, p)$, it must be the case that $p \cdot z \geqq p \cdot x$, which is a contradiction. Consequently, we conclude that $p \cdot y = p \cdot x$ for every $y \in D^*(X, \succ, x, p)$. □

The following result serves to confirm that the subdemand relations constitute an extension of the demand relations.

Proposition 4.10 (Subdemands and Demands)

Given a preference relation $(X, \succ)$, a commodity vector $x \in \mathbb{G}$, and a price vector $p \in \mathbb{R}^\ell$, the subdemand set for $(X, \succ)$ at (x, p) exhibits the following properties with regard to the demand set:

(1) $D(X, \succ, p \cdot x, p) \subset D^*(X, \succ, x, p)$.
(2) $D(X, \succ, p \cdot x, p) = D^*(X, \succ, x, p)$ *if* $p \cdot x \geqq \inf p \cdot X$ *and* $x \notin C_p(X)$.

Proof (1) Let $y \in D(X, \succ, p \cdot x, p)$, then $\succ(y) \subset H_>(p, p \cdot x)$ implying $\succ(y) \cap H_<(p, p \cdot x) = \emptyset$. Thus, $y \in D^*(X, \succ, x, p)$.

(2) To show $D^*(X, \succ, p \cdot x, p) \subset D(X, \succ, x, p)$, let $y \in D^*(X, \succ, x, p)$ for $x \notin C_p(X)$ and $p \cdot x \geqq \inf p \cdot X$. Clearly, $B(X, p \cdot x, p) \neq \emptyset$. Assume that for

some $z \in B(X, p \cdot x, p)$, we had $z \in \succ(y) \cap H_{\leqq}(p, p \cdot x)$. Since $z \notin H_{<}(p, p \cdot x)$, it must be $z \in H(p, p \cdot x)$. $x \notin C_p(X)$ implies that there is a sequence of cheaper points $(z_n)_n$ in $B(X, p \cdot x, p) \cap H_{<}(p, p \cdot x)$ such that $z_n \to_n z$. Consequently, $z_n \in \succ(y) \cap H_{<}(p, p \cdot x)$ for all n large enough in contradiction to $y \in D^*(X, \succ, x, p)$. Hence, one has $\succ(y) \cap H_{\leqq}(p, p \cdot x) = \emptyset$, which shows that $y \in D(X, \succ, x, p)$. □

Continuity Property of Subdemand Relations

In contrast to demand relations, which are defined only for wealth levels at least as much as minimum income levels and are closed only at points off the critical sets in the wealth space (cf. Proposition 4.6, p. 147), subdemand relations are defined even when budgets sets are empty, and the relations are closed at any points in their domains.

Proposition 4.11 (Closedness of Subdemand Relations)
The subdemand relation $D^* : \mathcal{P} \times \mathbb{G} \times \mathbb{R}^\ell \to \mathbb{G}$ *is closed.*

Proof Let $(X_n, \succ_n, x_n, p_n, y_n)_n$ be a sequence in $D^* \subset \mathcal{P} \times \mathbb{G} \times \mathbb{R}^\ell \times \mathbb{G}$ converging to $(X, \succ, x, p, y)$. Clearly, $x \in \mathbb{G}$, and $y \in X$ by Corollary 3.1, p. 99. To confirm $y \in D^*(X, \succ, x, p)$, we shall show:

$$\begin{cases} (1)\ y = x & \text{if } p \cdot x < \inf p \cdot X, \\ (2)\ \succ(y) \subset H_{\geqq}(p, p \cdot x) & \text{if } p \cdot x \geqq \inf p \cdot X. \end{cases}$$

(1) If $p \cdot x < \inf p \cdot X$, then $p_n \cdot x_n < \inf p_n \cdot X_n$ for all n large enough. Consequently, $y_n = x_n$ residually. Thus, $y_n \to_n y$ and $x_n \to_n x$ imply that $y = x$. It shows that $y \in D^*(X, \succ, x, p)$.

(2) If $p \cdot x \geqq \inf p \cdot X$, then clearly, $y \in B(X, p \cdot x, p)$. Suppose that $z \in \succ(y) \cap H_{<}(p, p \cdot x)$ for some $z \in X$. Since $X_n \to_n X$, there is a convergent sequence $(z_n)_n \to_n z$ such that $z_n \in X_n, n \in \mathbb{N}_+$, for which one has $z_n \in \succ(y) \cap H_{<}(p_n, p_n \cdot x_n)$ for n large enough in contradiction to the fact that $\succ(y) \cap H_{<}(p_n, p_n \cdot x_n) = \emptyset$ for $n \in \mathbb{N}_+$. Therefore, it must be $\succ(y) \subset H_{\geqq}(p, p \cdot x)$. □

Notes 4

Notes 4.1 [Section 4.1 , p. 131]

[Cheaper points and local cheaper points, p. 132] See [83] for the use of the concept of cheaper points in demand theory. It appears that L. McKenzie first used a sequence of local cheaper points in the general equilibrium analysis in his lecture at University of Rochester before 1960.
[Critical sets of consumption sets, p. 132] See [126, 127].
[Theorem 4.1, p. 135] See [126].
[Example, Fig. 4.4, p. 135] It should be noted that the symbols w_j, with $j = 1, 2, 3$, represent wealth levels rather than vectors in $\mathbb{R}^2$. Since the straight line $\mathbb{R}$ in the figure is drawn to be perpendicular to each hyperplane $H(p, w_j)$, the vectors z_j in $\mathbb{R}^2$, corresponding to w_j, $j = 1, 2, 3$, satisfy $\|z_j\| = w_j$, $z_j = w_j p$, $j = 1, 2, 3$. It is due to the fact that the norm of p is equal to one, which implies that $w_j = p \cdot z_j = \|p\| \|z_j\| \cos 0 = \|z_j\|$.
[Proposition 4.2, p. 137] See [127].
[Proposition 4.3, p. 139] See [127].

Notes 4.2 [Section 4.2 , p. 139]

[Example, Fig. 4.9, p. 142] This is a classical example due to [22, §4.8, p. 63].
[Example, p. 149] This example is due to [81, p. 445].

Notes 4.3 [Section 4.3 , p. 150]

[Subdemands, p. 151] The concept of subdemands represents a means of delineating what Koopmans [68, Section 2.7, p. 62] referred to as a theoretical "hard-boiled alternative" approach to the survival problem of consumers.

Chapter 5
Economies with a Measure Space of Consumers

Abstract This chapter commences with an introduction to two types of formulations of economies with a measure space of economic agents. The aggregate demand relation which summarizes the total behavior of economic agents in perfectly competitive markets is introduced. Following this we proceed to discuss *aggregation effects*, that is, some specific effects on the total demands generated by aggregation or "summing" over individual demands. Two distinct sources of the effects are identified: one is a straightforward effect resulting from the mere fact of summing individual demands over a "large" population of economic agents, the other coming from the distribution of varying characteristics in the population. Aggregation effects will play an important role in discussing equilibria of economies in subsequent chapters. In the appendix, quotations from some of the most prominent classical economic theorists of the 19th and early 20th centuries are presented, in which they express their perspectives on the aggregation effects.

5.1 Formulation of Trading and Consumption Economies

Markets are abstract or tangible settings in which the trading of commodities takes place among economic agents. An *economy* is a description of economic agents who participate in market transactions and engage in consumption activities. In this sense an economy is also called a *market economy*. General equilibrium analysis is essentially relevant to *perfectly competitive markets*. However, as is well-known in economic theory, the "perfectly competitive" behaviors of individuals are not directly articulated. Instead, the standard way of formulating the behaviors of individuals in perfectly competitive markets is passive rather than active. This assumes that individuals in markets regard prevailing prices as given, such that each one of the individuals alone is not in a position to influence market prices by its buying and selling amounts of various commodities. In this context, the competitive behaviors of individuals are defined as the demand relation representing their optimal choices, whose mathematical properties are examined in the previous chapter.

A. Yamazaki, *Mathematical Economics of General Equilibrium*, Monographs in Mathematical Economics 8, https://doi.org/10.1007/978-981-96-9497-6_5

Measure Space of Economic Agents

Population of Economic Agents

Let A be a non-empty set representing the *set of all economic agents*, or the *population of economic agents*. In everyday terms, each element a in A is identified with its individual name. Economic agents are accounted for by the description of their needs, tastes, and endowments (or wealth) as individuals pursuing trading and consumption activities. To be more precise, each economic agent $a \in A$ is associated with its preference relation $(X_a, \succ_a)$ and its endowments of various commodities $e_a \in \mathbb{G}$ (or wealth $w_a \in \mathbb{R}$). In accordance with this particularity of economies with trading and consumption activities, an *economy* is detailed in the form of a mapping $\mathscr{E} : A \to \mathcal{P} \times \mathbb{G}$ from the population of economic agents A to the space of economic agents' characteristics $\mathcal{P} \times \mathbb{G}$. It is acknowledged that the population of any economies in our world is in fact finite, while in a setting of abstract perfectly competitive markets where none of single individuals can influence the outcome of a whole economy, logical consistency requires an infinite population. Moreover, as will be substantiated later on from a statistical perspective, the population of economic agents is not constrained to a non-empty finite set.

A *measure space of economic agents* $(A, \mathscr{A}, \nu)$ is a finite measure space with $\nu(A) = 1$, representing a space of economic agents. That is, A is a population of economic agents, which is either a finite or infinite set. In general, $\mathscr{A}$ is a σ-algebra on A, and is the set of all subsets of A if it is a non-empty finite set, i.e., if $0 < \sharp A < \infty$. It is understood that $\mathscr{A}$ represents the family of all conceivable groups of members of an economy under consideration. The measure ν is the *normalized counting measure*, i.e., $\nu(C) = \sharp C / \sharp A$ if $\sharp A$ is finite; on the other hand, ν is assumed to be a *non-atomic* (or, equivalently, *atomless*) measure if $\sharp A$ is infinite. The assumption of non-atomicity or atomlessness is designed to portray economic agents in perfectly competitive markets in which the *actions of each individual are negligible*. In other words, market prices and the quantities of commodities obtained at prevailing prices are not affected by what a single individual buys or sells in markets unless a significant proportion of the other individuals change what they intend to buy or sell at the prevailing prices. This is a natural extension of a finite population with a counting measure to an infinite population case represented by a countably additive finite measure space (cf. Note 5.1, p. 171). Regardless of whether the number $\sharp A$ of population is finite or not, the measure $\nu(C)$ represents the proportion of individuals in C to the entire population.

Formulation of Economies without Production

Economies as Measurable Mappings

Definition 5.1 (Economies) For a mapping $\mathscr{E} : (A, \mathscr{A}, \nu) \to \mathcal{P} \times \mathbb{G}$ from a finite measure space $(A, \mathscr{A}, \nu)$, with $\nu(A) = 1$, to the space of the economic agents' characteristics $(\mathcal{P}, \mathbb{G})$, let $\mathscr{E} = (\mathscr{E}_1, \mathscr{E}_2)$ with $\mathscr{E}_i = \mathrm{pr}_i \circ \mathscr{E}, i = 1, 2$, for which pr_i

denotes the projection from $\mathcal{P} \times \mathbb{G}$ onto $\mathcal{P}$ and $\mathbb{G}$, respectively. An *economy* is defined as a measurable mapping $\mathcal{E} : (A, \mathcal{A}, \nu) \to \mathcal{P} \times \mathbb{G}$ such that the integral of the second component, $\int \mathcal{E}_2 d\nu$, with respect to the measure ν is finite. In other words, the *aggregate endowments per capita* (i.e., the *mean endowments*) are finite. A *Walrasian economy* is an economy defined on a non-atomic or atomless measure space of economic agents. It is also called an *atomless economy*, or an *economy with a continuum of agents or traders*.

In an economy $\mathcal{E} : (A, \mathcal{A}, \nu) \to \mathcal{P} \times \mathbb{G}$, each member $a \in A$ is an *economic agent* and is equally referred to as an *individual* or a *consumer*. The preference relation and the endowments of consumer a are $\mathcal{E}(a) = (X_{\mathcal{E}(a)}, \succ_{\mathcal{E}(a)}, e_{\mathcal{E}(a)})$. When the mapping $\mathcal{E}$ is clearly understood from the context, it is abbreviated as $(X_a, \succ_a, e_a)$. In this case, we refer to the mappings $X : A \to \mathcal{X}$ and $e : A \to \mathbb{G}$ given by $a \mapsto X_a$ and $a \mapsto e_a$, respectively, as the *consumption set mapping* and the *endowment mapping* of the economy.

In terms of an economy $\mathcal{E} : A \to \mathcal{P} \times \mathbb{G}$ without production, the budget set and the demand set of consumer a at $p \in \mathbb{R}^\ell$ are given by $B(X_a, p \cdot e_a, p)$ and $D(X_a, \succ_a, p \cdot e_a, p)$, respectively.

The *distribution* of an economy $\mathcal{E}$, denoted by $\mu_{\mathcal{E}}$, is the *image measure of* ν *under* $\mathcal{E}$. That is, $\mu_{\mathcal{E}} = \nu \circ \mathcal{E}^{-1}$. It is the measure on $\mathcal{B}(\mathcal{P} \times \mathbb{G})$ with $\mu_{\mathcal{E}}(\mathcal{P} \times \mathbb{G}) = 1$ that assigns a non-negative real number $\mu_{\mathcal{E}}(T) = \nu(\{a \in A \mid \mathcal{E}(a) \in T\})$ to each Borel subset T of the space of consumption characteristics $\mathcal{P} \times \mathbb{G}$. It reflects the proportion of individual consumers having consumption characteristics belonging to a subset $T \subset \mathcal{P} \times \mathbb{G}$. The distribution $\mu_{\mathcal{E}}$ of the economy $\mathcal{E}$ is also denoted by $\mu_{\succ,e}$, and is occasionally called the *preference–endowment distribution* or the *joint distribution of preference relations and endowments* of the economy $\mathcal{E}$. Two marginal distributions of $\mu_{\mathcal{E}}$ are denoted by $\mu_{\succ}$ and μ_e, respectively, and are referred to as the *preference distribution* and the *endowment distribution*, respectively. We have $\mu_{\succ} = \mu_{\mathcal{E}} \circ \mathrm{pr}_1$ and $\mu_e = \mu_{\mathcal{E}} \circ \mathrm{pr}_2$ where p_i, $(i = 1, 2)$, are projections on $\mathcal{P} \times \mathbb{G}$ such that $\mathrm{pr}_1 : (X, \succ, e) \mapsto (X, \succ)$ and $\mathrm{pr}_2 : (X, \succ, e) \mapsto e$.

Alternative Expression of Economies

In certain circumstances the space $\mathcal{P} \times \mathbb{R}$ is employed as the space of consumption characteristics instead of $\mathcal{P} \times \mathbb{G}$. This is due to the fact that individual demands depend upon the preferences and wealth levels of individual economic agents. In this respect, the strands of an economy could also be captured by way of a pair of measurable mappings, one of which is an assignment of preference relations and the other being an assignment of wealth functions to each economic agent. To be precise, in the case of an economy $\mathcal{E} = (\mathcal{E}_1, \mathcal{E}_2) \to \mathcal{P} \times \mathbb{G}$, a pair of measurable mappings $(\mathcal{E}_1, w)$, where $\mathcal{E}_1 : A \to \mathcal{P}$ is the first coordinate mapping of $\mathcal{E}$ and w is a *wealth (assignment) function* from $S \times A$ to $\mathbb{R}$ for a subset $S \subset \mathbb{R}^\ell$. Apart from environments with possible production activities for $\mathcal{E}$, the wealth function is defined by $w : (p, a) \mapsto w(p, a) = p \cdot e_a$. For a fixed price vector $p \in S \subset \mathbb{R}^\ell$, the function $w_p : A \to \mathbb{R}$ may be used to denote the wealth function $w(p, \cdot) : a \mapsto w(p, a), a \in A$. In the context of a given market price vector $p \in \mathbb{R}^\ell$, economic characteristics that directly influence the demands of economic agents are pairs of a preference relation and a wealth level. Consequently, in examining demands at $p \in S \subset \mathbb{R}^\ell$, references will be made on several occasions to economies $\mathscr{E}_p : A \to \mathcal{P} \times \mathbb{R}$, where $p \in S$. In the case of an economy $\mathscr{E} : A \to \mathcal{P} \times \mathbb{G}$, $\mathscr{E}_p$ is given by the pair $\mathscr{E}_p = (\mathscr{E}_1, w_p)$.

Economies as Distributions over Characteristics

The distribution of consumption characteristics in connection with an economy $\mathscr{E}_p = (\mathscr{E}_1, w_p), p \in \mathbb{R}^\ell$, is the distribution $\mu_{\mathscr{E}_p}$ of the mapping $\mathscr{E}_p, p \in \mathbb{R}^\ell$, that is, the image measure $\nu \circ \mathscr{E}_p^{-1}$ at each p. For each Borel set $U \subset \mathcal{P} \times \mathbb{R}$, we have that $\mu_{\mathscr{E}_p}(U) = \nu(\{a \in A \mid \mathscr{E}_p(a) \in U\})$, which shows the proportion of individuals who possess a pair of preferences and wealth within the set U. For the sake of brevity, the distribution $\mu_{\mathscr{E}_p}$ of an economy at $p \in \mathbb{R}^\ell$ is denoted by $\mu_{\succ, w_p}$, and is called the *preference–wealth distribution* or the *joint distribution of preference relations and wealth* at p. The two marginal distributions of the preference–wealth distribution, written as $\mu_\succ$ and μ_{w_p}, are respectively called the *preference* and the *wealth distribution*. The preference–endowment distribution $\mu_{\succ,e}$ as well as the preference–wealth distribution $\mu_{\succ,w_p}, p \in \mathbb{R}^\ell$, of an economy will be called a *distributional economy*. The individuality of economic agents is elided in a distributional economy, and consumption characteristics are taken into account alone.

In a distributional economy $\mu_{\succ,w_p}, p \in \mathbb{R}^\ell$, we let $\mu_{X,w_p} = \mu_{w_p} \circ \pi_{X,w}^{-1}$ and $\mu_X = \mu_{w_p} \circ \pi_X^{-1}$, where the mappings $\pi_{X,w} : \mathcal{P} \times \mathbb{R} \to \mathcal{X} \times \mathbb{R}$ and $\pi_X : \mathcal{P} \times \mathbb{R} \to \mathcal{X}$ are given, respectively, by $(X, \succ, w) \mapsto (X, w)$ and $(X, \succ, w) \mapsto X$. μ_{X,w_p} and μ_X are referred to as the *joint distribution of consumption sets and wealth*, and the *consumption set distribution* of $\mu_{\succ,w_p}$, respectively. μ_X is the marginal distribution of $\mu_{\succ,w_p}$ on $\mathcal{X}$. In the context of a *joint distribution of consumption sets and wealth*, $\mu_{X,w_p} = \mu_{w_p} \circ \pi_{X,w}^{-1}$, we employ the symbol $\mu_{w_p|X}$ to represent a conditional wealth distribution given consumption set X (cf. Mathematical Property 1.33, p. 58).

Economic Meanings of Some Mathematical Conditions

The following paragraphs will examine the economic implications of certain mathematical conditions.

Meaning of Measurability Requirements

Some of those who are pursuing the study of economic theory have expressed perplexity at the measurability assumption inherent to theoretical analysis with a measurable structure. This may be due to the perception that it may be a precarious mathematical assumption. In light of this, it is important to take time to comment on the meaning of the measurability requirement. The assumption of the measurability of mapping $\mathscr{E} : (A, \mathscr{A}, \nu) \to \mathcal{P} \times \mathbb{G}$ is not an essential restriction of the theory from an economic point of view. In fact, in a mathematical theory without the Axiom of Choice the imposition of the Axiom of Solvay, which states that "all the subsets of $\mathbb{R}^\ell$ are Lebesgue measurable," does not lead to an inconsistency of the theory. This demonstrates that arguments about measurability are concerned with problems related to the mathematical foundations that are as profound as those related to the Axiom of Choice. Thus, one cannot interpret the assumption of measurability as a restriction on economic theory. Should one desire to provide a justification about the measurability requirement on an intuitive and informal basis, then one could say, "If a restriction on economic analysis is understood to be restrictions to mappings that can be constructed by a computer, then the measurability assumption does not impose any restrictions."

Large Economies from Statistical Viewpoint

It is beneficial to briefly elucidate the distinction between the viewpoint of the measure theoretic approach in equilibrium analysis and the statistical viewpoint. In the analysis of an economy, its distribution μ of the population is given, and various properties of the economy are analyzed on that basis. In contrast, typical statistical inferences are about the properties of the distribution μ of the population, and are done on the basis of sampling (i.e., gathering of an actual data) from the population. In this context, the following statistical viewpoint has been proposed by W. Hildenbrand in economic analysis and has been proven to provide a fruitful way in equilibrium analysis:

> A distribution μ on the space of agents' characteristics $\mathcal{P} \times \mathbb{G}$ is a distribution of a "hypothetical" population. Then, the "actual" distribution is considered as a distribution of a *sample* of size $\sharp\bar{A}$ drawn from the population.

Here, empirical reality is the distribution $\bar{\mu}$ of a sample $\bar{A}$ of size $\sharp\bar{A}$ drawn from the hypothetically valid distribution μ of the population A associated with an atomless conceptual economy $\mathscr{E} : (A, \mathscr{A}, \nu) \to \mathcal{P} \times \mathbb{G}$. In this context an atomless economy is conceived as a conceptualization of an *ideal type* of economies with perfectly competitive markets.

5.2 Aggregate Demand of an Economy as Mean Demand

Given an economy $\mathscr{E} : (A, \mathscr{A}, \nu) \to \mathcal{P} \times \mathbb{G}$, how should its *aggregate demand* (or, *total demand*) be expressed? We consider this question in this section. Aggregate demand is an indispensable basis on which market prices are determined. It is intuitively clear that the aggregate demand, which is the total of individual demands, is the sum of all the individual demands of an economy when the population is finite. For a given market price vector $p \in \mathbb{R}^\ell$, the individual demand of a consumer $a \in A$ is expressed by its demand set $D(X_a, \succ_a, w_p(a), p)$, where $w_p(a) = w(p, a)$ is the wealth level of consumer a at p indicated by the wealth function $w : A \to \mathbb{R}$, which is associated with the economy $\mathscr{E} : A \to \mathcal{P} \times \mathbb{G}$. Accordingly, in the event that the economy in question is finite in population, i.e. $0 < \sharp A < \infty$, the sum of individual demands of all consumers in the economy is equal to

$$\sum_{a \in A} D(X_a, \succ_a, w_p(a), p). \tag{5.1}$$

In terms of an integral with the counting measure ν, (5.1) is equal to

$$(\sharp A) \int_A D(X_a, \succ_a, w_p(a), p) d\nu. \tag{5.2}$$

It is evident that the representation of aggregate demand of an economy as given by equations (5.1) and (5.2) is not mathematically sound unless the population size, $\sharp A$, is constrained to a finite value. However, by dividing the aforementioned expressions by the total population size, we obtain the following:

$$\begin{aligned} \frac{1}{\sharp A} \sum_{a \in A} D(X_a, \succ_a, w_p(a), p) &= \sum_{a \in A} D(X_a, \succ_a, w_p(a), p) \left(\frac{1}{\sharp A} \right) \\ &= \int_A D(X_a, \succ_a, w_p(a), p) d\nu, \end{aligned}$$

where the last expression is independent of the population size of an economy. This prompts us to represent the aggregate demand of an economy with any population size by an integral of its demand correspondence with respect to a normalized finite measure, which is understood to show the proportion of individuals in a group (i.e., a measurable set) to the entire population. Mathematically speaking, it amounts to representing the *aggregate demand* of an economy by its *mean demand*. However, from an economic perspective, it is admittedly preferable to interpret the expression of the integral of a demand correspondence as follows: The integral of the aggregate demand function, $D(X_a, \succ_a, w_p(a), p)$, with respect to the normalized finite measure, ν, $\int_A D(X_a, \succ_a, w_p(a), p) d\nu$, represents the total sum of individual demands of an economy wherein units of measurement of commodities are switched to units per capita of the entire population.

Indeed, if we consider a finite population A, and let C be a subpopulation of A, the integral

$$\int_C D(X_a, \succ_a, w_p(a), p) d\nu = \sum_{a \in C} D(X_a, \succ_a, w_p(a), p) \left(\frac{1}{\sharp A} \right)$$

is no longer the mean demand of the subpopulation C, but it still is the aggregate demand of the subpopulation C with units of measurement being the units of commodities per capita of the entire population of the economy. This provides a rationale for referring to the integral $\int \mathscr{E}_2 d\nu$ of an economy $\mathscr{E} = (\mathscr{E}_1, \mathscr{E}_2) : A \to \mathcal{P} \times \mathbb{G}$ *aggregate endowments per capita* in Definition 5.1.

Definition 5.2 (Aggregate Demand of Economies) The *aggregate demand* of an economy $\mathscr{E} : (A, \mathscr{A}, \nu) \to \mathcal{P} \times \mathbb{G}$ is a relation $\Phi_{\mathscr{E}}$ from $\mathbb{R}^\ell$ to $\mathbb{R}^\ell$ defined by

$$p \mapsto \Phi_{\mathscr{E}}(p) = \int D(X_a, \succ_a, w_p(a), p) d\nu,$$

where $w_p(a) = p \cdot e_a$ for each $a \in A$ and $p \in \mathbb{R}^\ell$.

In the case of a distributional economy given by its preference–wealth distribution $\mu_{\succ, w_p}$ at $p \in \mathbb{R}^\ell$, the *aggregate demand* of the distribution $\mu_{\succ, w_p}$ at $p \in \mathbb{R}^\ell$ is a relation $\Phi_{\succ, w_p}$ defined for each $p \in \mathbb{R}^\ell$ by

$$p \mapsto \Phi_{\succ, w_p}(p) = \int_{\mathcal{P} \times \mathbb{R}} \mathrm{co}\, D(X, \succ, w, p) d\mu_{\succ, w_p}.$$

A rationale will be provided at a later stage for defining the aggregate demand of a distributional economy as the integral of the *convex hull* of values of the demand relation instead of that of the demand relation itself.

5.3 Aggregation Effects on Demands

One intriguing topic in the field of measure theoretical analysis of general economic equilibrium is the phenomenon of *aggregation effects* on the total (mean) demand of an economy induced by aggregating economic agents' individual demands of commodities over all economic agents. A few classical perspectives in the literature on this topic are reviewed in the appendix to this chapter (cf. p. 172).

Historical Perspectives

Discontinuity vs. Continuity

It is challenging to ascertain whether individual demands are continuous with respect to prices and wealth levels from our own everyday experiences or from our observations of other consumers in markets. In this vein, Cournot [20] who perceived demands from a statistical viewpoint and called it "les loi des debit" recognized very clearly that individual demands do not change continuously (cf. p. 173). In addition to Cournot, prominent economic theorists from the late 19th century to the beginning of the 20th century such as Walras [122], Pareto [96], and Marshall [78], appear to have acknowledged as an observed fact that individual demands do not respond continuously to changes in market prices. This is in contrast with economic theorists including Hicks [49], Samuelson [101], Arrow–Debreu [3], Debreu [22], McKenzie [83, 84, 86], as well as Nikaido [95], who promoted general equilibrium analysis after Walras and developed the theory in frameworks that ensure the continuity or upper hemi-continuity of individual demands. It is noteworthy that several eminent economic theorists, including Cournot, Walras, Marshall, and Pareto, in the history of economic theory, were not reticent about assuming aggregate demands to be Lipschitzian or at least continuous in their analysis of total market demands. All of their astute justifications are based on the aggregation effects on demands due to the diversity and heterogeneity of consumption characteristics of individual consumers.

Modern Conjectures on Aggregation Effects

Conjectures in General Equilibrium Model with a Measurable Structure

The advent of the measure-theoretic model of Aumann, as presented in [6], appears to have motivated Debreu to propose the first workable and very specific conjectures on the aggregation effects in his presidential address to the Econometric Society in Barcelona in 1971 (cf. [26, p. 614]):

> "... Still another way to approach the study of regular differentiable economies is to postulate an atomless measure on the space A of economic agents defined as the cartesian product $P \times \mathcal{F}$, where the first factor is interpreted as the set of endowments, and $\mathcal{F}$ is the set of monotone, convex, continuous, complete preference relations on P, endowed with one of the metrics of Y. Kannai, G. Debreu, and W. Hildenbrand. One expects that if the measure ν is suitably diffused over the space A, integration over A of the demand *correspondences* of the agents will yield a total demand *function*, possibly even a total demand function of class C^1."

This was followed by a related conjecture of Hildenbrand [55, p. 111]:

> "There is another, probably deeper, reason why one should consider atomless distribution of agent's characteristics: the very fact that economic agents are not all alike — which means in our framework that the support of the preference-wealth distribution is "spread over" the set $\mathcal{P} \times \mathbb{R}$ — can give rise to properties, for example of the mean demand, which would not hold without a diversification of agent's characteristics. ... "

Both of these conjectures are focused on what will be called *economic effects of aggregation* in subsequent sections.

5.4 Direct Effects of Aggregation

Direct Effects vs. Economic Effects

It is beneficial to distinguish between two types of aggregation effects. One type is what we like to call the *direct effect of aggregation*, which is a consequence of aggregation itself and does not depend on the particularities of objects that are being aggregated. Put another way, the direct effect of aggregation is an intrinsic phenomenon to aggregation itself rather than an economic phenomenon. The other is an effect that is induced by the nature of objects being those of economic analysis; it is an aggregation effect as an economic phenomenon. Fittingly, it will be called the *economic effect of aggregation*. The direct effect of aggregation is verified by examining effects which are consequences of totaling values associated with many individuals. In terms of everyday language, it refers to a "law of large numbers" (not in the sense of statistics nor probability theory). In the present section the direct effects of aggregation on demands are substantiated. Subsequently, we return to the topic of economic effects of aggregation.

Example of Effects in Finite Populations

It is not necessarily the case that an individual economic agent's consumption characteristics $(X, \succ, w) \in \mathcal{P} \times \mathbb{R}$ result in a convex demand set $D(X, \succ, w)$ for that agent. The reason for this may be due to the existence of indivisible commodities or due to preferences that do not satisfy convexity even if all of the commodities are perfectly divisible. Nevertheless, the population size directly affects the geometric property of aggregated sets expressed as mean demand when these individual demand sets are aggregated over individual agents. To illustrate, consider mixtures $z = tx + (1 - t)y$, where $0 < t, 1 - t < 1$, of two consumption vectors x and y in the demand set $D(X, \succ, w)$. The status of these mixtures as demand vectors depends upon the population size. The following example illustrates this aspect.

Example [Finite Population Sizes Affecting Direct Effects of Aggregation]

Let $\ell = 2$, $\mathbb{G} = \mathbb{R}^2$, $X = \mathbb{R}^2_+$, $\succ = \{((x_1, x_2), (y_1, y_2)) \in X \times X \mid x_1^2 + x_2^2 > y_1^2 + y_2^2\}$, $\bar{e} = (20, 20)$, $p = \frac{1}{2}(1, 1)$. Then, we have $D(X, \succ, p \cdot \bar{e}, p) = \{(0, 40), (40, 0)\}$. (See Fig. 5.1.) For each positive integer $n \in \mathbb{N}_+$, let $A_n = \{\frac{k}{n} \mid k = 1, \ldots, n\}$, $\mathcal{A}_n = \mathbb{P}(A_n)$ (i.e., the power set of A_n), ν_n the normalized counting measure on $\mathcal{A}_n$, and define economies $\mathcal{E}^n : A_n \to \mathcal{P} \times \mathbb{G}$ by $a \mapsto \mathcal{E}^n(a) = (X, \succ, \bar{e})$. In other words, we consider economies with individual agents having an identical consumption characteristics and the population size n. It means that in terms of distributional economies, $\mu_{\mathcal{E}^n} = \mu^n_{\succ, e}$ one has $\mu^1_{\succ, e} = \mu^n_{\succ, e}$ for all $n \in \mathbb{N}_+$. The following calculations show that the mean demands of these economies vary in accordance with population size, despite their identical distribution of consumption characteristics:

$$\int D(X, \succ, p \cdot e, p) d\mu^2_{\succ, e} = \{(0, 40), (20, 20), (40, 0)\}$$

$$\vdots$$

$$\int D(X, \succ, p \cdot e, p) d\mu^n_{\succ, e} = \left\{ \frac{k}{n}(0, 40) + \left(1 - \frac{k}{n}\right)(40, 0) \,\middle|\, k = 0, 1, \ldots, n \right\}.$$

In the above example an increasing number of consumption vectors along the line segment between two consumption vectors $(0, 40)$ and $(40, 0)$ in the demand set,

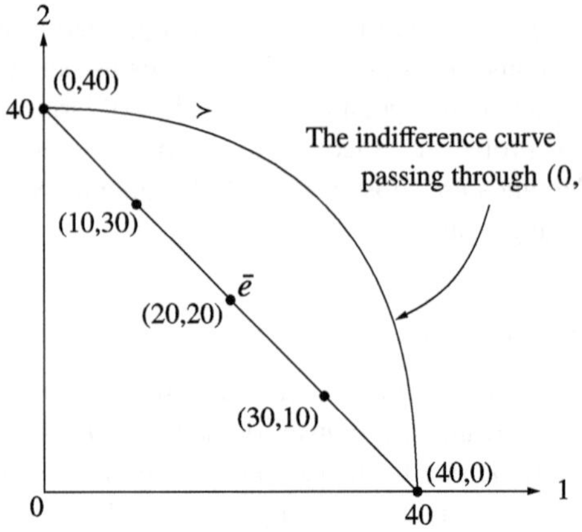

Aggregate Demand for $n = 4$

k	0	1	2	3	4
Demand Vector	(40,0)	(30,10)	(20,20)	(10,30)	(0,40)

Fig. 5.1 Direct effects of aggregation in finite population

as well as in the mean demand set for a single population size, belong to the mean demand set as the population size grows. This phenomenon is precisely the "*average phenomena and those involving large numbers*" noted by Pareto and what we like to associate with the *direct effects of aggregation*. Strictly speaking, this effect is different from the law of large numbers in statistics which states that as the sample size increases, the mean of the sample becomes closer to that of the population. Yet, given that mean demand sets represent the average of individual demand sets, Theorem 5.1 and Theorem 5.2 in finite and infinite economies, respectively, support that it is comparable to a law of large numbers except in a different but similar context. We refer to this as the *convexity approximation*, which is related to the Shapley–Folkman Theorem (cf. Lemma 5.1). Note, however, that it is distinct from the phenomenon Walras called the law of large numbers in the earlier quotation from his book.

Direct Effects

Two results concerning the direct effects of aggregation are presented below. The first result is stated in terms of *convexity approximations* in finite economies. In this context, the term "convexity approximation of the mean demand" means that points in the convex hull of the mean demand set can be approximated by points in the mean demand set arbitrarily close if the population size is sufficiently large.

Convexity Approximations

Theorem 5.1 (Convexity Approximations in Large Finite Economies)
Let $\mathscr{E} : (A, \mathscr{A}, \nu) \to \mathcal{P} \times \mathbb{G}$ be an economy for which $0 < \sharp A < \infty$. Then, for every compact set $K \subset \mathbb{R}^{\ell}_{++}$, $\varepsilon > 0$, and $\delta > 0$, there exists a positive integer n_0 such that whenever $\sharp A \geqq n_0$ and $\sup\{w_p(a) \mid a \in A, p \in K\} \leqq \delta$, we have $\operatorname{co}\Phi_{\mathscr{E}}(p) \subset N_{\varepsilon}(\Phi_{\mathscr{E}}(p))$ for all $p \in K$.

The following mathematical result, due to Shapley and Folkman, is employed in the proof of the theorem.

Lemma 5.1 [Shapley–Folkman Theorem] *Let $S_i, i = 1, \ldots, n$, be non-empty subsets of $\mathbb{R}^m$, and let $x \in \operatorname{co} \sum_{i=1}^n S_i$. Then, there exist points $x_i \in \operatorname{co} S_i, i = 1, \ldots, n$, satisfying $x = \sum_{i=1}^n x_i$ such that the number of indices i for which $x_i \notin S_i$, $\sharp\{i \mid x_i \notin S_i\}$, is at most m.*

Proof For $x \in \operatorname{co} \sum_{i=1}^n S_i$, since $\operatorname{co} \sum_{i=1}^n S_i = \sum_{i=1}^n \operatorname{co} S_i$, it is possible to write x as follows:

$$x = \sum_{i=1}^{n} x_i \text{ such that } x_i = \sum_{j=1}^{n_i} s_{ij} y_{ij} \in \operatorname{co} S_i, \tag{5.3}$$

$$\text{where } \sum_{j=1}^{n_i} s_{ij} = 1, y_{ij} \in S_i, s_{ij} \geqq 0 \text{ for all } ij,$$

for some n positive integers $n_i, i = 1, \ldots, n$.

We show first that if the sum, $\sum_{i=1}^n n_i$, of the integers in an expression (5.3) of x strictly exceeds $n + m$, then there is an alternative way to write x among these expressions with the sum of integers $\sum_{i=1}^n n_i$ not exceeding $n + m - 1$. Indeed, if $\sum_{i=1}^n n_i > n + m$, then we have $\sum_{i=1}^n (n_i - 1) > m$. It follows that the vectors $y_{ij} - y_{i1}, (j = 2, \ldots, n_i, i = 1, \ldots, n)$, must be linearly dependent. Consequently, there are real numbers t_{ij}, not all equal to zero, satisfying $\sum_{i=1}^n \sum_{j=1}^{n_i} t_{ij}(y_{ij} - y_{i1}) = 0$. By multiplying this equation by a number $s \neq 0$ and adding it to the expression for x, one obtains $x = \sum_{i=1}^n \sum_{j=1}^{n_i} s'_{ij} y_{ij}$ where $s'_{i1} = s_{i1} - s \sum_{j=2}^{n_i} t_{ij}$ and $s'_{ij} = s_{ij} + s t_{ij}$ for $i = 1, \ldots, n, j = 2, \ldots, n_i$. It is possible to select the number s in a way that $s'_{ij} = 0$ for at least one ij. Therefore, the sum of integers in this alternative expression of x in (5.3) does not exceed $n + m - 1$.

The preceding arguments demonstrate that among the expressions of x in (5.3), there exists one with the sum of integers, $\sum_{i=1}^n n_i$, at most equal to $n + m$. For such an expression of x, we have $n_i = 1$ except for at most m i's. This proves the assertion of the lemma. □

Proof of Theorem 5.1 The theorem is proven if the following property is established:

> *Let $K \subset \mathbb{R}^{\ell}_{++}$ be a compact set, $\varepsilon > 0$, and $\delta > 0$. Then, there is a point $y \in \Phi_{\mathscr{E}}(p)$ such that $\|x - y\| < \varepsilon$ for any $p \in K$ and for any $x \in \operatorname{co}\Phi_{\mathscr{E}}(p)$ if $\sup\{w_p(a) \mid a \in A, p \in K\} \leqq \delta$ and $\sharp A \geqq n_0$ for some positive integer n_0.*

Let $p \in K$ and $x \in \operatorname{co}\Phi_{\mathscr{E}}(p)$; then, $(\sharp A)x \in \operatorname{co}\sum_{a\in A} D(X_a, \succ_a, w_p(a), p)$. It follows from the Shapley–Folkman Theorem that there exists a subset $S \subset A$, and mappings $f : A \to \mathbb{G}$ and $g : S \to \mathbb{R}^\ell$ satisfying the following properties (a)–(d):

(a) $\sharp S \leqq \ell$;
(b) $f(a) \in D(X_a, \succ_a, w_p(a), p)$ for all $a \in A$;
(c) $g(a) \in \operatorname{co} D(X_a, \succ_a, w_p(a), p)$ for all $a \in S$;
(d) $(\sharp A)x = \sum_{a\in A} f(a) + \sum_{a\in S}(g(a) - f(a))$.

Since $X_a \in \mathscr{X}$ for all $a \in A$, for any a, $y = (y_1, \ldots, y_\ell) \in X_a$, and $i = 1, \ldots, \ell$, we have $y_i \geqq \underline{\beta}$. Let $\eta = \min\{p_i \mid p = (p_1, \ldots, p_\ell) \in K, i = 1, \ldots, \ell\}$. Then, for each $a \in A$ and $i = 1, \ldots, \ell$, we have $\underline{\beta} \leqq f_i(a) \leqq \delta/\eta$, $\underline{\beta} \leqq g_i(a) \leqq \delta/\eta$. It follows that $\|f(a) - g(a)\| \leqq \sqrt{\ell}(\delta - \eta\underline{\beta})/\eta$ for all a. Thus, by setting $y = \frac{1}{\sharp A}\sum_{a\in A} f(a)$, one has

$$\|x - y\| \leqq \sum_{a\in S} \frac{\|g(a) - f(a)\|}{\sharp A} \leqq \frac{\ell\sqrt{\ell}(\delta - \eta\underline{\beta})}{\eta(\sharp A)}.$$

By defining a positive integer n_0 so that $n_0 \geqq \ell\sqrt{\ell}(\delta - \eta\underline{\beta})/\varepsilon\eta$, one obtains $\|x - y\| < \varepsilon$. The integer n_0 depends only upon K, ε, and δ, but not upon x or y. This completes the proof of the theorem. □

Convexity Effects

Theorem 5.1 above attests that the mean demand sets of large but finite economies can be approximated by vectors in their convex hulls. The following theorem demonstrates the convexity of mean demands in infinite (atomless) economies.

Theorem 5.2 (Convexity Effects in Infinite Economies)
Let $\mathscr{E} : (A, \mathscr{A}, \nu) \to \mathcal{P} \times \mathbb{G}$ be an economy for which $\sharp A = \infty$. Then, the aggregate demand set $\Phi_{\mathscr{E}}(p)$ is convex for any $p \in \mathbb{R}^\ell_{++}$, and we have $\Phi_{\mathscr{E}}(p) = \Phi_{\mu_\succ, w_p}(p)$.

Proof By Proposition 4.5, p. 145, the demand relation D is measurable. Consequently, the measurability of the mapping $\mathscr{E} : (A, \mathscr{A}, \nu) \to \mathcal{P} \times \mathbb{G}$ implies that two correspondences $a \mapsto D(X_a, \succ_a, w_p(a), p)$ and $(X, \succ, w, p) \mapsto D(X, \succ, w, p)$ are measurable. Since all consumption sets $X \in \mathscr{X}$ are uniformly bounded below, these correspondences are integrably bounded below.

The measure space of economic agents $(A, \mathscr{A}, \nu)$ is atomless by the definition of an economy $\mathscr{E}$ with $\sharp A = \infty$. It follows from Mathematical Theorem 3.4, p. 119, that we have

$$\Phi_{\mathscr{E}}(p) = \int D(X_a, \succ_a, w_p(a), p)d\nu = \int \operatorname{co} D(X_a, \succ_a, w_p(a), p)d\nu.$$

This shows that $\Phi_{\mathscr{E}}(p)$ is convex. If we change variables of integration for the correspondence $a \mapsto D(X_a, \succ_a, w_p(a), p)$ using Mathematical Theorem 3.5, p. 122, then

$$\int_A \operatorname{co} D(X_a, \succ_a, w_p(a), p) d\nu = \int_{\mathcal{P}\times\mathbb{R}} \operatorname{co} D(X, \succ, w, p) d(\nu \circ \mathscr{E}_p{}^{-1})$$
$$= \int_{\mathcal{P}\times\mathbb{R}} \operatorname{co} D(X, \succ, w, p) d\mu_{\mathscr{E}_p}.$$

Therefore, one obtains that

$$\Phi_{\mathscr{E}}(p) = \int_{\mathcal{P}\times\mathbb{R}} \operatorname{co} D(X, \succ, w, p) d\mu_{\succ, w_p}$$
$$= \Phi_{\mu_{\succ, w_p}}(p).$$

□

The rationale behind defining the mean demand of distributional economies to be the integral of the *convex hull* of values of individual demand relations instead of individual demand relations themselves is twofold. Firstly, it allows for the convexity effects in infinite economies. Secondly, it represents the convexity approximation in large but finite economies.

5.5 Economic Effects of Aggregation

The direct effects of aggregation in the previous section are contingent upon the aggregation of a substantial number of objects and are not contingent upon the distribution of economic characteristics associated with the objects. The "large numbers effects," which classical economic theorists such as Cournot, Walras, and Marshall had in mind, are aggregation effects attributable to the diversity and heterogeneity of characteristics of economic agents. In this section, we present one of the key aggregation effects on total demands induced by aggregation over individual demands with diversified individual consumption characteristics.

Diversified Wealth Distribution

In our concern with the continuity property of mean demand relations, it is necessary to consider the possible continuity and discontinuity properties of individual demand relations (cf. Theorem 4.2, p. 148) in conjunction with the fundamental property of critical sets of consumption sets in the wealth space (cf. Theorem 4.1, p. 135). This necessitates the existence of a "well-diversified" wealth distribution, as this is a prerequisite for the aggregation effects on mean demand relations to be relied upon. This is consistent with the verbal arguments put forth by some theoretical economists as previously cited.

Hemi-Continuity Effects of Wealth Distribution

In accordance with Theorem 4.2, an individual demand relation may fail to be upper hemi-continuous if its wealth level w belongs to the critical set $J_p(X)$ of its consumption set X in the wealth space. However, since the critical set $J_p(X)$ is countable (Theorem 4.1), it is possible that the aggregate demand relation of individual demand relations having the same consumption set X may exhibit continuity properties in the sense of upper hemi-continuity if the wealth distribution is sufficiently dispersed among them. To illustrate, if the endowments of all individuals with the consumption set X are uniformly distributed over the set S as depicted in Fig. 5.2, then the induced wealth distribution is sufficiently spread for any $p \in \mathbb{R}^{\ell}_{++}$.

The preceding considerations lead us to introduce the notion of a *well-diversified wealth distribution* in a natural way.

Definition 5.3 (Dispersion of Wealth Distribution) In a distributional economy $\mu_{\succ,w_p}$, the wealth distribution is said to be *dispersed* at $p \in \mathbb{R}^{\ell}$ if, for any X with $(X, \succ) \in \operatorname{supp} \mu_{\succ}$, the conditional wealth distribution $\mu_{w_p|X}$ given consumption set X is *atomless*; that is, if it assigns zero weight to any wealth level $w \in \mathbb{R}$, for μ_X-almost every X, where μ_X is the consumption set distribution of the economy $\mu_{\succ,w_p}$ at p. The *wealth distribution of an economy* $\mathcal{E} : (A, \mathcal{A}, \nu) \to \mathcal{P} \times \mathbb{G}$ is said to be *dispersed* at $p \in \mathbb{R}^{\ell}$ if the wealth distribution of the distributional economy $\mu_{\succ,w_p}$, associated with $\mathcal{E}$, is dispersed at $p \in \mathbb{R}^{\ell}$.

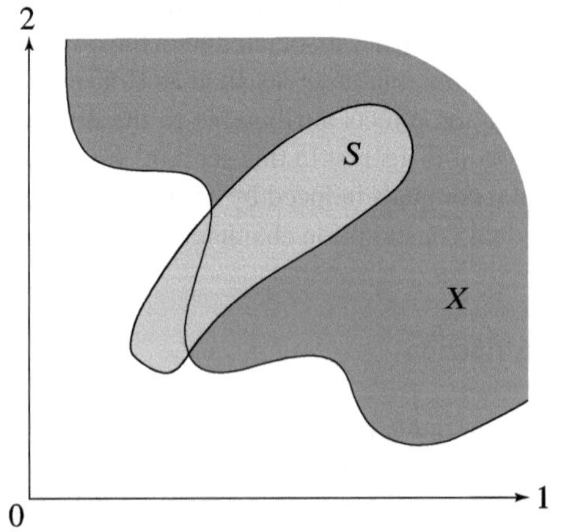

Fig. 5.2 Dispersion of endowments

Examples [Economies with Dispersed Wealth Distributions]

(1) Let $\mathbb{G} = \mathbb{N}^{\ell-1} \times \mathbb{R}$, and $\mathscr{E} : (A, \mathscr{A}, \nu) \to \mathcal{P} \times \mathbb{G}$ be an economy satisfying the following conditions:

(1.1) $X_a = \mathbb{G} \cap \mathbb{R}^\ell_+$ for all $a \in A$.
(1.2) The endowment distribution of $\mathscr{E}$ is the uniform distribution on $L = [e_1, e_2]$ where $e_1 = (2, \ldots, 2, 2)$ and $e_2 = (2, \ldots, 2, 10)$.

Then, the economy $\mathscr{E}$ has a dispersed wealth distribution at any price vector $p \in \mathbb{R}^\ell_+$ with $p^\ell > 0$.
(2) (See Fig. 5.3) Let $\mathbb{G} = \mathbb{R}^2_+ \times \mathbb{N}$ and $(A, \mathscr{A}, \nu) = ([0, 1]^2, \mathcal{B}([0, 1]^2), \lambda^2)$. Define $X_t = \{x \in \mathbb{G} \mid x_i \geqq t \text{ for all } i\}$ for each $t \in [0, 1]$. Let an economy $\mathscr{E} : A \to \mathcal{P} \times \mathbb{G}$ be defined by setting for each $a = (a_1, a_2) \in A = [0, 1]^2$,

(2.1) $(X_a, \succ_a) = (X \succ)$ for some $(X \succ) \in \mathcal{P}$ such that $X = X_t$ with $t = a_1$;
(2.2) $e(a) = (a_1 + a_2 + 1, a_2 + 1.5, 2)$.

Put $e_I(t) = (t + 1, 1.5, 2)$ and $e_{II}(t) = (t + 2, 2.5, 2)$. Then, the conditional endowment distribution given consumption set X_t for the economy $\mathscr{E}$ is the uniform distribution on the line segment $L_t = [e_I(t), e_{II}(t)]$. Let $p = (p_1, p_2, p_3) \in \mathbb{R}^3_+$ be a price vector with $\max\{p_1, p_2\} > 0$. Since the set $H(p, w) \cap L_t$ is either a singleton or empty for $w = p \cdot e, e \in L(t)$, the wealth distribution of the economy $\mathscr{E}$ is dispersed at p.

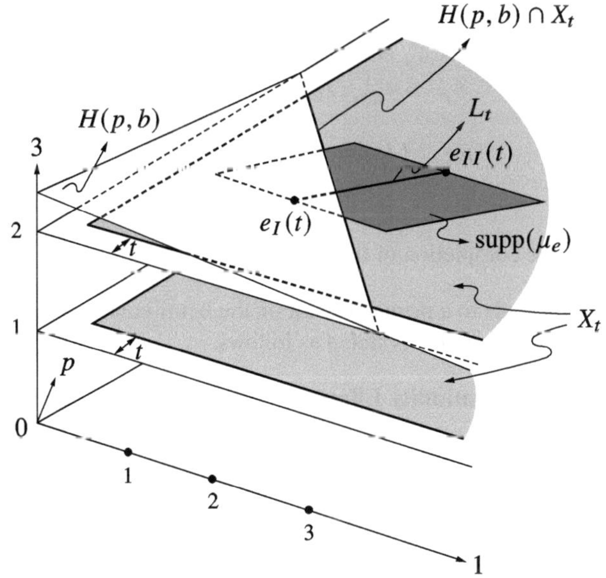

Fig. 5.3 Dispersed wealth distribution

The dispersion of wealth sets the stage for the economic effects of aggregation to operate, as the next result shows.

Proposition 5.1 (Dispersion of Wealth and Null Critical Sets)

(1) *If the wealth distribution of a distributional economy* $\mu_{\succ,w_p}$ *is dispersed at* $p \in \mathbb{R}^\ell$, *then we have* $\bar{\mu}_{X,w_p}(J_p) = 0$, *where* $\bar{\mu}_{X,w_p}$ *is the completion of the joint distribution of consumption sets and wealth of the distributional economy* $\mu_{\succ,w_p}$.
(2) *Let* $\mathscr{E} : (A, \mathscr{A}, \nu) \to \mathcal{P} \times \mathbb{G}$ *be an economy with a dispersed wealth distribution at* $p \in \mathbb{R}^\ell$. *Then, the set* A_p *of all economic agents whose wealth levels belong to the critical sets in the wealth space, i.e.* $A_p = \{a \in A \mid p \cdot e_a \in J_p(X_a)\}$, *is a* ν*-null set.*

Proof (1) The relation $J_p = \{(X, v) \in \mathcal{X} \times \mathbb{R} \mid v \in J_p(X)\}$ is universally measurable by Proposition 4.3, p. 139, and thus J_p belongs to the completion of $\mathcal{B}(\mathcal{X} \times \mathbb{R})$. We have

$$\begin{aligned}\bar{\mu}_{X,w_p}(J_p) &= \int_{\mathcal{X}} \mu_{w_p|X}(J_p(\bar{X}), \bar{X}) d\mu_X(\bar{X}) \\ &= \int_{\mathcal{X}} \sum_{w \in J_p(\bar{X})} \mu_{w_p|X}(\{w\}, \bar{X}) d\mu_X(\bar{X}) \\ &= 0,\end{aligned}$$

where $\mu_{w_p|X}$ is the conditional wealth distribution given consumption set X, and the last two equalities are consequences of Theorem 4.1, p. 135, that $J_p(\bar{X})$ is countable for any set $\bar{X} \in \mathcal{X}$ and that $\mu_{w_p|X}(\cdot, \bar{X})$ is an atomless measure μ_X-a.e. in $\mathcal{X}$ since the wealth distribution of $\mu_{\succ,w_p}$ is dispersed at p.

(2) As we have $A_p \subset \mathscr{E}_p^{-1} \circ \pi_{X,w}^{-1}(J_p)$, it follows from (1) that we have

$$\begin{aligned}\bar{\mu}_{X,w_p}(J_p) &= \bar{\nu}(\mathscr{E}_p^{-1} \circ \pi_{X,w}^{-1}(J_p)) \\ &\geqq \bar{\nu}(A_p) = 0,\end{aligned}$$

where $\bar{\nu}$ denotes the completion of the measure ν. □

Proposition 5.1 leads to a primary result on the hemi-continuity effects of aggregation on total demands which is stated as follows.

Theorem 5.3 (Hemi-Continuity Effects of Aggregation on Total Demands)
If $\mathscr{E} : (A, \mathscr{A}, \nu) \to \mathcal{P} \times \mathbb{G}$ *is an economy whose wealth distribution is dispersed at* $p \in \mathbb{R}^\ell_{++}$, *then the total demand* $\Phi_{\mathscr{E}} : \mathbb{R}^\ell_{++} \to \mathbb{R}^\ell$ *is upper hemi-continuous at* p.

Proof The conclusion of the theorem follows from Proposition 5.1 (2) and Theorem 4.2, p. 148, by taking account of Mathematical Proposition 1.1, p. 36, 3.18 and 3.19, p. 127. □

Notes 5

Notes 5.1 [Section 5.1, p. 155]

[A measure space of economic agents $(A, \mathscr{A}, \nu)$, p. 156] An analytical framework representing perfect competition in markets, where each individual cannot influence the overall outcome of an economy, requires an infinite number of individuals without any influence on the overall outcome. The concept of an atomless economy was first proposed by Aumann [6], where, as in some classical general equilibrium models with a finite population in the 1950s, the commodity space is a finite-dimensional Euclidean space in which all economic agents have identical consumption sets given by the non-negative orthant of the commodity space.

Nevertheless, some readers may consider that a very natural extension of a finite number of individuals is the set of positive integers, i.e. $A = \mathbb{N}_+$. Moreover, [76] shows that an example of a measure defined on $\mathbb{N}_+$ that is a natural extension of the counting measure is one that assigns a zero measure to each element. However, this is a finitely additive measure. For this reason, if one wishes to stay within a framework of measure spaces with countably additive measures, a finite non-atomic measure space is a very good analytical framework for representing a set of individuals in perfectly competitive markets.

It should be noted that there are those in probability theory and statistics who do not consider constructing theories by assuming countable additivity of measures. See, for example, [29, 30] and related literature. If one assumes only finite additivity of measures, the claim in [6] that the population of economic agents must have a continuum of individuals in order to have individuals without any influence on the overall outcome does not hold in this case. In fact, the following finite additive measure space, which is perhaps most familiar as an extended notion of a finite population of individuals, is atomless.

Let $A = \mathbb{N}_+$ and $\mathscr{A} = \mathbb{P}(A)$ to define a measure space $(A, \mathscr{A}, \nu)$. Define a positive set function ν on $\mathscr{A}$ by the following steps. First, let $\mathcal{F}$ be a class of subsets of A given by

$$\mathcal{F} = \left\{ S \in \mathbb{P}(A) \middle| \lim_n \frac{\sharp(S \cap \{1, \ldots, n\})}{n} \text{ exists.} \right\}.$$

Then let $m : \mathcal{F} \to [0, 1]$ be a set function on $\mathcal{F}$ defined by $m(S) = \lim_n \frac{\sharp(S\cap\{1,\ldots,n\})}{n}$ for each $S \in \mathcal{F}$. ($m(S)$ is known as the *asymptotic density* of S.) This set function $m : \mathcal{F} \to [0, 1]$ can be extended to a finitely additive measure on $\mathscr{A}$ (e.g., see [91, p. 119]). Let us write the measure with the symbol ν. (ν is called the *asymptotic density measure*.) ν is a finitely additive atomless measure and satisfies $\nu(\{a\}) = 0$ for every $a \in A$.

[Reference to conditional distributions, p. 158] As references to Mathematical Property 1.33, p. 58, one can see [11, Chapter 10, Theorem 10.3.5, Problems 1–2, p. 334] and [97, Chapter V, Section 8, pp. 146–150].

[Axiom of Solvay, p. 159] Readers interested in this issue could refer to [41] to see that the measurability issue is related to the foundations of mathematics.

Notes 5.2 [Section 5.4, p. 162]

[Shapley–Folkman Theorem, p. 165] The proof of Lemma 5.1 is due to [18, pp. 435–436].

Notes 5.3 [Section 5.5, p. 167]

[Example, p. 169] (1) is from [81], and (2) is from [127].

Appendix to Chapter 5: Some Classical Perspectives on the Aggregation Effects

Reference to Large Numbers Effect by Walras

The following quotes are from [122, English translation, Section 52, p. 95]:

> "... There is nothing to indicate that the individual demand curves ..., or the individual demand equations ... are continuous, in other words, that an infinitesimally small increase in p_a produces an infinitesimally small decrease in d_a. *On the contrary, these functions are often discontinuous.* In the case of oats, for example, surely our first holder of wheat will not reduce his demand gradually as the price rises, but he will do it in some intermittent way every time he decides to keep one horse less in his stable. *His individual demand curve will, in reality, take the form of a step curve .... All the other individual demand curves will take the same general form. And yet the aggregate demand curve ... can, for all practical purposes, be considered as continuous by virtue of the so-called law of large numbers.* In fact, whenever a very small increase in price takes place, at least one of the holders of (B), *out of a large number of them,* will then reach the point of being compelled to keep one horse less, and thus a very small diminution in the total demand for (A) will result. ..." *(Mostly my italics.)*

Pareto's Intuitive Arguments about Economic Effects of Aggregation

The following quotations are from [96, English translation, pp. 122–123)]:

> "65. Continuous variations and discontinuous variations. The indifference curves and the paths could be discontinuous, and they are in reality. That is, the variations in the quantities occur in a discontinuous fashion. An individual passes from a state in which he has 10 handkerchiefs to a state in which he has 11, and not through intermediate states in which he would have, for example, 10 and 1/100 handkerchiefs, 10 and 2/100 handkerchiefs, etc.
>
> In order to come closer to reality, we would have to consider finite variations, but there is a technical difficulty in doing so.
>
> Problems concerning quantities which vary by infinitely small degrees are much easier to solve than problems in which the quantities undergo finite variations. Hence, every time it is possible, we must replace the latter by the former; this is done in all the physiconatural sciences. We know that an error is thereby committed; but it can be neglected either when it is small absolutely, or when it is smaller than other inevitable errors which make it useless to seek a precision which eludes us in other ways. *This is precisely so in political economy, for there we consider only average phenomena and those involving large numbers. We speak of the individual, not in order actually to investigate what one individual consumes or produces, but only to consider one of the elements of a collectivity and then add up the consumption and the production of a large number of individuals.*
>
> 66. When we say that an individual consumes one and one-tenth watches, it would be ridiculous to take those words literally. A tenth of a watch is an unknown object for which we have no use. Rather these words simply signify that, for example, one hundred individuals consume 110 watches.
>
> When we say that equilibrium takes place when an individual consumes one and one-tenth watches, we simply mean that equilibrium takes place when 100 individuals consume — some one, others two or more watches and some even none at all — in such a way that all together they consume about 110, and the average is 1.1 for each ..." *(Mostly my italics.)*

Perceptive Arguments about Continuity Effects of Aggregation by Cournot

Here are quotes from [20, English translation, Section 22, pp. 49–50]:

> "... We will assume that the function $F(p)$, which expresses the law of demand or of the market, is a *continuous* function, i.e. a function which does not pass suddenly from one value to another, but which takes in passing all intermediate values. *It might be otherwise if the number of consumers were very limited:* thus in a certain household the same quantity of firewood will possibly be used whether wood costs 10 francs or 15 francs the stere, and the consumption may suddenly be diminished if the price of the stere rises above the latter figure. *But the wider the market extends, and the more the combinations of needs, of fortunes, or even of caprices, are varied among consumers, the closer the function $F(p)$ will come to varying with p in a continuous manner.* However little may be the variation of p, there will be some consumers so placed that the slight rise or fall of the article will affect their consumptions, and will lead them to deprive themselves in some way or to reduce their manufacturing output, or to substitute something else for the article that has grown dearer, as, for instance, coal for wood or anthracite for soft coal. ..." *(Mostly my italics).*

Marshall's Insightful Account of Regularizing Effects of Aggregation by Diversity of Consumption Characteritcs

[78, Book III, Chapter III, Section 5, pp. 82–83] give an explicit reference to the regularizing effects of aggregation due to the diversity of consumers' characteristics:

> "... *But even among those things which are in constant use, there are many for which the demand on the part of any single individual cannot vary continuously with every small change in price,* but can move only by great leaps. For instance, a small fall in the price of hats or watches will not affect the action of every one; but it will induce a few persons, who were in doubt whether or not to get a new hat or a new watch, to decide in favour of doing so.
>
> There are many classes of things the need for which on the part of any individual is inconstant, fitful, and *irregular.* There can be no list of individual demand prices for wedding-cakes, or the services of an expert surgeon. But the economist has little concern with particular incidents in the lives of individuals. He studies rather "the course of action that may be expected under certain conditions from the members of an industrial group," in so far as the motives of that action are measurable by a money price, *and in these broad results the variety and the fickleness of individual action are merged in the comparatively regular aggregate of the action of many.*
>
> *In large markets, then — where rich and poor, old and young, men and women, persons of all varieties of tastes, temperaments and occupations are mingled together, — the peculiarities in the wants of individuals will compensate one another in a comparatively regular gradation of total demand. Every fall, however slight in the price of a commodity in general use, will, other things being equal, increase the total sales of it; just as an unhealthy autumn increases the mortality of a large town, though many persons are uninjured by it. And therefore if we had the requisite knowledge, we could make a list of prices at which each amount of it could find purchasers in a given place during, say, a year. (Mostly my italics.)*

Chapter 6
Walrasian Equilibria of Economies

Abstract This chapter is concerned with the Walrasian equilibria of economies. It is focused on a most foundational question in economic theoretical analysis: the existence of market equilibria of economies. In order to address this question, the chapter abstracts economies from their production possibilities. This allows the redistribution of existing and produced commodities to be brought to the fore in the analysis. The production possibilities of economies will be taken into consideration in the following chapter.

6.1 Concepts of Equilibrium in Perfectly Competitive Markets

Economic Transactions in Markets

Let $(A, \mathscr{A}, \nu)$ be a measure space of economic agents and let $\mathscr{E} : (A, \mathscr{A}, \nu) \to \mathcal{P} \times \mathbb{G}$ be an economy. Economic transactions taking place in markets are described by an assignment of commodity vectors to each agent in the economy. These commodity vectors are constrained by total quantities of various commodities available for transactions in markets. It should be recalled that the total availability of commodities in $\mathscr{E}$ is given by the aggregate or the mean endowments $\int \mathscr{E}_2 d\nu$ of $\mathscr{E} = (\mathscr{E}_1, \mathscr{E}_2) : A \to \mathcal{P} \times \mathbb{G}$. The mapping $\mathscr{E}_2 : A \to \mathbb{G}$ is the endowment mapping, and the notation $e : A \to \mathbb{G}$ is used when the mapping $\mathscr{E}$ is readily apparent from the context. Note that by the assumption placed on economies the measurable mapping $e : A \to \mathbb{G}$ is integrable, so that $\| \int e d\nu \| < \infty$.

Allocations and Suballocations

In an economy $\mathscr{E} : (A, \mathscr{A}, \nu) \to \mathcal{P} \times \mathbb{G}$, a *suballocation* f for $\mathscr{E}$ is an integrable mapping f from the set A of economic agents to the commodity space $\mathbb{G}$ such that $f(a) \in X_a \cup \{e_a\}$, a.e. $a \in A$. A suballocation f is called an *allocation* for $\mathscr{E}$ if it satisfies $f(a) \in X_a$, a.e. $a \in A$. A primary aspect of allocations and suballocations is

A. Yamazaki, *Mathematical Economics of General Equilibrium*, Monographs in Mathematical Economics 8, https://doi.org/10.1007/978-981-96-9497-6_6

that they function to specify an assignment or a distribution of commodities among economic agents. Another functional aspect is that they are useful for delimiting transactions in markets in various ways. In particular, an allocation or suballocation f for an economy $\mathscr{E}$ is called *attainable* if the aggregate quantities of all commodities required to realize f (i.e., $\int f dv$) are equal to those available in the economy (i.e., $\int e dv$). Accordingly, f is defined as *attainable* if it satisfies the equation $\int f dv = \int e dv$.

In the context of economies without production, alternative concepts of attainability of allocations that are less restrictive are often applied. Notably, an allocation or suballocation f for an economy $\mathscr{E}$ is called *attainable with disposal of commodities* if the aggregate quantities of all commodities required by f do not exceed those available in the economy. That is, f is said to be *attainable with disposal of commodities* if it satisfies the inequality $\int f dv \leqq \int e dv$. The inequality may be rephrased as an equality, namely, $\int f dv + \delta(f, e) = \int e dv$, where $\delta(f, e) = \int e dv - \int f dv \geqq 0$ represents the amounts of commodities to be disposed of.

Allocations as well as suballocations may well be specified without any reference to an economy when it is clear from the context. It is also important to note that there is a fundamental distinction between allocations and suballocations. The assignment or distribution of commodities by allocations ensures the survival of economic agents, whereas that of suballocations may not do so.

Walrasian Equilibrium

Walrasian equilibrium is a classical equilibrium concept that is variously termed a *market equilibrium* or a *competitive equilibrium*. It is the main equilibrium concept in general equilibrium theory and is called *Walrasian equilibrium* in this book.

Definition 6.1 (Walrasian Equilibrium) A *Walrasian equilibrium* for an economy $\mathscr{E} : (A, \mathscr{A}, \nu) \to \mathcal{P} \times \mathbb{G}$ is a pair (p, f) composed of a price vector $p\ (\neq 0) \in \mathbb{R}^{\ell}$ and an allocation f for $\mathscr{E}$ satisfying the following conditions:

(a) For a.e. a in A, $f(a)$ belongs to the demand set $D(X_a, \succ_a, p \cdot e_a, p)$, i.e., $f(a)$ is a maximal element for the preference relation $(X_a, \succ_a)$ in the budget set $B(X_a, p \cdot e_a, p)$.

(b) The allocation $f : A \to \mathbb{G}$ is attainable, i.e. $\displaystyle\int f dv = \int e dv.$

When the condition (b) is replaced by the condition:

(b') The allocation $f : A \to \mathbb{G}$ is *attainable with free disposal of commodities*, i.e. $\int f dv + \delta(f, e) = \int e dv$ where $\delta(f, e) = \int e dv - \int f dv \geqq 0$ and $p \cdot \delta(f, e) = 0$;

then, (p, f) is called a *Walrasian equilibrium with free disposal* for $\mathscr{E}$. The price vector p of (p, f) is an *equilibrium price vector*, and the allocation $f : A \to \mathbb{G}$ is a *Walrasian allocation with free disposal* or an *equilibrium allocation with free disposal*. The condition $p \cdot \delta(f, e) = 0$ in (b') signifies that the commodity bundle $\delta(f, e)$ to be disposed of has a zero market value.

Institutional Presumption of Walrasian Equilibrium

The concept of Walrasian equilibrium, as is well-known, presumes a transaction institution which is usually called "markets." However, markets do not simply mean "physical marketplaces" that we have known historically in our daily life. Rather, they represent "abstract fields" where each and every conceivable commodity is bought and sold not only by direct negotiations but also by means of all sorts of communication devices, including telephones, telegrams, facsimiles, or computer networks. Moreover, in recent years, an increasing number of transactions have been conducted online using personal computers and mobile communication tools. In our analysis of market transactions, prices for various commodities are uniquely quoted in the form of a price vector $p = (p_1, \ldots, p_\ell)$, and commodities are voluntarily bought and sold under these prices. That is, a unit of the i-th commodity may be freely bought in exchange for the value of p_i/p_j units of the j-th commodity. This is an underlying institutional presumption of Walrasian equilibrium.

Assuming that all the economic agents are able to participate in the institution, let us consider the behavior of economic agent a and its consequences in markets. The consumption characteristics of agent a are given by $(X_a, \succ_a, e_a)$ in an economy $\mathscr{E}$. Agent a selects a most desirable consumption vector (i.e., a maximal element for $(X_a, \succ_a)$) in the budget set $B(X_a, p \cdot e_a, p) = \{x \in X_a \mid p \cdot x \leqq p \cdot e_a\}$. This decentralized decision-making process is thus made by each agent in the economy. If the resulting total demands are equal to the total supplies of each and every commodity, then the price vector and the consumption vectors are called the Walrasian equilibrium for the economy $\mathscr{E}$. A set of market prices is the only device that coordinates demands and supplies of individual agents who make their own decision in a completely independent way. Let us refer to an economy with a Walrasian equilibrium as a *Walrasian economy*. In our view, a natural interpretation of institutional arrangements with regard to transaction contracts in a Walrasian economy is that buying or selling contracts of individual agents are made with the markets and not directly with other individual agents. This is because market prices are taken as given, and that transaction partners and possible negotiations among them are not specified.

Preconditioned Minimum Income Levels for Walrasian Equilibrium

Given a Walrasian equilibrium (p, f) for $\mathscr{E} : (A, \mathscr{A}, \nu) \to \mathcal{P} \times \mathbb{G}$, budget sets $B(X_a, p \cdot e_a, p)$ must be non-empty for almost every agent $a \in A$. This implies that almost every agent owns endowed commodities with their values no less than its minimum income level. Otherwise, the budgets of some agents are insufficient to enable them to participate in market transactions, obtaining commodity vectors in their consumption sets, and survive. This is certainly a circumstance not envisioned by the definition of Walrasian equilibrium. In this regard, Koopmans [68, p. 59] points out in a somewhat overemphasized manner that the "hardest part" of the Walrasian model is to "*make sure that each consumer can both survive and participate in the market, without anticipating in the postulates what specific prices will prevail in an equilibrium.*" Indeed, historical evidence demonstrates that starvation has posed a significant challenge to market economies. This suggests that a market equilibrium may not preclude the possibility of starvation among some economic agents in markets. From this perspective, it is only appropriate to introduce a market equilibrium concept based on suballocations and subdemands (cf. p. 151).

Market Subequilibrium

Definition 6.2 (Market Subequilibrium) A *market subequilibrium*, or simply a *subequilibrium*, for an economy $\mathscr{E} : (A, \mathscr{A}, \nu) \to \mathcal{P} \times \mathbb{G}$ is a pair (p, f) composed of a price vector $p\ (\neq 0) \in \mathbb{R}^\ell$ and a suballocation f for $\mathscr{E}$ satisfying the following conditions:

(a) For a.e. a in A, $f(a)$ belongs to the subdemand set $D^*(X_a, \succ_a, e_a, p)$.

(b) The suballocation $f : A \to \mathbb{G}$ is attainable, i.e. $\displaystyle\int f d\nu = \int e d\nu$.

The price vector p of a market subequilibrium (p, f) is a *subequilibrium price vector*, and the suballocation $f : A \to \mathbb{G}$ is a *subequilibrium allocation*. When the condition (b) of the attainability of suballocations is replaced by (b') of Definition 6.1, the adjective "*with free disposal (of commodities)*" will be added to these terms in a similar manner.

Since $f(a)$ might be a subdemand vector without being a demand vector in the above condition (a), $f(a)$ is a consumption vector in the budget set $B(X_a, p \cdot e_a, p)$ to which no preferred consumption vectors are obtainable with strictly less expenditure if $p \cdot e_a \geqq \inf p \cdot X_a$. However, we have $f(a) = e(a) \notin B(X_a, p \cdot e_a, p)$ if $p \cdot e_a < \inf p \cdot X_a$. This means that a subequilibrium allocation f may fail to secure the survival of some economic agents in the market.

6.2 Existence of Subequilibria in Perfectly Competitive Markets

It is evident that environments in which a market subequilibrium exists are much more general than those in which a Walrasian equilibrium can be shown to exist. There are two basic reasons for this. One obvious reason is that a market subequilibrium allocation may ignore the survival issue for all the agents, whereas a Walrasian allocation must ensure it. The other reason is that the subdemand relation exhibits a continuity property even at critical points of the consumption sets, whereas the demand relations may fail to have a continuity property at the critical points of consumption sets.

Theorem 6.1 (Existence of Subequilibria)

Let $\mathscr{E} : (A, \mathscr{A}, \nu) \to \mathcal{P}_{lns} \times \mathbb{G}$ be an atomless economy. There exists a subequilibrium with free disposal (p, f) for $\mathscr{E}$ with $p(\neq 0) \in \mathbb{R}^\ell_+$.

Proof To begin, we define

$$B(a, p) = B(X_a, p \cdot e_a, p) \text{ and } D^*(a, p) = D^*(X_a, \succ_a, e_a, p).$$

For each $a \in A$ and $p \in \mathbb{R}^\ell_{++}$, the budget set $B(a, p)$ is compact when $p \cdot x \geqq \inf p \cdot X$. By Proposition 4.8, p. 152, we have $D^*(a, p) \neq \emptyset$ for each $a \in A$ and $p \in \mathbb{R}^\ell_{++}$. Now, given a price vector $\overline{p} \in \mathbb{R}^\ell_{++}$, let $U \subset \mathbb{R}^\ell_{++}$ be a compact neighborhood of $\overline{p}$. Define two real numbers, $\underline{m}$, $\overline{m}$, by $\underline{m} = \min\{p_i \mid p = (p_1, \dots, p_i, \dots, p_\ell) \in U, i = 1, \dots, \ell\}$ and $\overline{m} = \max\{p_i \mid p = (p_1, \dots, p_i, \dots, p_\ell) \in U, i = 1, \dots, \ell\}$. If we also define two mappings $m : A \to \mathbb{R}$ and $M : A \to \mathbb{R}$ by $m(a) = \max\{\min p \cdot X_a \mid p \in U\}$ and $M(a) = \max\{\overline{m} \sum_{i=1}^\ell |e_{ai}|, m(a)\}$ for each $a \in A$, it follows from Mathematical Proposition 3.6, p. 108, that the mapping $M : A \to \mathbb{R}$ is $\mathscr{A}_\nu$-measurable as the mappings $X : A \to \mathscr{X}$ and $e : A \to \mathbb{G}$ are measurable, and thus, without loss of generality we can take M itself as measurable. Let $h : A \to \mathbb{R}$ be a measurable mapping defined by $a \mapsto (\sqrt{\ell}/\underline{m})M(a)$. Then, since all the consumption sets in $\mathscr{X}$ have the uniform lower bound, h is integrable by the integrability of the mapping e. Moreover, we have for a.e. $a \in A$ that

$$\|x\| \leqq h(a) \text{ for all } x \in D^*(a, p) \text{ and for all } p \in U. \tag{6.1}$$

Since the correspondence D^* from A to $\mathbb{G}$ is measurable by Proposition 4.11, p. 153, one obtains from the Measurable Selection Theorem and (6.1) that $\int D^*(\cdot, \overline{p}) d\nu \neq \emptyset$. This shows that we have

$$\int D^*(\cdot, p) d\nu \neq \emptyset \text{ for any } p \in \mathbb{R}^\ell_{++}. \tag{6.2}$$

Next, let us introduce *truncated price simplices* $PS_k \subset PS, (k \in \mathbb{N}_+)$ given by $PS_k = \{p \in PS \mid p_i \geqq 1/(2\ell k) \text{ for all } i = 1, \dots, \ell\}$ where $PS = \{p \in \mathbb{R}^\ell_+ \mid \sum_{i=1}^\ell p_i = 1\}$. Then, define correspondences F_k from PS_k to $\mathbb{R}^\ell$, $(k \in \mathbb{N}_+)$, by $F_k : p \mapsto \int \{D^*(\cdot, p) - e\} d\nu$. We will show that the correspondence $F_k : PS_k \to \mathbb{R}^\ell$ has the following properties for each $k \subset \mathbb{N}_+$:

$$F_k(p) \neq \emptyset \text{ for all } p \in PS_k; \tag{6.3}$$

$$F_k(p) \subset K \text{ for all } p \subset PS_k \text{ for some compact set } K \subset \mathbb{R}^\ell; \tag{6.4}$$

$$F_k(p) \subset H_{\leqq}(p, 0) \text{ for all } p \in PS_k; \tag{6.5}$$

$$F_k : PS_k \to \mathbb{R}^\ell \text{ is closed-valued and upper hemi-continuous;} \tag{6.6}$$

$$F_k(p) \text{ is a convex set in } \mathbb{R}^\ell \text{ for all } p \in PS_k. \tag{6.7}$$

(6.3) is immediate from (6.2). (6.4) follows from $\int e d\nu < \infty$ and $D^*(a, p) \subset X_a \cup \{e_a\}$ for each $p \in PS_k$ and every $a \in A$. To check the property (6.5), let $z \in F_k(p)$. Then, there is an integrable mapping $g : A \to \mathbb{G}$ such that we have $z = \int (g - e) d\nu$ and $g(a) \in D^*(a, p)$, a.e. $a \in A$. It follows that $g(a) \in H_{\leqq}(p, p \cdot e_a)$, a.e. $a \in A$, which in turn shows that $z \in H_{\leqq}(p, 0)$. To affirm (6.6), recall that on account of Proposition 4.11, p. 153, the correspondence

$F_k : PS_k \to \mathbb{R}^\ell$ is closed for every $k \in \mathbb{N}_+$. Thus, F_k is closed-valued, and the upper hemi-continuity of F_k is a consequence of Mathematical Proposition 1.1, p. 36, and (6.4). The last property (6.7) follows from the atomlessness of the measure space of economic agents $(A, \mathscr{A}, \nu)$ and Richter's Proposition (cf. Mathematical Proposition 3.14, p. 117) which is a consequence of the Liapounoff convexity property of purely non-atomic vector measures (cf. Mathematical Proposition 3.12, p. 112).

By applying the Equilibrium Fixed Point Theorem (cf. Mathematical Property 1.22, p. 43) to each F_k, $k \in \mathbb{N}_+$, one obtains a pair (p_k, f_k) of a price vector $p_k \in PS_k$ and an integrable mapping $f_k : A \to \mathbb{G}$ satisfying for each $k \in \mathbb{N}_+$:

$$f_k(a) \in D^*(a, p), \text{a.e. } a \in A; \tag{6.8}$$

$$\int (f_k - e)d\nu \subset H_{\leqq}(p, 0) \text{ for all } p \in PS_k. \tag{6.9}$$

By setting $p = (1/\ell, \ldots, 1/\ell)$, we get from (6.9) that

$$\int \sum_{i=1}^{\ell} (f_k)_i d\nu \leqq \int \sum_{i=1}^{\ell} e_i d\nu. \tag{6.10}$$

Therefore, $\int f_k d\nu$ belongs to a compact set in $\mathbb{R}^\ell$ given by

$$\left\{ z \in \mathbb{R}^\ell \,\middle|\, \sum_{i=1}^{\ell} z_i \leqq \int \sum_{i=1}^{\ell} e_i d\nu \text{ and } z_i \geqq \underline{\beta} \text{ for each } i = 1, \ldots, \ell \right\}.$$

(Remember that $\underline{\beta}$ is the infimum of the components of all the consumption vectors (cf. p. 75).) Thus, we can assume without loss of generality that $(\int f_k d\nu)_k \to_k \overline{z}$. By Mathematical Theorem 3.6 (cf. Multidimensional Fatou's Lemma, p. 123) there is an integrable mapping $f : A \to \mathbb{G}$ such that

$$f(a) \in \text{Ls}\,(f_k(a)), \text{a.e. } a \in A, \text{ and} \tag{6.11}$$

$$\int f d\nu \leqq \overline{z}. \tag{6.12}$$

Since we have $p \cdot \overline{z} \leqq p \cdot \int e d\nu$ for all $p \in \cup_{k \in \mathbb{N}_+} PS_k = \mathbb{R}^\ell_{++} \cap PS$ from (6.9), it must be that $\overline{z} \leqq \int e d\nu$. Then, we get from (6.12) that

$$\int f d\nu \leqq \int e d\nu. \tag{6.13}$$

Seeing that $(p_k)_k$ is a sequence in the compact set PS, there is a convergent subsequence, still denoted by $(p_k)_k$, with $p_k \to_k p$. This price vector p is fixed from now on. Moreover, in view of (6.11), there is a subsequence of $(p_k, f_k)_k$, again denoted by $(p_k, f_k)_k$, for which one has $p_k \to_k p \in PS$ and $f_k(a) \to_k f(a)$, a.e. $a \in A$. Since the relation D^* is closed by Proposition 4.11, p. 153, one verifies that

$$f(a) \in D^*(a, p), \text{a.e. } a \in A. \tag{6.14}$$

Accordingly, we have

a.e. $a \in A$,

$$p \cdot f(a) \leq p \cdot e_a, \text{ and} \tag{6.15}$$

$$\begin{cases} p \cdot x \geqq p \cdot f(a) \text{ for all } x \succ_a f(a) & \text{if } p \cdot e_a \geqq \inf p \cdot X_a, \\ p \cdot f(a) = p \cdot e_a & \text{if } p \cdot e_a < \inf p \cdot X_a. \end{cases} \tag{6.16}$$

By the local non-satiation of preference relations, (6.16) together with (6.15) implies that $p \cdot \int f dv = p \cdot \int e dv$. Hence, if we define the amounts of disposal of commodities $\delta(f, e)$ by $\delta(f, e) = \int e dv - \int f dv \geqq 0$, then

$$\int f dv + \delta(f, e) = \int e dv \text{ and } p \cdot \delta(f, e) = 0, \tag{6.17}$$

showing that $\delta(f, e)$ is a free disposal of commodities with respect to the suballocation f and the price vector p.

Therefore, we can conclude from (6.14) and (6.17) that the pair (p, f) with $p(\neq 0) \in PS \subset \mathbb{R}^\ell_+$ is a subequilibium with free disposal for $\mathscr{E}$. □

Theoretical Basis of Survival of Economic Agents at Equilibrium

Subequilibrium allocations are well defined despite the fact that some or even all the economic agents are unable to participate in market transactions due to shortfalls of their income or wealth. This is because some or all of the agents in an economy may not have sufficient endowments to support their lives under a subequilibrium allocation. Despite the significant economic challenge of survival in times of famine in certain regions of our world, standard general equilibrium analysis typically imposes theoretical conditions to circumvent concerns about survival. (See the appendix to this chapter, p. 187, for further remarks on survival issues.)

First, what is required is that the economic agents as a collective possess a sufficient quantity of endowed commodities to ensure the survival of all. This is expressed by the condition termed the *social positive income condition* for an economy that will guarantee economic agents as a whole to have sufficient income to obtain an adequate quantity of commodities for their survival if commodities are distributed among them in a proper way. The second condition to be introduced is the condition termed the *endowments interdependence among preferred commodities*. This condition ensures that all economic agents receive income that strictly exceeds their minimum income levels in a market economy.

Definition 6.3 (Social Positive Income Condition for an Economy) An economy $\mathscr{E} : (A, \mathscr{A}, \nu) \to \mathcal{P} \times \mathbb{G}$ is said to satisfy the *social positive income condition for an economy* if the mean endowments are in the interior of the mean consumption sets. This is to say that we have $\int e d\nu \in \text{int} \int X d\nu$ for the endowment mapping $e : A \to \mathbb{G}$ and the consumption set mapping $X : A \to \mathscr{X}$.

When we note that the aggregate (or mean) consumption set of $\mathscr{E}$, $\int X d\nu$, reflects the social or overall consumption needs of the entire economic agents, we can identify two aspects of the condition "$\int e d\nu \in \text{int} \int X d\nu$." One aspect is that for each of the ℓ types of commodities, a non-negligible fraction of individuals express a need for the commodity. The second aspect is that a non-negligible proportion of individuals possess the commodity in question in quantities that strictly exceed their own necessary consumption. However, it is possible that some non-trivial proportion of individuals lack an adequate quantity of commodities to survive. For this reason, another condition is required to serve as a way for the total excess of commodities over the minimal quantities necessary for the survival of whole economic agents to be distributed throughout the economy by way of market transactions. This condition will be called the *endowments interdependence among preferred commodities*.

Definition 6.4 (Endowments Interdependence among Preferred Commodities) An economy $\mathscr{E} : (A, \mathscr{A}, \nu) \to \mathcal{P} \times \mathbb{G}$ is said to satisfy the *endowments interdependence among preferred commodities* if, for any given suballocation $f : A \to \mathbb{G}$ with $\int f d\nu \leqq \int e d\nu$, and a non-trivial two-part measurable partition $\{A_1, A_2\}$ of A, each element having a positive measure, there is a suballocation $g : A \to \mathbb{G}$ with $\int g d\nu \leqq \int e d\nu$ satisfying

(a) a.e. a in A_1, $g(a) \in X_a$;
(b) a.e. a in A_1, $g(a) \succ_a f(a)$ whenever $f(a) \in X_a$;
(c) $g(a) \neq f(a)$ on a subset of A_2 with a positive measure.

To rephrase the above conditions (a)–(c), for a suballocation g with $\int g d\nu = \int f d\nu$, set $\Delta z = \int_{A_2} (f - g) d\nu$. One can write

$$\int g d\nu = \int_{A_1} g d\nu + \int_{A_2} g d\nu$$
$$= \left(\int_{A_1} f d\nu + \Delta z \right) + \left(\int_{A_2} f d\nu - \Delta z \right) \tag{6.18}$$
$$\text{where } \int_{A_1} g d\nu = \int_{A_1} f d\nu + \Delta z. \tag{6.19}$$

Conditions (a) through (c) on the suballocation $g : A \to \mathbb{G}$ can be replaced by those of a variation Δz of aggregate allocated commodities to the agents in A_1 and A_2. In this context, the property of the *endowments interdependence among preferred commodities* is expressed as follows:

For any given suballocation $f : A \to \mathbb{G}$ with $\int f d\nu \leqq \int e d\nu$, and a non-trivial two-part measurable partition $\{A_1, A_2\}$ of A, each set A_i, $i = 1, 2$, having a positive measure, there is a variation Δz of aggregate allocated commodities to the agents in A_1 and A_2 that satisfies:

(a') $\int_{A_1} f d\nu + \Delta z$
$\in \int_{A_1} \Big\{ \{x \in \mathbb{G} \mid x \in X_a\} \cap \{x \in X_a \mid x \succ_a f(a) \text{ if } f(a) \in X_a\} \Big\} d\nu;$
(b') $\int_{A_2} f d\nu + \Delta z \in \int_{A_2} \{x \in \mathbb{G} \mid x \in X_a \cup \{e_a\}\} d\nu.$

The interdependence of endowments among preferred commodities can be understood in terms of a supplementary allotment of commodities to one set of agents from the other set, which is made possible by a probable sacrifice of allocated commodities to the other set of agents.The following interpretation is suggested by conditions (a') through (c'). For any suballocation of commodities for consumption such that the total quantities do not exceed those available from the endowed commodities in the economy, if all economic agents are divided into two non-trivial sets, it will be possible for each set of agents to provide the other set of agents with additional commodities by changing parts of what they would have received and by contributing to an improvement in the commodity bundles received by the other set of agents.

Subequilibrium Allocations and Survival of Agents

Take notice of the fact that the conditions (a) through (c) in the stipulation of the endowments interdependence among preferred commodities are somewhat vague with regard to the commodity bundles of economic agents $a \in A_2$ to be in its consumption set X_a or not. It is therefore unclear whether the modification of commodity allocation by the suballocation g among the second set of agents impedes its own members from reaching at least their survival levels. We will confirm below that, as a consequence of the two conditions placed on an economy, all economic agents are able to engage in market transactions to reach their survival levels of consumption. Therefore, the social positive income condition, in conjunction with the endowments interdependence among preferred commodities, provides a safeguard for all economic agents in terms of survival.

In particular, in addition to Theorem 6.1, we obtain the following proposition, which shows the survival of all economic agents under the subequilibrium allocation when the conditions of social positive in income and endowments interdependence among preferred commodities are met.

Proposition 6.1 (Allocations with Survival) *Let $\mathscr{E} : (A, \mathscr{A}, \nu) \to \mathcal{P}_{Ins} \times \mathbb{G}$ be an atomless economy that satisfies the social positive income condition and the endowments interdependence among preferred commodities. Then, a subequilibrium (p, f) for $\mathscr{E}$ with free disposal of commodities satisfies the following properties:*

(a) $f(a) \in X_a$ *for* a.e. $a \in A$;
(b) $f(a) \in D^*(X_a, \succ_a, e_a, p)$ *for* a.e. $a \in A$;
(c) $f(a) \in D(X_a, \succ_a, p \cdot e_a, p)$ *in (b) if* $p \cdot e_a \notin J_p(X_a)$.

Proof Since (p, f) is a subequilibrium for $\mathscr{E}$ with free disposal of commodities, the suballocation f satisfies (b). Let a subset $\widetilde{S} \subset A$ be defined by $\widetilde{S} = \{a \in A \mid \inf p \cdot X_a \leqq p \cdot e_a\}$. As $\widetilde{S} \in \mathscr{A}_\nu$ by Mathematical Proposition 3.6, p. 108, there is a measurable set $\overline{S}$ such that $\overline{S} \Delta \widetilde{S}$ is a ν-null set. By the social positive income condition for an economy, i.e. $\int e d\nu \in \operatorname{int} \int X d\nu$, we have $\inf p \cdot \int X d\nu < p \cdot \int e d\nu$. This implies $\nu(\overline{S}) > 0$. Let $\underline{S} = A - \overline{S}$. The suballocation f satisfies (a) if we can confirm that $\nu(\underline{S}) = 0$. Suppose $\nu(\underline{S}) > 0$. Then, by the condition of the endowments interdependence among preferred commodities, there is a suballocation $g : A \to \mathbb{G}$ satisfying

(i) a.e. a in $\overline{S}$, $g(a) \in X_a$ and $g(a) \succ_a f(a)$;
(ii) $g(a) \neq f(a)$ on a subset of $\underline{S}$ with a positive measure;
(iii) $\int g d\nu \leqq \int e d\nu$.

As f satisfies (b), we have $f(a) = e_a$, a.e. $a \in \underline{S}$. This implies that $p \cdot e_a \leqq p \cdot g(a)$, a.e. $a \in \underline{S}$. Furthermore, by (ii) it must be $p \cdot e_a < p \cdot g(a)$ on a subset of $\underline{S}$ with a positive measure. Thus, we must have

$$\int_{\underline{S}} (p \cdot e - p \cdot g) d\nu = p \cdot \int_{\underline{S}} (e - g) d\nu < 0. \tag{6.20}$$

Since (i) and (b) imply that $p \cdot g(a) \geqq p \cdot e_a$, a.e. $a \in \overline{S}$, (iii) and (6.20) lead to a contradiction that

$$0 \leqq p \cdot \int_{\overline{S}} (g - e) d\nu \leqq p \cdot \int_{\underline{S}} (e - g) d\nu < 0.$$

Therefore, we conclude that $\nu(\underline{S}) = 0$. This proves (a). Finally, (a) and (b) together with Proposition 4.10 (2) (cf. p. 152) confirm (c). □

6.3 Existence of Walrasian Equilibria

Walrasian and Subequilibrium Allocations

We will now summarize relations between Walrasian and the subequilibrium allocations.

Proposition 6.2 (Walrasian and Subequilibrium Allocations)
Let $\mathscr{E} : (A, \mathscr{A}, \nu) \to \mathcal{P} \times \mathbb{G}$ be an atomless economy.

(1) A Walrasian allocation $f : A \to \mathbb{G}$ for $\mathscr{E}$ is a subequilibrium allocation for $\mathscr{E}$.
(2) Conversely, a subequilibrium allocation $f : A \to \mathbb{G}$ for $\mathscr{E}$ with a subequilibrium price vector p is a Walrasian allocation for $\mathscr{E}$ if the set $A_p = \{a \in A \mid p \cdot e_a \in J_p(X_a)\}$ is a ν-null set and $\mathscr{E}$ satisfies the social positive income condition and the endowments interdependence among preferred commodities.

Proof Both (1) and (2) are immediate consequences of Proposition 4.10, p. 152, and Proposition 6.1. □

A Case of Non-Existence of Walrasian Equilibrium

The following example illustrates an economy without Walrasian equilibria due to the absence of upper hemi-continuity of individual demand correspondences. This phenomenon arises from a non-convex commodity space and/or consumption sets.

Example [Non-Existence of Walrasian Equilibria due to Discontinuity of Demand Correspondences]

Let $\mathbb{G} = \mathbb{R} \times \{\ldots, -2, -1, 0, 1, 2, \ldots\}$, $X = \mathbb{G} \cap \mathbb{R}^2_+$, and $e = (2, 2)$. A preference relation $(X, \succ)$ is defined by the intersections of the dotted curves in Fig. 6.1 and the consumption set X. $(A, \mathscr{A}, \nu)$ is any given atomless probability space. Define an economy $\mathscr{E} : (A, \mathscr{A}, \nu) \to \mathcal{P} \times \mathbb{G}$ by setting $\mathscr{E}(a) = (X, \succ, e)$ for every $a \in A$. In this economy, all individuals have the same consumption characteristics, namely $(X, \succ, e)$. The price consumption curve (or the offer curve) of each individual in this economy is as depicted in Fig. 6.1. The individual demand correspondence lacks the upper hemi-continuity at a price vector $p = (p_1, p_2)$ with $p_1/p_2 = 1$. As a result, the total (or the mean) demand excess demand correspondence does not contain the vector 0. Since the only possible Walras equilibrium allocation is the no-trade allocation, the economy $\mathscr{E}$ does not have any Walrasian equilibria.

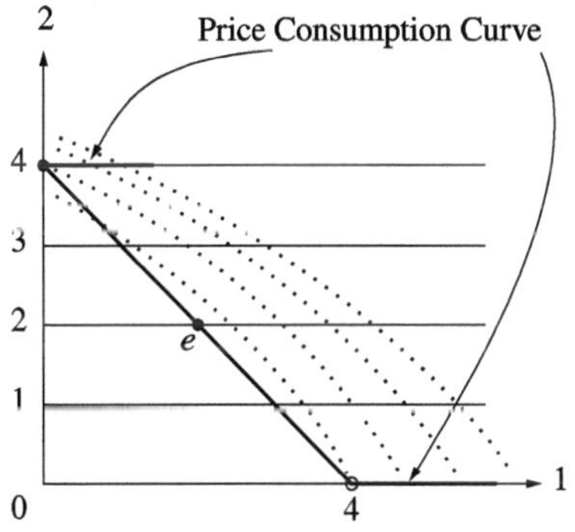

Fig. 6.1 An example of non-existence of Walrasian equilibrium

Subequilibria and Walrasian Equilibria

In light of the proof of the existence of subequilibria presented in Theorem 6.1, Proposition 6.2 (2) delineates the circumstances under which subequilibrium allocations can be considered Walrasian allocations. The formation of these environments is influenced by two key factors. The initial factor is the safeguarding of adequate resources for individual agents, which enables their involvement in market transactions. The second factor is the wealth distribution to be sufficiently diversified so as to keep off from the critical sets except for a negligible fraction of individuals.

Proposition 6.3 (Dispersion of Wealth Distribution and Subequilibria)

Let an atomless economy $\mathscr{E} : (A, \mathscr{A}, \nu) \to \mathcal{P}_{Ins} \times \mathbb{G}$ *satisfy the social positive income condition and the endowments interdependence among preferred commodities. Then, the subequilibria for* $\mathscr{E}$ *are Walrasian equilibria provided that the wealth distribution of* $\mathscr{E}$ *is dispersed at each subequilibrium price vector.*

Proof On account of Proposition 5.1 (2), p. 170, the conclusion of the proposition follows from Proposition 6.2 (2). □

Remark 6.1 (Subequilibria with Sufficient Budgets in Convex Environments) In the literature of the existence of general equilibria, economies are frequently defined in convex environments in such a way that the commodity space $\mathbb{G}$ is the entire Euclidean ℓ-space $\mathbb{R}^\ell$ and all the consumption sets X are identical to $\mathbb{R}^\ell_+$. In this context, if a special condition of the *strict individual survival* such that $e_a \in \mathbb{R}^\ell_{++}$ a.e. $a \in A$ for an economy $\mathscr{E}$ defined on $(A, \mathscr{A}, \nu)$, is postulated, we have that $p \cdot e_a > \inf p \cdot \mathbb{R}^\ell_{++}$ a.e. $a \in A$ for any subequilibria (f, p) for $\mathscr{E}$. It then follows from Proposition 4.10 (2), p. 152, and Proposition 4.1, p. 133, that subequilibria for $\mathscr{E}$ are Walrasian equilibria.

The preceding results obtained in this chapter provide the foundation for the main existence theorem of Walrasian equilibria in atomless economies with non-convex environments.

Theorem 6.2 (Existence of Walrasian Equilibrium in Non-Convex Environments)

Let $\mathscr{E} : (A, \mathscr{A}, \nu) \to \mathcal{P}_{Ins} \times \mathbb{G}$ *be an atomless economy that satisfies the social positive income condition and the endowments interdependence among preferred commodities. Then, a Walrasian equilibrium for* $\mathscr{E}$ *exists if the wealth distribution of* $\mathscr{E}$ *is dispersed at each price vector* $p \in \mathbb{R}^\ell_+$.

Proof Given an atomless economy $\mathscr{E} : (A, \mathscr{A}, \nu) \to \mathcal{P}_{Ins} \times \mathbb{G}$, a subequilibrium (p, f) for $\mathscr{E}$ with $p(\neq 0) \in \mathbb{R}^\ell_+$ exists by Theorem 6.1. Since the social positive income condition and the endowments interdependence among preferred commodities are satisfied, it follows from Proposition 6.3 that (p, f) is a Walrasian equilibrium for $\mathscr{E}$ with $p(\neq 0)$ as the wealth distribution of $\mathscr{E}$ is dispersed at each price vector $p \in \mathbb{R}^\ell_+$. □

Notes 6

Notes 6.1 [Section 6.2, p. 178]

[Endowments interdependence among preferred commodities, p. 182] The condition stated here is the Hildenbrand adaptation of McKenzie's irreducibility assumption to a continuum of agents model (see [52, Assumption 2.3, p. 85] and [86, p. 58]). Note that automatic survival of individual agents is not warranted in an economy since it is not assumed that $e(a) \in X(a)$ a.e. in A. Even within the framework of economies with no production, the removal of the survival condition is of importance here, since even if we have $e(a) \in \operatorname{co} X(a) \cap \mathbb{G}$, it is possible that $e(a) \notin X(a)$. This is in contrast to the models of [6], [107], and [81].

Notes 6.2 [Section 6.3, p. 184]

[Existence of Walrasian equilibria in atomless economies, p. 184] The existence proof of Walrasian equilibrium in an economy with a finite-dimensional Euclidean space as the commodity space and an atomless measure space as the space of economic agents was first given by [8], which showed that the convexity assumption of preference relations is unnecessary due to the Liapounoff convexity theorem of non-atomic vector measures. [107] extended the theorem in [8] by giving an existence proof that does not require the completeness of preference relations. [52, 53] showed that the monotonicity requirement for preference relations in both papers can be weakened to local non-satiation. [81] extended the proof of the existence of Walras equilibrium in an atomless economy to the case where all consumption sets are non-negative orthants of a commodity space composed of a finite number of all perfectly divisible commodities except one type of purely indivisible commodity. A general equilibrium model that essentially imposes no restrictions on the indivisibility of commodities and the non-convex shapes of consumption sets, was introduced in [126, 127], and a proof of the existence of the Walras equilibrium was given.

Appendix to Chapter 6: Backgrounds to the Survival Issue in General Equilibrium Model

The following paragraphs introduce some of the contentions in the literature related to the survival issue as backgrounds for this topic.

Survival of Consumers in a Competitive Equilibrium

Conditions for the Survival of Consumers

Koopmans [68, p. 59] elaborates on the way Walrasian models are constructed in order to guarantee that each consumer can both survive and participate in the markets without anticipating what specific prices will prevail in an equilibrium as the postulates.

> To achieve this the authors assume first of all that the aggregate supply set contains a point which supplies just a little more of every commodity than is necessary, as indicated by some point in the aggregate consumption set, for every consumer to survive. Secondly, they assume that each consumer can, if necessary, survive on the basis of the resources he holds and the direct use of his own labor, without engaging in exchange, and still have something to spare of some type of labor which is sure to meet with a positive price in any equilibrium. The latter question, in any situation that can conceivably be reached within the limitations of resources, technology, and survival of all consumers, has a positive marginal productivity with regard to at least one out of a certain class of commodities, each of which, again

in any situation reachable within the same limitations, can still add to the satisfaction for every consumer. From these assumptions, the authors prove the existence of a competitive equilibrium.

These important results illustrate vividly the great difficulties to be overcome, and the extent of rethinking of supposedly familiar problems needed, if even only the more classical portions of economic theory are to be rebuilt on a rigorous basis. In particular, most authors have ignored the analytical difficulty of formulating a model that ensues the possibility of survival, blithely admitting any nonnegative rates of consumption as sustainable. Arrow and Debreu face this issue, and find it to be a complicated one. . . .

Market Equilibria without Survival

Koopmans acknowledges that in "modern society few of us can indeed survive without engaging in exchange." He proposes one *"hard-boiled"* alternative, which would entail assuming the

instantaneous elimination by starvation of those whose resource prove insufficient for survival, and to look for conditions ensuring existence of an "equilibrium" involving survival of some consumers. An alternative more realistic for highly industrialized private enterprise societies would be to recognize the existence of income transfers through taxation and social insurance, and to look for conditions, including tax and benefit schedules, ensuring general survival in an equilibrium. ([68, p. 62])

Debreu's Interpretation of Survival in a General Equilibrium Model

Koopmans describes Debreu's interpretation of survival in the Arrow–Debreu model of a general equilibrium (cf. [3]) as follows:

It has been pointed out to me orally by Debreu that the treatment of time adopted by Arrow and Debreu in which each choice is a plan extending over a sequence of future periods permits more subtle version of what was called above the "hard-boiled" interpretation of the model in regard to the survival problem. The plan formed by each consumer in response to the price system includes a specification of the length of life compatible with his present resources, his ability to do remunerative work or shift for himself, and other aspects of his life plan. All that is to be assured by the postulates is survival of every consumer at least into the first period. The amount and initial distribution of resources and skills in the population determine the pace and extent, if any, of starvation.

Granted the inadequacies of any model unable to recognize the element of uncertainty in individual survival, this interpretation has an appeal of descriptive accuracy with regard to societies or phases of development in which income transfers are quantitatively unimportant. It neatly clarifies a group of questions in regard to which the liberalist and Marxist schools of economic thought have never quite achieved communication. ([68, pp. 62–63])

Difficulty in Debreu's Interpretation

The interpretation of survival in the Arrow–Debreu model proposed by Debreu, as articulated by Koopmans, appears to present an obstacle to formalization within the model. To illustrate this, consider the following example: let $(X, \succ, e) \in \mathcal{P} \times \mathbb{G}$ be given. For the sake of simplicity, assume that a sequence of periods is two, with periods numbered 1 and 2. Bearing in mind that consumption vectors that assure survival into the first period are distinct from those that ensure survival into the second period, the consumer's needs and preferences in general differ between these two categories of consumption vectors. Accordingly, rather than a single specification of a pair of needs and preferences $(X, \succ) \in \mathcal{P}$, a distinct pair, $(X_t, \succ_t)$, must be provided for each period $t = 1, 2$ with $X_1 \neq X_2, \succ_1 \neq \succ_2$ and $X_t \neq X, t = 1, 2$. In this context, it is less innocuous to postulate that the consumer survives at least into the first period with its endowment, i.e., $e \in X_1 \cup X_2$. In addition, it is anticipated that $p \cdot e \leqq \min p \cdot X_1 < \min p \cdot X_2$. Although the budget sets $B(X_1, p \cdot e, p), B(X_2, p \cdot e, p), B(X_1 \cup X_2, p \cdot e, p)$ are non-empty, it is not indisputable how individual demands of the consumer should be defined in this framework. Moreover, it is not at all clear how the universal preference set could be defined.

Survival Issue in Relation to Availability of Foods

The issue of survival has been inextricably linked to the availability of food supplies in our world. However, it should be clearly understood from discussions of existence of market equilibrium that the survival issue is rooted in the functioning of market prices in relation to the distribution of endowments, preferences and needs.

Availability of Foods and Non-Survival

Famines have been observed throughout our history. Sen emphasized that, in many cases, the aggregate availability of food is not the primary cause of starvation deaths. The following quotations from one of Sen's works [110] will be presented here.

> On a conservative estimate, about half a billion people in the world—about one person in eight—suffer on a regular basis from significant malnutrition. In addition, severe famine conditions develop in different parts of the world with distressing regularity, slaying many and seriously debilitating many more. Several authors have recently argued that there is evidence of increasing starvation for the world as a whole, and there has been quite an outburst of alarm about the 'food crisis of the world'. . . . . .
>
> . . . . . . While a variety of causes have been considered in understanding the so-called "world wide famine," the most common approach has been that of "too many mouths and too little food". The statistical basis of the observation of a declining food availability per capita is, however, rather weak. Not only has the trend rate of growth of population not exceeded that of food production for the world as a whole—it has been significantly lower—but there has been no such excess in general for the world of underdeveloped countries either. Nor has there been such an excess for any major region of the world, with the possible exception of Africa. ([110, p. 33])

Sen places great emphasis on the necessity of comprehending the market mechanisms of exchange economies, which could potentially result in widespread starvation.

> The traditional—and in some ways the most obvious—approach to famines is in terms of variation of food availability: "a sudden, sharp reduction in the food supply in any particular

> geographic locale has usually resulted in widespread hunger and famine." This approach of "food availability decline" I shall call for brevity (not premature disrespect) the FAD view.
>
> FAD works best when much of the food eaten by a family happens to be grown by it without being acquired through exchange. In an exchange economy, however, the terms of exchange constitute a factor of some importance of its own, and a family's ability to buy food depends on the rates at which its labour and commodity possessions can be exchanged into food. While food availability will clearly be an important influence on these terms of exchange, other forces are also involved, and famines can thus arise from causes other than food availability decline. In an exchange economy, whether a family will starve or not will depend on what it has to sell, whether it can sell them, and at what prices, and also on the price of food. An economy in a state of comparative tranquillity may develop a famine if there is a sudden shake-up of the system of rewards for exchange of labour, commodities and other possessions, even without a "sudden, sharp reduction in the food supply." ([110, p. 34])

A Further Remark on Survival

A prominent religious figure has made a well-known pronouncement on the subject of economic survival in market economies. This pronouncement offers a religious perspective on economic life in the context of a transition period from socio-communistic systems in Eastern Europe to free market economies.

> It would appear that, on the level of individual nations and of international relations, the free market is the most efficient instrument for utilizing resources and effectively responding to needs. But this is true only for those needs which are "solvent", insofar as they are endowed with purchasing power, and for those resources which are "marketable", insofar as they are capable of obtaining a satisfactory price. But there are many human needs which find no place on the market. It is a strict duty of justice and truth not to allow fundamental human needs to remain unsatisfied, and not to allow those burdened by such needs to perish. It is also necessary to help these needy people to acquire expertise, to enter the circle of exchange, and to develop their skills in order to make the best use of their capacities and resources. Even prior to the logic of a fair exchange of goods and the forms of justice appropriate to it, there exists something which is due to man because he is man, by reason of his lofty dignity. Inseparable from that required "something" is the possibility to survive and, at the same time, to make an active contribution to the common good of humanity. ([62, Section 34])

Chapter 7
Production in Economies

Abstract This chapter formalizes the supply side of economies in terms of production, apart from the initial endowments of commodities, in competitive markets. Previous chapters have developed a general equilibrium analysis that abstracts from the production possibilities open to economic agents. After introducing the product space, which is distinct from the commodity space faced by individual agents, production is defined as technological transformations in the product space. After discussing two types of production, namely Hicksian production and Walrasian production, the chapter proceeds with the general equilibrium analysis of perfectly competitive markets with Walrasian production.

7.1 Production as Technological Transformations

To extend the analysis of market economies in which only consumption activities and exchange transactions of individuals are explicitly considered, we proceed to examine the possibility that individuals may also obtain additional or new commodities through technological transformations that detail existing production processes.

Production and Product Space

Production, when considered in isolation from its institutional context, is defined by its technological transformation of commodities through production processes or activities. This raises the question of whether the space in which production takes place may differ from the commodity space $\mathbb{G}$ faced by individuals in their consumption activities. In particular, while the quantities of commodities consumed by individuals are expressed in units of commodities of the space $\mathbb{G}$, produced amounts of commodities must be in units of total or mean demand in order for the attainability or feasibility of allocations in atomless economies to be coherently

A. Yamazaki, *Mathematical Economics of General Equilibrium*, Monographs in Mathematical Economics 8, https://doi.org/10.1007/978-981-96-9497-6_7

expressed in combination with mean endowments. This will be clarified below by introducing the *product space*.

Product Space

To substantiate the point raised in the preceding paragraph, let $f : A \to \mathbb{G}$ be an allocation or a suballocation for an economy $\mathscr{E} : (A, \mathscr{A}, \nu) \to \mathcal{P} \times \mathbb{G}$. f requires the total amount of commodities $\int f d\nu \in \mathbb{R}^\ell$. The commodities allocated to each individual economic agent $a \in A$ are in the commodity space $\mathbb{G}$. This means that units are those of the space $\mathbb{G}$, whereas its total amount $\int f d\nu$ of commodities allocated in the economy is measured by units per capita (of economic agents). It follows that $\int f d\nu$ belongs to $\mathbb{R}^\ell$, but is not required to belong to the space $\mathbb{G}$. In fact, when the space of economic agents $(A, \mathscr{A}, \nu)$ is atomless and the space $\mathbb{G}$ is non-convex, it is possible that $\int f d\nu$ does not belong to $\mathbb{G}$.

Technological Transformation of Commodities in Product Space

A technological transformation of commodities is delineated by a *production process or activity*. This is expressed by a vector in the product space, which is called a *production vector or plan*. Given a production vector in the product space, negative components represent inputs and positive components are outputs of the process. Thus, on the basis of the product space a production technology is represented by the possibilities of technological transformation of commodities open to it.

Definition 7.1 (Product Space and Production Set) In economies with ℓ commodities, the *product space* is given by the Euclidean ℓ-space $\mathbb{R}^\ell$. A point $y = (y_i)_{i=1,\dots,\ell}$ in the product space $\mathbb{R}^\ell$ is referred to as a *production vector or plan*. A commodity i is an *(net) input* (resp. *(net) output*) if $y_i < 0$ (resp. $y_i > 0$). A *production set* or a *production possibility set* is defined as a subset Y of the product space $\mathbb{R}^\ell$. A production vector $y \in Y$ represents a feasible or possible technological transformation of commodities by the production set Y.

In the following section, we will examine some of the traditional expressions for describing probable properties of technological transformation of various commodities.

7.2 Expression for Technological Transformations

Traditional Description of Technological Properties

Returns to Scale

There are well-known standard expressions of technological transformation properties which relate production activities to production scales. These are known as *constant returns to scale, decreasing returns to scale, and increasing returns to scale*. In this context, *scale* refers to the degree of a proportional shift of amounts of inputs of a reference production vector, while *returns* refer to the resulting effect on outputs. Similarly, the term *constant returns to scale* in production indicates that a proportional change in all inputs induces the same proportional change in all outputs. In contrast, *decreasing returns to scale* (resp. *increasing returns to scale*) means that an increasing proportional change in all the inputs brings about a decreased (resp. increased) proportional change in all the outputs.

Production Set and Returns to Scale

In the context of technological transformations of production processes expressed by a production set $Y \subset \mathbb{R}^\ell$, constant returns and non-constant returns to scale are technically specified as follows. The production process of Y is said to satisfy *constant returns to scale* if, for any $y \in Y$ and $t \in \mathbb{R}_+$, we have $ty \in Y$. In other words, Y satisfies constant returns to scale if and only if Y is a cone with vertex at the origin 0. Y is said to satisfy *non-increasing returns to scale* if for any $y \in Y$ we have $ty \in Y$ for any $0 \leqq t \leqq 1$. It is said to satisfy *non-decreasing returns to scale* if, for any $y \in Y$, we have $ty \in Y$ for any $t \geqq 1$.

Properties of Production Sets

A list of the properties of production sets that are frequently cited in the literature is provided below.

List of Properties of Production Sets:

(a) [Reality of Production Possibilities] The set of production possibilities, Y, is not empty. There exists at least one production process that can be carried out.
(b) [Continuity of Production Process] *Y is closed.* If $y \in \mathbb{R}^\ell$ is a vector that has production vectors $y_n \in Y$, $(n = 1, 2, \dots)$, arbitrarily close to y for some n, then y is a possible production vector $y \in Y$.
(c) [Irreversibility of Production] $Y \cap -Y = \emptyset$. It is not possible to recover inputs from outputs by reversing the production process.
(d) [No Free Production and Possibility of No Production] $Y \cap \mathbb{R}^\ell_+ = \{0\}$. No outputs are produced unless inputs are provided. Furthermore, productive inaction or shutdown of production is possible, that is, $0 \in Y$.
(e) [Convex Production] Y is a convex subset of $\mathbb{R}^\ell$; that is, for any two production vectors $y, z \in Y$ and $t \in [0, 1]$, the vector $ty + (1 - t)z$ is also a production vector in Y.

An interpretation of property (e) is that the balancing of inputs of two production processes, represented by y and z, by a proportion t and $(1 - t)$ results in identically balanced outputs of the two processes. Thus, the convexity of production set Y is a consequence of the presumption that separate production processes do not interfere with each other in a positive or negative way.

Let us now turn to the specification of production sets in general equilibrium analysis.

7.3 Competition in Production Sector

In general equilibrium analysis, there are essentially two different formulations of the production sector. One is the Walrasian formulation of perfect competition in production and the other is the Hicksian formulation of competition in production. The difference in the formulation of the production sector arises from how producing economic agents are modeled in competitive environments. An economic agent engaged in production is referred to as an *entrepreneur*, and a production unit is an *enterprise* or a *firm*.

Hicksian Production

The representation of production in a competitive economy by a fixed number of firms maximizing profit over production possibilities exhibiting decreasing or non-increasing returns to scale dates back, at least, to Hicks [49]. Arrow and Debreu [3] followed Hicks by using convex production sets containing the origin.

Definition 7.2 (Hicksian Production Economy) *Hicksian production* for an economy $\mathscr{E} : (A, \mathscr{A}, \nu) \to \mathcal{P} \times \mathbb{G}$ is a fixed number of production sets $Y_k \subset \mathbb{R}^\ell$, $(k = 1, \ldots, n)$, which satisfy the properties (a) through (e) listed above. A pair $(\mathscr{E}, \{Y_k\}_{k=1,\ldots,n})$ of an economy and its Hicksian production is called a *Hicksian production economy*.

In the preceding chapters, the analysis of market economies has focused on consumption activities and exchange transactions of individuals. The concept of an economy (denoted by the symbol $\mathscr{E}$) is abstracted from the production possibilities open to economic agents. We are now in a position to consider the possibility that they may also obtain additional or new commodities through production, thereby extending the analysis to this point. In the context of Hicksian production, entrepreneurs are defined as individual economic agents who secure commodities through technical transformation, rather than through the sale of their own productive services. An entrepreneur acquires factors of production or some commodities with the objective of fully exploiting its productive opportunities and not for the purpose of its own consumption of those commodities. The amount of commodities it acquires depends solely upon its production plan. For this reason, an enterprise of the entrepreneur is naturally regarded as a separate economic unit from its consumption activity unit. Differently put, the account of the entrepreneur is set apart from its private account as a consumer.

Institutional Setting of Hicksian Production

The Hicksian production sector is constituted by a *fixed number of enterprises* with technologies exhibiting non-increasing returns to scale. This view of the production sector is admittedly based on short-term production plans. Entrepreneurs do not carry out production plans unless it is more profitable to do so. It is to the advantage of entrepreneurs to transform tradable commodities into other commodities by production only if the commodities produced give a higher market net value than the traded commodities. In other words, the shutdown or creation of an enterprise is inherently an economic judgment. When a production set containing the origin is convex, in other words, when a production technology exhibiting non-increasing returns to scale is used, positive profits may be earned by an enterprise at a competitive equilibrium. In such circumstances, whenever the same production technology is available to any economic agents, market conditions will prompt further entries of those who desire to set themselves up as entrepreneurs. In this manner, different options will be open to different economic agents under differing market conditions. Accordingly, the number of entrepreneurs may vary depending on market conditions. Furthermore, the concept of perfect competition has historically been interpreted to imply the freedom of entry into any productive activities. In light of these considerations, the Hicksian production model, which assigns production sets with non-increasing returns to scale to a fixed number of economic agents, is, from an economic perspective, only suitable for understanding a restricted class of competitive economies without free entries of enterprises.

Walrasian Production

Walrasian production incorporates the idealized concept of perfect competition into the theoretical framework of market economies, abstracting from the specific institutional organization of the production sector. The aspect of perfect competition on the production side of economies is characterized by the free entry of any economic agent into productive activities. Free entry to productive activities is based upon the presumption that there are no impediments to the utilization and transfer of all available technological knowledge and productive resources.

Free Entries of Production Units and Linear Production Technologies

The assumption of free entries of production units essentially implies a focus on production processes rather than enterprises. Alternatively stated, the Walrasian production sector is composed of constant returns to scale or linear technologies. In this regard, Walras (cf. [122, Section 188: p. 225]) posited that "in a state of equilibrium in production, entrepreneurs neither make a profit nor a loss ['les entrepreneurs ne font ni bénéfice ni perte']." McKenzie (cf. [86, 87, 88]) formalized Walrasian production as a convex cone representing a *total or social production set*. From an economic perspective, Walrasian production describes the ideal type of perfectly competitive production sector.

Economies with Walrasian Production

Definition 7.3 (Walrasian Production Economy) The *Walrasian production* of an economy $\mathscr{E} : (A, \mathscr{A}, \nu) \to \mathcal{P} \times \mathbb{G}$ is defined as a production set $Y \subset \mathbb{R}^\ell$ which is a convex cone in $\mathbb{R}^\ell$ satisfying the properties (a) through (d) in the above list of properties of production sets. A pair $(\mathscr{E}, Y)$ of an atomless economy and its Walrasian production is called a *Walrasian production economy.*

Allocations and Suballocations for a Walrasian Production Economy

In a Walrasian production economy $(\mathscr{E}, Y)$ with $\mathscr{E} : (A, \mathscr{A}, \nu) \to \mathcal{P} \times \mathbb{G}$ and a Walrasian production $Y \subset \mathbb{R}^\ell$, an *allocation* (resp., *suballocation*) f for $(\mathscr{E}, Y)$ is defined as an allocation f (resp., suballocation) for $\mathscr{E}$. An allocation or suballocation f for $(\mathscr{E}, Y)$ is said to be *attainable* if the total amounts of all commodities called for in order to materialize the allocation or suballocation f (i.e., $\int f d\nu$) that exceed initially available amounts (i.e., $\int e d\nu$) can be produced by the given production technology. In more precise terms, an allocation or suballocation f for $(\mathscr{E}, Y)$ is said to be *attainable* if it satisfies the condition $\int f d\nu - \int e d\nu \in Y$, that is, if $\int f d\nu - \int e d\nu = y$ for some $y \in Y$.

7.4 Equilibria and Subequilibria for Walrasian Production Economies

We now turn our attention to equilibrium concepts in Walrasian production economies. As in Chapter 6 for economies without production, two types of equilibrium concepts are introduced for economies with Walrasian production.

Equilibrium for Walrasian Production Economies

Definition 7.4 (Walrasian Equilibrium for Production Economies) A *Walrasian equilibrium* for a Walrasian production economy $(\mathscr{E}, Y)$ with $\mathscr{E} : (A, \mathscr{A}, \nu) \to \mathcal{P} \times \mathbb{G}$ is a triple (p, f, y) composed of a price vector p $(\neq 0) \in \mathbb{R}^\ell$, an allocation f and a production vector $y \in Y$ for $(\mathscr{E}, Y)$ satisfying the following conditions:

(a) a.e. a in A, $f(a)$ belongs to the demand set $D(X_a, \succ_a, p \cdot e_a, p)$, i.e., $f(a)$ is a maximal element for $(X_a, \succ_a)$ in the budget set $B(X_a, p \cdot e_a, p)$;
(b) the production vector $y \in Y$ in equilibrium earns zero profit, i.e., $p \cdot y = 0$, and no production plan can earn positive profit, i.e., $p \cdot z > 0$ implies $z \notin Y$; and
(c) the allocation $f : A \to \mathbb{G}$ is attainable, i.e., $\int f d\nu - \int e d\nu = y$.

The price vector p of a Walrasian equilibrium (p, f, y) is an *equilibrium price vector* and the allocation $f : A \longrightarrow \mathbb{G}$ is called a *Walrasian allocation*. The set of all Walrasian allocations for $(\mathscr{E}, Y)$ is denoted by $\mathscr{W}(\mathscr{E}, Y)$.

The concept of suequilibria for Walrasian production economies is defined by the substitution of suballocations for allocations in the definition of equilibria, in a manner analogous to that for economies without production.

Definition 7.5 (Production Subequilibrium for Walrasian Production Economies) A *production subequilibrium* for a Walrasian production economy $(\mathscr{E}, Y)$ with $\mathscr{E} : (A, \mathscr{A}, \nu) \longrightarrow \mathcal{P} \times \mathbb{G}$ is a triple (p, f, y) composed of a price vector $p\ (\neq 0) \subset \mathbb{R}^\ell$, a suballocation f and a production vector $y \in Y$ for $(\mathscr{E}, Y)$ that satisfies the following conditions:

(a) a.e. a in A, $f(a)$ belongs to the subdemand set $D^*(X_a, \succ_a, e_a, p)$, that is, $p \cdot x \geqq p \cdot f(a)$ for all $x \in X_a$ such that $x \succ f(a)$ if $p \cdot f(a) \geqq \inf p \cdot X_a$, and $f(a) = e_a$ if $p \cdot f(a) < \inf p \cdot X_a$;
(b) the production vector $y \in Y$ in subequilibrium earns zero profit, i.e., $p \cdot y = 0$, and no production vector can earn positive profit, i.e., $p \cdot z > 0$ implies $z \notin Y$; and
(c) the suballocation $f : A \longrightarrow \mathbb{G}$ is attainable, i.e., $\int f d\nu - \int e d\nu = y$.

The price vector p of a production subequilibrium (p, f, y) is a *subequilibrium price vector* and the suballocation $f : A \longrightarrow \mathbb{G}$ is called a *production subequilibrium allocation*. The set of all production subequilibium allocations for $(\mathscr{E}, Y)$ is denoted by $\mathscr{W}_s(\mathscr{E}, Y)$.

7.5 Existence of Equilibria for Walrasian Production Economies

As provision for existence of equilibria of production economies, we introduce two definitions at the outset. The first of these is a property analogous to one already introduced for economies without production in Chapter 6, technically termed "*social positive income condition*" for production economies. This condition posits that economic agents as a whole are capable of attaining sufficient income to procure adequate quantities of commodities for their survival, contingent upon the utilization of production technologies and the distribution of commodities among them in an appropriate manner. This condition is considerably more tenable in a production economy than in an economy without production.

Definition 7.6 (Social Positive Income Condition for Production Economy) A production economy $(\mathscr{E}, Y)$ with $\mathscr{E} : (A, \mathscr{A}, \nu) \longrightarrow \mathcal{P} \times \mathbb{G}$ and a production set $Y \subset \mathbb{R}^\ell$ is said to satisfy the *social positive income condition for a production economy* if its endowment mapping $e : A \longrightarrow \mathbb{G}$ and consumption set mapping $X : A \longrightarrow \mathscr{X}$ satisfy the condition that the relative interior of the set of points $\int (X - e) d\nu$ that lie within the relative interior of Y is not empty with respect to the production set Y; that is, $\text{rel.int} \int (X - e) d\nu \cap \text{rel.int}\, Y \neq \emptyset$.

The social positive income condition for a production economy ensures that all economic agents are able to participate in market transactions and that a non-trivial fraction of them have positive income. This is achieved by using production technologies to protect their wealth levels to strictly exceed their minimum income levels. The condition "rel.int $\int(X - e)d\nu \cap \text{rel.int}\, Y \neq \emptyset$" indicates that consumers as a whole are able to engage in market transactions to obtain what they need for their survival with the help of available production technologies.

An additional condition introduced is the free disposal of consumption commodities in production economies. The property of free disposal of consumption commodities is satisfied if the complete depletion of any surplus aggregate amounts of consumption commodities left over after their allotments to economic agents is always possible. This is defined as follows:

Definition 7.7 (Free Disposal of Consumption Commodities in Production Economy) A production economy $(\mathscr{E}, Y)$ with $\mathscr{E} : (A, \mathscr{A}, \nu) \to \mathcal{P} \times \mathbb{G}$ and a production set $Y \subset \mathbb{R}^\ell$ is said to satisfy the *free disposal of consumption commodities in production* if we have $x - y \in Y$ for any commodity vectors $x, y \in \mathbb{G}$ with $x \leqq y$ and $x, y \in \int X d\nu$.

Existence of Equilibria in Production Economies

The initial step is to demonstrate the existence of production subequilibria for Walrasian production economies. The existence of Walrasian equilibria is discussed on the basis of the existence of production subequilibria for Walrasian production economies.

Existence of Production Subequilibria

Theorem 7.1 (Existence of Production Subequilibria)

Let $(\mathscr{E}, Y)$ be a Walrasian production economy with $\mathscr{E} : (A, \mathscr{A}, \nu) \to \mathcal{P} \times \mathbb{G}$ and a Walrasian production $Y \subset \mathbb{R}^\ell$. If $(\mathscr{E}, Y)$ satisfies the social positive income condition for a production economy and the free disposal of consumption commodities in production, then there exists a production subequilibrium (p, f, y) for $(\mathscr{E}, Y)$ with $p(\neq 0) \in Y^\circ = \{p \in \mathbb{R}^\ell \mid p \cdot z \leqq 0 \text{ for all } z \in Y\}$.

Proof Let us first introduce truncation of consumption sets to cope with their potential for unboundedness. For each agent $a \in A$ and integer $k \in \mathbb{N}_+$, the truncated consumption set $X_{a,k}$ is defined by $X_{a,k} = X_a \cap k[\{x = (x_i)_{i=1,\dots,\ell} \in \mathbb{G} \mid x_i \leqq \|e(a)\| + 1 \text{ for all } i\}]$. Recall that for any consumption set $X \in \mathscr{X}$, there is an element $z = (z_i)_{i=1,\dots,\ell} \in X$ such that $z_i \leqq \overline{\beta}$ for each i. Moreover, we have $x_i \geqq \underline{\beta}$ for each i and every $x = (x_i)_{i=1,\dots,\ell} \in X$ (cf. Definition 2.4, p. 75). By the former property we can assume without loss of generality that we have $X_{a,k} \neq \emptyset$ for all a and k. Furthermore, consumption sets $X \in \mathscr{X}$ are uniformly bounded below by the latter property, thereby ensuring that truncated consumption sets are bounded. To facilitate the notation, we write $B(a, p) =$

$B(X_a, p \cdot e_a, p), D(a, p) = D(X_a, \succ_a, p \cdot e_a, p)$, $D^*(a, p) = D^*(X_a, \succ_a, e_a, p)$, and $B^*(a, p) = B(X_a^*, p \cdot e_a, p)$, where $X_a^* = X_a \cup \{e_a\}$. Similarly for truncated consumption sets, let $B_k(a, p) = B(X_{a,k}, p \cdot e_a, p), D_k(a, p) = D(X_{a,k}, \succ_a, p \cdot e_a, p)$, $D_k^*(a, p) = D^*(X_{a,k}, \succ_a, e_a, p)$, and $B_k^*(a, p) = B(X_{a,k}^*, p \cdot e_a, p)$ with $X_{a,k}^* = X_{a,k} \cup \{e_a\}$, given that the preferece relation $\succ_a$ restricted to the truncated consumption set $X_{a,k}$ is again denoted by $\succ_a$. The truncated production set Y_k is also defined for each $k \in \mathbb{N}_+$ by $Y_k = \{y \in Y \mid -k \leqq y_i \leqq k\}$.

Now, since rel.int $\int (X - e)dv \cap$ rel.int$Y \neq \emptyset$, there is a point y^* in the relative interior of the mean consumption sets and the relative interior of the production set, so $y^* \in$ rel.int $\int X dv \cap$ rel.int$Y$. Then, define the price set PS by $PS = \{p = (p_i)_{i=1,\ldots,\ell} \in Y^\circ \mid p \cdot y^* = -1\}$. Here, $Y^\circ \subset \mathbb{R}^\ell$ is the dual (or the polar) of the cone Y, that is, $Y^\circ = \{p \in \mathbb{R}^\ell \mid p \cdot y \leqq 0 \text{ for all } y \in Y\}$. The convexity and the closedness of PS are immediate. It is compact if it is shown to be bounded. Indeed, suppose to the contrary that there is an unbounded sequence $(p_n)_{n \in \mathbb{N}_+}$ in PS, and consider the associated sequence $(q_n)_{n \in \mathbb{N}_+}$ in the unit sphere $S = \{p \in \mathbb{R}^\ell \mid \|p\| = 1\}$ where $q_n = p_n / \|p_n\|$. The compactness of S implies that there is a limit point $q^* \in S$ of $(p_n)_n$. One may assume that $(q_n)_n \to q^*$. As $q_n \cdot y^* = -1/\|p_n\| \to_n 0$, one must have $q^* \cdot y^* = 0$ in contradiction to $y^* \in$ rel.int $\int (X - e)dv \cap$ rel.int$Y$. Hence, PS is bounded.

We begin by showing the existence of subequilibria (p_k, f_k, y_k), $(k \in \mathbb{N}_+)$, for the truncated economies of $(\mathscr{E}, Y)$, each of which satisfies the following conditions:

(a') a.e. a in A, $f_k(a)$ belongs to the subdemand set $D_k^*(X_a, \succ_a, e_a, p_k)$;
(b') no production plan y in Y_k earns profit strictly exceeding the profit earned by the production plan $y_k \in Y$ in subequilibrium, i.e., $p_k \cdot y_k \geqq p_k \cdot y$ for any $y \in Y_k$; and
(c') the suballocation $f_k : A \to \mathbb{C}$ is attainable, i.e., $\int f_k dv - \int e dv = y_k$.

Let F_k be the mean excess subdemand for the truncated economy, defined as the mapping from PS to $\mathbb{R}^\ell$ given by $F_k(p) = \int \{D_k^*(\cdot, p) - e\}dv$. It is shown that the correspondence $F_k : PS \to \mathbb{R}^\ell$ satisfies the following properties for each $k \in \mathbb{N}_+$:

$$F_k(p) \neq \emptyset \text{ for all } p \in PS; \tag{7.1}$$

$$F_k(p) \subset H_{\leqq}(p, 0) \text{ for all } p \in PS; \tag{7.2}$$

$$F_k(p) \subset K \text{ for all } p \in PS \text{ for some compact set } K \subset \mathbb{R}^\ell; \tag{7.3}$$

$$F_k : PS \to \mathbb{R}^\ell \text{ is closed-valued and upper hemi-continuous;} \tag{7.4}$$

$$F_k(p) \text{ is a convex set in } \mathbb{R}^\ell \text{ for all } p \in PS. \tag{7.5}$$

Since the measurability of the correspondence $D_k^*(\cdot, p)$ from A to $\mathbb{C}$ at any $p \in PS$ follows from that of $D^*(\cdot, p)$ as a consequence of Proposition 4.11, p. 153, by the Measurable Selection Theorem, p. 106, and the uniform boundedness of the truncated consumption sets, we have

$$\int D_k^*(\cdot, p)dv \neq \emptyset \text{ for any } p \in PS. \tag{7.6}$$

(7.1) is immediate from (7.6). To confirm the validity of (7.2), let $z \in F_k(p)$. Then, there is an integrable mapping $g : A \longrightarrow \mathbb{G}$ for which one has $z = \int (g - e)dv$ and $g(a) \in D^*_k(a, p)$, a.e. $a \in A$. This implies that $g(a) \in H_{\leqq}(p, p \cdot e_a)$, a.e. $a \in A$. Consequently, $z \in H_{\leqq}(p, 0)$.

If we put

$$K = \left\{x = (x_i)_{i=1,\dots,\ell} \in \mathbb{R}^\ell \;\middle|\; -k(1 + |\underline{\beta}|) \leqq x_i \leqq k\left(1 + \left\|\int edv\right\|\right) \text{ for each } i\right\},$$

(7.3) follows from $\int edv < \infty$ and $D^*_k(a, p) \subset K$ for each $p \in PS$ and every $a \in A$. Since the correspondence $F_k : PS \longrightarrow \mathbb{R}^\ell$ is closed for every $k \in \mathbb{N}_+$ by Proposition 4.11, p. 153, F_k is closed-valued, and the upper hemi-continuity of F_k is a consequence of Mathematical Proposition 1.1, p. 36 and (7.3). This confirms the property (7.4). The last property (7.5) follows from the atomlessness of the measure space of economic agents $(A, \mathscr{A}, v)$ and Richter's Proposition (cf. Mathematical Proposition 3.14, p. 117).

We now apply the Equilibrium Fixed Point Theorem, p. 43, to each correspondence $F_k : PS \longrightarrow K$ for $k \in \mathbb{N}_+$. The properties (7.1) through (7.5) imply that there exists for each k, a price vector $p_k \in PS$ such that $F_k(p_k) \cap PS^\circ \neq \emptyset$. Thus, there is an integrable mapping $f_k : PS \longrightarrow K$ and a production vector $y_k \in Y$ for each $k \in \mathbb{N}^+$ satisfying

(i) *a.e. a in A, $f_k(a)$ belongs to the subdemand set $D^*_k(X_a, \succ_a, e_a, p_k)$, and*
(ii) $\int (f_k - e)dv = y_k \in Y$,

since we have that $PS^\circ = (Y^\circ)^\circ = Y$. The sequence of price vectors $(p_k)_k$ in PS is bounded. To show that the sequences $(\int f_k dv)_k$ and $(y_k)_k$ are bounded, let us initially verify that the set $\int (X - e)dv \cap Y$ is bounded. Indeed, suppose that there is a sequence $(z_n)_n$ in $\int (X - e)dv \cap Y$ such that $\|z_n\| \longrightarrow_n \infty$. Set $x_n = (1/\|z_n\|)z_n$ for each n. As $(x_n)_n$ is a sequence in the unit sphere which is compact, there is a limit point $\overline{x}$ of the sequence. We may assume $x_n \longrightarrow_n \overline{x}$. Since $\int Xdv$ is bounded below, it must be that $\overline{x} \geqq 0$ and $\overline{x} \neq 0$. As the production set Y is a closed cone, we must have $\overline{x} \in Y$ in contradiction to the no free production property $Y \cap \mathbb{R}^\ell_+ = \{0\}$. Thus, $\int (X - e)dv \cap Y$ is bounded. It then follows from $\int f_k dv - \int edv = y_k \in \int (X - e)dv \cap Y$ for every k that $(\int f_k dv)_k$ and $(y_k)_k$ are bounded. Without loss of generality, we assume that the sequences $(p_k)_k$, $(y_k)_k$, and $(\int f_k dv)_k$ are convergent. Let $(p_k)_k \longrightarrow_k \overline{p}$, $(y_k)_k \longrightarrow_k \overline{y}$, and $(\int f_k dv)_k \longrightarrow_k \overline{z}$. We have $\overline{p}(\neq 0) \in PS \subset Y^\circ$ and $\overline{y} \in Y$. Moreover, it must be that $\overline{p} \cdot \overline{y} \geqq \overline{p} \cdot y$ for every $y \in Y$. Otherwise, since $Y_k \subset Y_{k+1}$ for each k and $Y = \cup_k Y_k$, there is a production vector $\widetilde{y} \in Y$ for which we have $\overline{p} \cdot \widetilde{y} > \overline{p} \cdot \overline{y}$, and $\widetilde{y}, \overline{y} \in Y_k$ for all k large enough. This leads to a contradiction, as for such large k, we have $p_k \cdot y \leqq p_k \cdot y_k$ for every $y \in Y_k$. Hence, the production vector $\overline{y}$ maximizes profits (i.e., the value of the production vector) over the production set Y. It also implies that $\overline{p} \cdot \overline{y} = 0$. If this is not the case, it must be that $\overline{p} \cdot \overline{y} > 0$. In this case one would have $\overline{p} \cdot 2\overline{y} > \overline{p} \cdot \overline{y}$ and $2\overline{y} \in Y$, given that Y is a cone in $\mathbb{R}^\ell$. This is a contradiction to the fact that $\overline{y}$ maximizes profits over Y.

Since the sequence $(f_k)_k$ of integrable functions is uniformly bounded below, it follows from the Multidimensional Fatou's Lemma, p. 123, that there is an integrable mapping $\overline{f} : A \to \mathbb{G}$ satisfying $\overline{f}(a) \in \mathrm{Ls}\,(f_k(a)$, a.e. $a \in A$, and $\int \overline{f} dv \leqq \overline{z}$. Define a vector $\widehat{y} \in \mathbb{R}^\ell$ by $\widehat{y} = \int \overline{f} dv - \int e dv$. We are going to show that the triple $(\overline{p}, \overline{f}, \widehat{y})$ represents a production subequilibrium for $(\mathscr{E}, Y)$. Since the subdemand relation D^* is closed by Proposition 4.11, p. 153, it follows from property (i) that

a.e. $a \in A$, $\overline{f}(a)$ *belongs to the subdemand set* $D^*(X_a, \succ_a, e_a, \overline{p})$.

As a consequence, by Proposition 4.9, p. 152, we obtain $\overline{p} \cdot \int \overline{f} dv = \overline{p} \cdot \int e dv$. To verify that $\widehat{y} \in Y$, set $y_\delta = \int \overline{f} dv - \int e dv - \overline{y}$. Then, $\widehat{y} = \overline{y} + y_\delta$. y_δ is the excess subdemand to be supplied in equilibrium. Since $\int f_k dv - \int e dv = y_k$ for all k implies $\lim_k \int f_k dv - \int e dv - \overline{y} = 0$, we have

$$\begin{aligned} y_\delta &= \left(\int \overline{f} dv - \lim_k \int f_k dv\right) + \left(\lim_k \int f_k dv - \int e dv - \overline{y}\right) \\ &= \int \overline{f} dv - \lim_k \int f_k dv = \int \overline{f} dv - \overline{z}, \end{aligned}$$

where $\int \overline{f} dv, \overline{z} \in \int X dv$, and $\int \overline{f} dv \leqq \overline{z}$. It follows from the free disposal of consumption commodities in production that we have $y_\delta \in Y$. Consequently, we see that $\widehat{y} = \overline{y} + y_\delta \in Y$, as Y is a convex cone.

The argument in the preceding paragraph shows that

$\overline{f} : A \to \mathbb{G}$ *is a suballocation, and attainable*, i.e., $\int \overline{f} dv - \int e dv = \widehat{y} \in Y$.

To complete the proof that $(\overline{p}, \overline{f}, \widehat{y})$ is a production subequilibrium for $(\mathscr{E}, Y)$, it remains to verify that $\overline{p} \cdot \widehat{y} \geqq \overline{p} \cdot y$ for every $y \in Y$ and that $p \cdot \dot{y} = 0$. Indeed, since $\overline{p} \cdot y_\delta = \overline{p} \cdot (\int \overline{f} dv - \int e dv) - \overline{p} \cdot \overline{y} = 0$, we must have $\overline{p} \cdot \widehat{y} = \overline{p} \cdot \overline{y} = 0 \geqq \overline{p} \cdot y$ for every $y \in Y$. □

Production Subequilibria and Survival of Agents

As in the case of Walrasian economies without production, subequilibrium allocations are well defined even if some or all economic agents are unable to participate in market transactions due to shortfalls in their income or wealth. Broadly speaking, in order to allow almost all agents to participate in market transactions, any non-trivial set of them must be able to provide the remaining set of agents with commodities that they prefer to receive in addition. In a production economy, this may be done by using possible production technologies together with some of what they have as endowments. A property required in this respect in economies with a finite number of agents is known as *irreducibility* (cf. [86, 87, 88]) or *resource relatedness* (cf. [4]). It is expressed in terms of a complementary allocation of commodities to one set of agents from the other set, made possible by production technologies and a probable sacrifice of allocated commodities to the other set. In production economies with an atomless measure space of economic agents it is convenient to define an "irreducibility" condition by directly specifying a possible suballocation that satisfies a required property. Accordingly, in what follows, irreducibility is defined as the "endowments and production interdependence among preferred commodities" in terms of attainable suballocations that satisfy a specific property.

Definition 7.8 (Endowments and Production Interdependence among Preferred Commodities) A production economy $(\mathscr{E}, Y)$ with $\mathscr{E} : (A, \mathscr{A}, \nu) \to \mathcal{P} \times \mathbb{G}$ and a production set $Y \subset \mathbb{R}^\ell$ is said to satisfy the *endowments and production interdependence among preferred commodities* if, for any given attainable suballocation $f : A \to \mathbb{G}$ with $\int f d\nu - \int e d\nu = y \in Y$, and a non-trivial two-part measurable partition $\{A_1, A_2\}$ of A, each set $A_i, i = 1, 2$, having a positive measure, there is a suballocation $g : A \to \mathbb{G}$ with a production vector $\widetilde{y} \in Y$ satisfying

(a) a.e. a in A_1, $g(a) \in X_a$;
(b) a.e. a in A_1, $g(a) \succ_a f(a)$ whenever $f(a) \in X_a$;
(c) $g(a) \neq f(a)$ on a subset of A_2 with a positive measure;
(d) $\int g d\nu - \int e d\nu = \widetilde{y}$.

Using condition (d) above along with the attainability of f, we can write

$$\int_{A_1} (g - f) d\nu = \widetilde{y} - y + \int_{A_2} (f - g) d\nu.$$

Then, by setting $\Delta y = \widetilde{y} - y$ and $\Delta z = \int_{A_2} (f - g) d\nu$, we get $\int_{A_1} g d\nu = \int_{A_1} f + (\Delta y + \Delta z)$, which allows us to reformulate the endowments and production interdependence among preferred commodities in terms of a complementary allocation of commodities to one set of agents from the other set, made possible by a production variation and a probable sacrifice of allocated commodities to the other set of agents, as follows:

> For any given attainable suballocation $f : A \to \mathbb{G}$ with $\int f d\nu - \int e d\nu = y \in Y$, and a non-trivial two-part measurable partition $\{A_1, A_2\}$ of A, each set $A_i, i = 1, 2$, having a positive measure, there is a variation Δz of allocated commodities to the agents in A_2 coupled with a variation Δy of the production plan y such that (cf. Notes 7.1, p. 205)
>
> (a') $\int_{A_1} f d\nu + (\Delta y + \Delta z) \in \int_{A_1} \{x \in X_a \mid x \succ_a f(a)\} d\nu$;
>
> (b') $\Delta z \neq 0$ and $\int_{A_2} f d\nu - \Delta z \in \int_{A_2} \{x \in \mathbb{G} \mid x \in X_a \cup \{e_a\}\} d\nu$;
>
> (c') $y + \Delta y \in Y$.

The reformulation of conditions (a)–(d) into (a')–(c') suggests a way of understanding the conditions of endowments and production interdependence among preferred commodities. It is this: given any attainable suballocation of commodities for consumption in conjunction with outputs and inputs in production, if the total set of economic agents is divided into two non-trivial sets, it will be possible for each set of agents, by changing parts of what they would have received, to provide the other set of agents with additional commodities, including parts of outputs and inputs from a possible production variation, thereby contributing to an improvement in the commodity vectors received by the other set of agents.

This condition has much more realistic validity than in the case of economies without production, for the reason that when production technologies are used between two non-trivial sets of partitioned whole economic agents, it is much easier for one set of agents to offer the other set of agents additional trades that improve the welfare of its members.

The following proposition about allocations with survival in production economies is based on Theorem 7.1.

Proposition 7.1 (Suballocations with Survival in Production Economies)

Let $(\mathscr{E}, Y)$ be a Walrasian production economy with $\mathscr{E} : (A, \mathscr{A}, \nu) \to \mathcal{P} \times \mathbb{G}$ and a Walrasian production Y satisfying the same conditions (including the social positive income condition for production economy) as in Theorem 7.1. Furthermore, if $(\mathscr{E}, Y)$ satisfies the endowments and production interdependence among preferred commodities, then there exists a production subequilibrium (p, f, y) for $(\mathscr{E}, Y)$ with $p(\neq 0) \in \mathbb{R}^\ell$ such that

(a) $f(a) \in X_a$ *for* a.e. $a \in A$;
(b) $f(a) \in D^*(X_a, \succ_a, e_a, p)$ *for* a.e. $a \in A$;
(c) $f(a) \in D(X_a, \succ_a, p \cdot e_a, p)$ *in (b) if* $p \cdot e_a \notin J_p(X_a)$.

Proof Given a Walrasian production economy $(\mathscr{E}, Y)$ satisfying the above conditions, by Theorem 7.1 there exists a production subequilibrium (p, f, y) for $(\mathscr{E}, Y)$ with $p(\neq 0) \in \mathbb{R}^\ell$. Since the suballocation $f : A \to \mathbb{G}$ is attainable with the production vector y, we have $\int f d\nu - \int e d\nu = y \in Y$, and (b) holds.

Let a subset $\widetilde{S} \subset A$ be defined by $\widetilde{S} = \{a \in A \mid \inf p \cdot X_a \leqq p \cdot e_a\}$. Since $\widetilde{S} \in \bar{\mathscr{A}}_\nu$ by Mathematical Proposition 3.6, p. 108, there is a measurable set $\overline{S}$ such that $\overline{S} \Delta \widetilde{S}$ is a ν-null set. Put $\underline{S} = A - \overline{S}$. (a) holds if $\nu(\underline{S}) = 0$. Suppose $\nu(\underline{S}) > 0$; then, it follows from the condition of endowments and production interdependence among preferred commodities that there is a suballocation $g : A \to \mathbb{G}$ with a production vector $\widetilde{y} \in Y$ that satisfies

(i) a.e. a in $\overline{S}$, $g(a) \in X_a$ and $g(a) \succ_a f(a)$;
(ii) $g(a) \neq f(a)$ on a subset of $\underline{S}$ with a positive measure;
(iii) $\int g d\nu - \int e d\nu = \widetilde{y}$.

Since f is a suballocation, it follows from (i) that $p \cdot g(a) \geqq p \cdot e_a$, a.e. $a \in \overline{S}$. Considering (ii) and (iii) above, we get the contradiction that

$$0 \leqq p \cdot \int_{\overline{S}} (g - e) d\nu = p \cdot \widetilde{y} + p \cdot \int_{\underline{S}} (e - g) d\nu < 0,$$

where the last strict inequality is due to the fact that $p \cdot \widetilde{y} \leqq p \cdot y$, $p \cdot e_a \leqq p \cdot g(a)$, a.e. $a \in \underline{S}$, and $p \cdot e_a < p \cdot g(a)$ on a subset of $\underline{S}$ with a positive measure. Therefore, we conclude that $\nu(\underline{S}) = 0$, proving (a).

Finally, (a) and (b) together with Proposition 4.10 (2), p. 152, confirm (c). □

Walrasian Allocations and Subequilibrium Allocations in Production Economies

Let us now turn to Walrasian allocations in production economies by way of production subequilibrium allocations in Walrasian production economies. As in the case of no-production economies, we first relate Walrasian allocations to production subequilibrium allocations.

Proposition 7.2 (Walrasian Allocations and Production Subequilibrium Allocations)

Let $(\mathscr{E}, Y)$ be a Walrasian production economy with $\mathscr{E} : (A, \mathscr{A}, \nu) \to \mathcal{P} \times \mathbb{G}$ and a Walrasian production $Y \subset \mathbb{R}^{\ell}$.

(1) A Walrasian allocation $f : A \to \mathbb{G}$ for $(\mathscr{E}, Y)$ is a production subequilibrium allocation for $(\mathscr{E}, Y)$.
(2) Conversely, a production subequilibrium allocation $f : A \to \mathbb{G}$ for $(\mathscr{E}, Y)$ with a subequilibrium price vector p is a Walrasian allocation for $(\mathscr{E}, Y)$, provided that $(\mathscr{E}, Y)$ satisfies the social positive income condition for production economy, the endowments and production interdependence among preferred commodities, and that the set $A_p = \{a \in A \mid p \cdot e_a \in J_p(X_a)\}$ is a ν-null set.

Proof Both (1) and (2) are immediate consequences of Proposition 4.10, p. 152, and Proposition 7.1. □

Proposition 7.3 (Dispersion of Wealth Distribution and Production Subequilibria)

Let $(\mathscr{E}, Y)$ be a Walrasian production economy with $\mathscr{E} : (A, \mathscr{A}, \nu) \to \mathcal{P} \times \mathbb{G}$ and a Walrasian production Y that satisfies the social positive income condition for production economy, and the endowments and production interdependence among preferred commodities. Then the production subequilibria for $(\mathscr{E}, Y)$ are Walrasian equilibria for $(\mathscr{E}, Y)$, provided that the wealth distribution of $\mathscr{E}$ is dispersed at each subequilibrium price vector.

Proof Based on Proposition 5.1 (2), p. 170, the conclusion of the proposition follows from Proposition 7.2 (2). □

The following main existence theorem established for Walrasian production economies with a consumption sector composed of an atomless measure space of economic agents in non-convex consumption environments is reached by integrating the results presented in this chapter.

Theorem 7.2 (Existence of Equilibrium in Walrasian Production Economies)

Let $(\mathscr{E}, Y)$ be a Walrasian production economy with $\mathscr{E} : (A, \mathscr{A}, \nu) \to \mathcal{P} \times \mathbb{G}$ and a Walrasian production Y. Assume that $(\mathscr{E}, Y)$ satisfies the free disposal of consumption commodities in production and the social positive income condition for production economy in conjunction with the endowments and production interdependence among preferred commodities. Then, a Walrasian equilibrium for $(\mathscr{E}, Y)$ exists if the wealth distribution of $\mathscr{E}$ is dispersed at each price vector $p(\neq 0) \in Y^{\circ} \subset \mathbb{R}^{\ell}$.

Proof Let $(\mathscr{E}, Y)$ be a Walrasian production economy satisfying the conditions of the theorem to be proved; then a production subequilibrium (p, f, y) for $(\mathscr{E}, Y)$ with $p(\neq 0) \in Y^\circ \subset \mathbb{R}^\ell$ exists by Theorem 7.1. Thus it follows from Proposition 7.3 that (p, f, y) is a Walrasian equilibrium provided that the wealth distribution of $\mathscr{E}$ is dispersed at p. □

Notes 7

Notes 7.1
[Rephrasing of conditions (a)–(d) by (a')–(c'), p. 202] The conditions (a')–(c') correspond to irreducibility conditions in finite economies. See, for example, [87, 88]. In a model of atomless economies we find it easier to express the condition of endowments and production interdependence among preferred commodities by directly stating that the properties (a)–(d) are satisfied by a modified suballocation.

Chapter 8
Edgeworth Equilibria as the Core of Economies

Abstract The Walrasian equilibrium of perfectly competitive economies is the standard classical formalization of market transactions and the valuation of goods and services that take place in perfectly competitive markets. This chapter examines another classical equilibrium concept, originally introduced by Edgeworth, in relation to Walrasian equilibria. The Edgeworth equilibrium is based on perfect competition in market transactions without imposing a market institutional framework that generates uniform prices for goods and services traded in markets.

8.1 Edgeworth's Concept of Equilibrium in Perfectly Competitive Markets and the Core of an Economy

Competition and Perfect Competition in Edgeworth's Sense

Edgeworth expresses competition among market participants in trade in terms of "contracts" and "recontracts" for traded items.

Edgeworth Contracts and Recontracts in Market Transactions

An *Edgeworth contract*, or simply a *contract*, among a set of participants or traders is an agreement on the details of trading among themselves. Given a contract among a set of traders, an *Edgeworth recontract*, or simply a *recontract*, is another contract among a different set of traders involving a subset of the traders who made the earlier contract.

Edgeworthian Field of Perfect Competition

An *Edgeworthian field of competition*, or simply a *field of competition*, with respect to a contract consists of all individual trading participants who are willing and able to recontract over the particulars of the contract. A field of competition is a *perfect*

A. Yamazaki, *Mathematical Economics of General Equilibrium*, Monographs in Mathematical Economics 8, https://doi.org/10.1007/978-981-96-9497-6_8

field of competition if any individual participant of trades is free to contract and recontract with any unspecified number of participants at the same time without the consent required of any third party.

An *Edgeworthian settlement of market transactions*, or simply a *settlement of market transactions*, is a contract that cannot be changed with the consent of all parties to it. An *Edgeworthian final settlement of market transactions*, or simply a *final settlement of market transactions*, is a settlement that cannot be modified by recontracts within the field of competition. The purpose of defining an Edgeworthian final settlement of market transactions was to explain Walrasian perfect competition from the point of view of Edgeworthian perfect competition, which does not involve a market institution of setting prices of traded commodities. It offers an advantage of the Edgeworthian formulation of competition in markets, since the Walrasian functioning of demand through uniform prices of various commodities is tied to an institutional framework of price-setting in markets. Another advantage is that it is also applicable to various cases of imperfect competition. Let us now see how Edgeworthian perfect competition is formalized in general equilibrium analysis.

Edgeworth Competition as the Core Competition

In general equilibrium analysis the Edgeworth competition concepts of the previous paragraph have been formalized as the "core" competition concepts.

Possible Improvements in Allocations: Principle of Edgeworthian Recontracts

Let $(\mathscr{E}, Y)$ be a Walrasian production economy with $\mathscr{E} : (A, \mathscr{A}, \nu) \to \mathcal{P} \times \mathbb{G}$ and a Walrasian production $Y \subset \mathbb{R}^\ell$. It is assumed that all non-trivial subsets of economic agents with positive measures have access to technologies represented by the same production set Y.

Given a Walrasian production economy $(\mathscr{E}, Y)$, a suballocation f for $\mathscr{E}$ is said to be *feasible for a set of economic agents* $S \in \mathscr{A}$ if there is a production vector $y \in Y$ satisfying $\int_S f d\nu - \int_S e d\nu = y$. A set $S \in \mathscr{A}$ with $\nu(S) > 0$ is often referred to in the literature as a *coalition of economic agents*.

Definition 8.1 (Possibility of Improvement) Let f be a suballocation for a Walrasian production economy $(\mathscr{E}, Y)$. A non-trivial set of economic agents $S \in \mathscr{A}, \nu(S) > 0$, is said to *improve upon the suballocation* f if there is a suballocation g for $\mathscr{E}$ that satisfies

(a) a.e. $a \in S$, $g(a) \succ_a f(a)$ if $f(a) \in X_a$, and $g(a) \in X_a$ if $f(a) \notin X_a$;
(b) a.e. $a \notin S$, $g(a) = e_a$;
(c) g is attainable for $(\mathscr{E}, Y)$, i.e. there is $y \in Y$ for which $\int g d\nu - \int e d\nu = y$.

In this context, the set S is called an *improving set (or coalition) of economic agents*.

Conditions (b) and (c) above imply $\int_S g d\nu - \int_S e d\nu = y$, which ensures that g is feasible for S. If a set $S \in \mathscr{A}$ can improve upon a suballocation f, sometimes S is said to *block* f, and S is said to be a *blocking coalition*. When this is the case, care must be taken to note that "S blocks f" *does not mean* that S can actually block any particular action taken by economic agents who are not members of S.

Definition 8.2 (Core of an Economy) Given a Walrasian production economy $(\mathscr{E}, Y)$, an attainable allocation f for $(\mathscr{E}, Y)$ is called a *core allocation* if it cannot be improved upon by any of the non-trivial sets in $\mathscr{A}$. The set of all core allocations for $(\mathscr{E}, Y)$ is called the *core of the (production) economy* $(\mathscr{E}, Y)$, and is denoted by $\mathscr{C}(\mathscr{E}, Y)$.

From a standard game theoretic point of view, the core of a game is understood as a *cooperative solution concept*. However, in view of the Edgeworthian concept of competition, which allows free contracts and recontracts among economic agents, improving sets or "coalitions" of economic agents are characterized so as to produce a field of competition. In other words, each member of an improving set of economic agents, considering only its own gain, voluntarily joins the set rather than forming it through the cooperative efforts of the members concerned. In an environment of free contracts and recontracts, an attainable allocation that can be improved upon does not fail to be so.

8.2 Subcore of Economies

Indivisibility of Commodities and Subcore Allocations

The core of an economy is introduced to formalize the concept of Edgeworthian perfect competition, which is intended to provide a basis for price-taking perfect competition. Edgeworth explicitly stated that a perfect field of competition requires the possibility of contracts and recontracts for each participant with an infinite number of market participants, for which infinite divisibilities of commodities are essential. Given the fact that the possibility of commodity indivisibilities is subsumed in the commodity space $\mathbb{G}$, it seems natural to "enlarge" the scope of core allocations by making the improvement by sets of economic agents *stricter* so as to require a margin for improvement. As done in Section 6.1, p. 178, which introduces subequilibrium allocations free from the assumption of survival of all economic agents, *subcore allocations* are introduced below.

Definition 8.3 (Strict Improvement) Given a Walrasian production economy $(\mathscr{E}, Y)$ with $\mathscr{E} : (A, \mathscr{A}, \nu) \to \mathcal{P} \times \mathbb{G}$ and a Walrasian production $Y \subset \mathbb{R}^\ell$, let f be a suballocation for $(\mathscr{E}, Y)$. A non-trivial set of economic agents $S \in \mathscr{A}$, $\nu(S) > 0$, is said to *strictly improve upon the suballocation* f if there is a suballocation g for $(\mathscr{E}, Y)$ that satisfies

(a) a.e. $a \in S$, $g(a) \in X_a$, and $g(a) \succ_a f(a)$ if $f(a) \in X_a$;
(b) a.e. $a \notin S$, $g(a) = e_a$;

(c) g is *strictly attainable* for $(\mathscr{E}, Y)$, i.e. there is $\widehat{y} \in \text{rel.int}\, Y$ for which $\int g dv - \int e dv = \widehat{y}$.

S is called a *strictly improving set (or coalition) of economic agents.*

Considering a relationship with a subequilibrium allocation for a production economy, we introduce the following concept of a *subcore allocation.*

Definition 8.4 (Subcore of a Production Economy) Given a Walrasian production economy $(\mathscr{E}, Y)$ with $\mathscr{E} : (A, \mathscr{A}, v) \to \mathcal{P} \times \mathbb{G}$ and a Walrasian production $Y \subset \mathbb{R}^\ell$, an attainable suballocation f for $(\mathscr{E}, Y)$ is called a *subcore allocation* if it cannot be strictly improved upon by any non-trivial sets in $\mathscr{A}$. The set of all subcore allocations for $(\mathscr{E}, Y)$ is called the *subcore of the (production) economy* $(\mathscr{E}, Y)$, and is denoted by $\mathscr{C}_s(\mathscr{E}, Y)$.

Core and Subcore Allocations

We can immediately check an inclusion relationship between core and subcore allocations.

Proposition 8.1 (Core and Subcore Allocations)

Let $(\mathscr{E}, Y)$ be a Walrasian production economy with $\mathscr{E} : (A, \mathscr{A}, v) \to \mathcal{P} \times \mathbb{G}$ and a Walrasian production $Y \subset \mathbb{R}^\ell$. Then core allocations for the economy $(\mathscr{E}, Y)$ are subcore allocations for $(\mathscr{E}, Y)$, that is, $\mathscr{C}(\mathscr{E}, Y) \subset \mathscr{C}_s(\mathscr{E}, Y)$.

Proof Let $f : A \to \mathbb{G}$ be a core allocation for $(\mathscr{E}, Y)$. Suppose that a suballocation $g : A \to \mathbb{G}$ for $(\mathscr{E}, Y)$ strictly improves upon the allocation f; then g satisfies

(i) a.e. $a \in S$, $g(a) \in X_a$, and $g(a) \succ_a f(a)$;
(ii) a.e. $a \notin S$, $g(a) = e_a$; and
(iii) $\int g dv - \int e dv = \widehat{y}$ for some $\widehat{y} \in \text{rel.int}\, Y \subset Y$.

Consequently, the suballocation g improves upon the allocation f in contradiction to f being a core allocation for $(\mathscr{E}, Y)$. □

Possibility of Non-Equivalence and Subequivalence

A celebrated equivalence theorem between core and Walrasian allocations given by Aumann [6] was presented in an atomless measure space of economic agents having consumption sets $\mathbb{R}^n_+$ in the commodity space $\mathbb{R}^n$. Considering that the commodity space $\mathbb{G}$ in this book embraces non-convexity due to indivisibility and other factors, it is natural to discuss the relationship between core and Walrasian allocations after clarifying the possibility of an equivalence relation between subcore and subequilibrium allocations.

Remark 8.1 If the commodity space $\mathbb{G} \subset \mathbb{R}^\ell$ is not convex, it is possible to have $\mathscr{W}(\mathscr{E}, Y) = \emptyset$ while $\mathscr{C}(\mathscr{E}, Y) \neq \emptyset$. To illustrate, let us take an economy $\mathscr{E}$ given in the example on p. 185. Let $\mathbb{G} = \mathbb{R} \times \{\ldots, -2, -1, 0, 1, 2, \ldots\}$, $X = \mathbb{G} \cap \mathbb{R}^2_+$, and $e = (2, 2)$.

Let a preference relation $(X, \succ)$ be defined by intersections of the dotted curves in Fig. 6.1 and the consumption set X. $(A, \mathscr{A}, \nu)$ is an arbitrary atomless probability space. Define an economy $\mathscr{E} : (A, \mathscr{A}, \nu) \to \mathcal{P} \times \mathbb{G}$ by setting $\mathscr{E}(a) = (X, \succ, e)$ for each $a \in A$. The economy $\mathscr{E}$ is such that everyone has the same consumption characteristics $(X, \succ, e)$. For simplicity, we also assume that a production set $Y \subset \mathbb{R}^\ell$ is given by $Y = \{0\}$, so that there are no technological transformations of commodities. The price-consumption curve (or offer curve) of each individual in this economy is as shown in Fig. 6.1, p. 185, and the individual demand correspondence lacks the upper hemi-continuity at a price vector $p = (p_1, p_2)$ with $p_1/p_2 = 1$. As a result, we can see that the aggregate (or mean) excess demand correspondence does not contain the vector 0. Since the only possible Walrasian equilibrium allocation is the no-trade allocation, the economy $\mathscr{E}$ has no Walrasian equilibria, i.e. $\mathscr{W}(\mathscr{E}, Y) = \emptyset$.

On the other hand, let $\{A_1, A_2\}$ with $\nu(A_1) = \nu(A_2) = 1/2$ be a measurable partition of A, and define an allocation $f : A \to X$ by $f(a) = (0, 4), a \in A_1$ and $f(a) = (4, 0), a \in A_2$. Then we can confirm that no trivial subset $S \in \mathscr{A}, \nu(S) > 0$, can improve upon f, so $\mathscr{C}(\mathscr{E}, Y) \neq \emptyset$. Note also that if in this example we had $\mathbb{G} = \mathbb{R}^2$, the allocation f would be improved upon by the subset A_2, and consequently we have $f \notin \mathscr{C}(\mathscr{E}, Y)$. Thus, the fact that f is not improved upon by any non-trivial subset of A in this example is interpreted as due to the indivisibility of commodity 2.

8.3 Subequivalence Theorem

In this section, we look at the equivalence relation between subequilibrium and subcore allocations. We like to call it the "subequivalence relation."

Subequilibrium and Subcore Allocations

Proposition 8.2 (Subequilibrium and Subcore Allocations)

Let $(\mathscr{E}, Y)$ be a Walrasian production economy with $\mathscr{E} : (A, \mathscr{A}, \nu) \to \mathcal{P} \times \mathbb{G}$ and a Walrasian production $Y \subset \mathbb{R}^\ell$. Then the production subequilibrium allocations for $(\mathscr{E}, Y)$ are the subcore allocations for the economy $(\mathscr{E}, Y)$, that is, $\mathscr{W}_s(\mathscr{E}, Y) \subset \mathscr{C}_s(\mathscr{E}, Y)$.

Proof Given a subequilibrium allocation f for $(\mathscr{E}, Y)$, there exists a price vector $p\ (\neq 0) \in Y^\circ$ and a production vector $y \in Y$ such that (p, f, y) is a subequilibrium for $(\mathscr{E}, Y)$, and satisfies $f(a) \in D^*(X_a, \succ_a, e_a, p)$, a.e. $a \in A$, and $\int f d\nu - \int e d\nu = y$. Suppose f is not in the subcore of $\mathscr{E}$. Then there is a strictly improving set of economic agents $S \in \mathscr{A}, \nu(S) > 0$ such that, for some suballocation g for $(\mathscr{E}, Y)$, one has

(i) a.e. $a \in S$, $g(a) \in X_a$, and $g(a) \succ_a f(a)$ if $f(a) \in X_a$;
(ii) a.e. $a \notin S$, $g(a) = e_a$;
(iii) $\int g d\nu - \int e d\nu = \widehat{y}$ for some production vector $\widehat{y} \in \text{rel.int}\, Y$.

But it leads to a contradiction that

$$
\begin{aligned}
0 &\leqq \int \big(p \cdot f(a) - p \cdot e(a)\big) d\nu \\
&\leqq \int \big(p \cdot g(a) - p \cdot e(a)\big) d\nu = p \cdot \widehat{y} < p \cdot y = 0,
\end{aligned}
$$

where the strict inequality in the last line above is a consequence of $p \in Y^{\circ}$ and $\widehat{y} \in \mathrm{rel.int}\, Y$. □

Subcore and Subequilibrium Allocations

Proposition 8.3 (Subcore and Subequilibrium Allocations)

Let $(\mathscr{E}, Y)$ be a Walrasian production economy with $\mathscr{E} : (A, \mathscr{A}, \nu) \to \mathcal{P}_{lns} \times \mathbb{G}$ and a Walrasian production $Y \subset \mathbb{R}^{\ell}$. Then the subcore allocations for the economy $(\mathscr{E}, Y)$ are the production subequilibrium allocations for $(\mathscr{E}, Y)$, that is, $\mathscr{C}_s(\mathscr{E}, Y) \subset \mathscr{W}_s(\mathscr{E}, Y)$.

Proof Given a subcore allocation f for $(\mathscr{E}, Y)$, there is a production vector $y \in Y$ with $\int f d\nu - \int e d\nu = y$. For each $a \in A$ define

$$
\succ_f(a) = \big\{x \in X_a \mid x \succ_a f(a) \text{ if } f(a) \in X_a\big\} \cup \big\{x \in \mathbb{G} \mid x \in X_a \text{ if } f(a) \notin X_a\big\},
$$

and $T(a) = \succ_f(a) - e_a$. By the local non-satiation of preference relation $(X_a, \succ_a)$, we have $\succ_f(a) \neq \emptyset$, and consequently $T(a) \neq \emptyset$ for every $a \in A$. Note that the correspondence T as a subset of $A \times \mathbb{G}$ is $T = \{(a, \hat{x}) \in A \times \mathbb{G} \mid \hat{x} + e_a \in \succ_f(a)\}$, and let $G \subset \mathcal{P} \times \mathbb{G} \times \mathbb{G}$ be a subset given by

$$
\begin{aligned}
G = \{(X, \succ, x, y) \in \mathcal{P} \times \mathbb{G} \times \mathbb{G} \mid x \succ y \text{ if } y \in X,& \\
\text{or } x \in X \text{ if } y \notin X\}&.
\end{aligned}
$$

It follows from Proposition 3.1, p. 100, and Proposition 3.3, p. 101, that G is a Borel set. Furthermore, one has $T = t^{-1}(G)$ by defining a map $t : A \times \mathbb{G} \to \mathcal{P} \times \mathbb{G} \times \mathbb{G}$ to be $(a, \hat{x}) \mapsto (X_a, \succ_a, \hat{x} + e_a, f(a))$. Since the measurability of the map $\mathscr{E}$ and the suballocation f implies that of the map t, T is a measurable correspondence from A to $\mathbb{G}$.

To show that $\int T d\nu \neq \emptyset$, let $\overline{T}$ be the correspondence from A to $\mathbb{G}$ given by $\overline{T}(a) = \mathrm{cl}\, T(a)$ for each $a \in A$. Also define another correspondence $Q : A \to \mathbb{G}$ related to T by setting $Q(a) = T(a) \cap U(f(a) - e_a, 1)$ for every $a \in A$, where $U(f(a) - e_a, 1)$ is the open neighborhood of point $f(a) - e_a$ with radius 1. Since the local non-satiation of the preference relations of economic agents implies that $(f(a) - e_a) \in \overline{T}(a)$, a.e. $a \in A$, we have $Q(a) \neq \emptyset$, a.e. $a \in A$. The measurability of the correspondence Q follows from that of the correspondence T and the mapping $(f - e)$. Therefore, by Measurable Selection Theorem, p. 106, there exists a measurable mapping $q : A \to \mathbb{G}$ such that $q(a) \in Q(a)$, a.e. $a \in A$. q is integrable due to the integrability of the mapping $(f - e)$. Consequently, we get $\int q d\nu \in \int T d\nu$, so $\int T d\nu \neq \emptyset$.

The next step is to consider a subset $\mathbb{T}$ of $\mathbb{R}^\ell$, given by

$$\mathbb{T} = \cup_{S \in \mathscr{A}} \left\{ \int_S T d\nu \subset \mathbb{R}^\ell \mid \nu(S) > 0 \right\},$$

and confirm that

$$\mathbb{T} \cap \text{rel.int}\, Y = \emptyset. \tag{8.1}$$

If we have $\mathbb{T} \cap \text{rel.int}\, Y \neq \emptyset$, there is an integrable mapping $g : A \to \mathbb{G}$ such that for some $S \in \mathscr{A}, \nu(S) > 0$, we have $g(a) - e_a \in T(a)$, a.e. $a \in S$, and $\int_S (g - e) d\nu = \widehat{y}$ for some $\widehat{y} \in \text{rel.int}\, Y$. Then consider the integrable mapping $\widehat{g} : A \to \mathbb{G}$, obtained by adjusting the values of g outside the set S, if necessary, so that $\widehat{g}(a) = e_a$ for $a \notin S$ and $\widehat{g}(a) = g(a)$ for $a \in S$. $\widehat{g} : A \to \mathbb{G}$ is a suballocation for $(\mathscr{E}, Y)$ and strictly improves upon the subcore allocation f, a contradiction. This confirms (8.1).

We now examine $\mathbb{T}$ and Y in the smallest linear subspace L that contains both $\mathbb{T}$ and Y. $\mathbb{T}$ is a convex set in $\mathbb{R}^\ell$ by Vind's Proposition, p. 118, since $(A, \mathscr{A}, \nu)$ is an atomless measure space, and rel.int$Y$ is a convex set as the relative interior of a convex set Y. Applying Minkowski's Separating Hyperplane Theorem, p. 41, to the convex sets $\mathbb{T}$ and rel.int$Y$, we obtain a vector $p (\neq 0) \in L$ such that for some $w \in \mathbb{R}$ we have

$$\mathbb{T} \subset H_{\geqq}(p, w), \quad \text{and} \quad \text{rel.int}\, Y \subset H_{\leqq}(p, w), \tag{8.2}$$

where the second inclusion implies that Y is also a subset of $H_{\leqq}(p, w)$. It implies $w = 0$ due to the fact that Y is a cone with its vertex at the origin. It also implies that $p \in Y^\circ$, and in particular $p \cdot y \leqq 0$. Noticing that $\int T d\nu \neq \emptyset$, it follows from the first inclusion of (8.2) that for every $S \in \mathscr{A}$ we have $p \cdot x \geqq 0$ for each $x \in \int_S T d\nu$, and therefore

$$\inf \left\{ p \cdot \overline{x} \mid \overline{x} \in \int_S T d\nu \right\} \geqq 0 \text{ for each } S \in \mathscr{A}. \tag{8.3}$$

Let us show that (8.3) leads to

$$\text{a.e. } a \in A, \; p \cdot z \geqq 0 \text{ for any } z \in T(a), \text{ and so}$$

$$p \cdot x \geqq p \cdot e_a \text{ for any } x \text{ such that } \begin{cases} x \succ_a f(a) & \text{if } f(a) \in X_a, \\ x \in X_a & \text{if } f(a) \notin X_a. \end{cases} \tag{8.4}$$

In fact, because of Mathematical Proposition 3.16, p. 118, we have

$$\inf \left\{ p \cdot \overline{x} \mid \overline{x} \in \int_S T d\nu \right\} = \int_S (\inf p \cdot T) d\nu.$$

So by (8.3) we get

$$\int_S (\inf p \cdot T) d\nu \geqq 0 \text{ for each } S \in \mathscr{A}.$$

This implies (8.4).

We are now ready to show that (p, f, y) is a subequilibrium for $(\mathscr{E}, Y)$. First we check that

$$p \cdot f(a) \leqq p \cdot e_a, \text{a.e. } a \in A, \tag{8.5}$$

so that $f(a)$ is in the budget set $B(X_a, p \cdot e_a, p)$, a.e. $a \in A$. By local non-satiation of preference relations it follows from (8.4) that $p \cdot f(a) \geqq p \cdot e_a$, a.e. $a \in A$. However, there is no set $S \in \mathscr{A}$ with $\nu(S) > 0$ for which we have $p \cdot f(a) > p \cdot e_a$, a.e. $a \in S$, because it would lead to a contradiction that

$$0 < p \cdot \left(\int f d\nu - \int e d\nu \right) = p \cdot y \leqq 0.$$

This confirms (8.5).

(8.4) and (8.5) verify that $f(a) \in D^*(X_a, \succ_a, e_a, p)$, a.e. $a \in A$. Thus, from the local non-satiation of preference relations and Proposition 4.9, p. 152, it follows that

$$p \cdot f(a) = p \cdot e_a, \text{a.e. } a \in A. \tag{8.6}$$

It only remains to show that $p \cdot y$ achieves a maximum profit of zero over the production set Y. But it is an immediate consequence of the attainability of f with respect to y and (8.6) that

$$p \cdot y = p \cdot \left(\int f d\nu - \int e d\nu \right) = 0.$$

This completes the proof that (p, f, y) is a subequilibrium for $(\mathscr{E}, Y)$. □

Subequivalence

By synthesizing Proposition 8.2 and Proposition 8.3, we obtain the following subequivalence theorem:

Theorem 8.1 (Subequivalence Theorem)

Let $(\mathscr{E}, Y)$ be a Walrasian production economy with $\mathscr{E} : (A, \mathscr{A}, \nu) \to \mathcal{P}_{Ins} \times \mathbb{G}$ and a Walrasian production $Y \subset \mathbb{R}^\ell$. Then the set of all subcore allocations of the economy $(\mathscr{E}, Y)$ is identical to the set of all production subequilibrium allocations for $(\mathscr{E}, Y)$, i.e. $\mathscr{C}_s(\mathscr{E}, Y) = \mathscr{W}_s(\mathscr{E}, Y)$.

It is now straightforward to discuss relationships between Walrasian and core allocations in non-convex consumption environments on the basis of the above subequivalence theorem.

8.4 Equivalence Theorem

Walrasian and Core Allocations

The fact that Walrasian allocations are core allocations is rudimentary and does not depend on whether the commodity space is convex or not.

Proposition 8.4 (Walrasian Allocations and Core)

Let $(\mathscr{E}, Y)$ *be a Walrasian production economy with* $\mathscr{E} : (A, \mathscr{A}, \nu) \to \mathcal{P} \times \mathbb{G}$ *and a Walrasian production* $Y \subset \mathbb{R}^\ell$. *Then the Walrasian allocations for* $(\mathscr{E}, Y)$ *are the core allocations of the economy* $(\mathscr{E}, Y)$, *i.e.* $\mathscr{W}(\mathscr{E}, Y) \subset \mathscr{C}(\mathscr{E}, Y)$.

Proof Given a Walrasian allocation f for $(\mathscr{E}, Y)$, there exists a price vector p $(\neq 0) \in Y^\circ$ and a production vector $y \in Y$ such that (p, f, y) is a Walrasian equilibrium for $(\mathscr{E}, Y)$, satisfying $f(a) \in D(X_a, \succ_a, e_a, p)$, a.e. $a \in A$, and $\int f d\nu - \int e d\nu = y$. Suppose f is not a core allocation for $\mathscr{E}$. Then there is an improving set of economic agents $S \in \mathscr{A}$, $\nu(S) > 0$ such that, for some suballocation g for $(\mathscr{E}, Y)$, one has

(i) a.e. $a \in S$, $g(a) \in X_a$, and $g(a) \succ_a f(a)$;
(ii) a.e. $a \notin S$, $g(a) = e_a$;
(iii) $\int g d\nu - \int e d\nu = \overline{y}$ for some production vector $\overline{y} \in Y$.

But this leads to a contradiction that

$$\begin{aligned} 0 &\leqq \int (p \cdot f(a) - p \cdot e(a)) d\nu \\ &< \int (p \cdot g(a) - p \cdot e(a)) d\nu = p \cdot \overline{y} \leqq 0, \end{aligned}$$

where the second inequality (which is strict) is due to (i), and the last inequality is due to $p \in Y^\circ$. □

Core and Walrasian Allocations

As we discussed in Remark 8.1, core allocations may not be Walrasian in some cases of non-convex commodity space. Nevertheless, we can verify that core allocations are *nearly* competitive equilibrium allocations in the sense that they are subequilibrium allocations with survival.

Core and Subequilibrium Allocations

Proposition 8.5 (Core and Subequilibrium Allocations with Survival)

Let $(\mathscr{E}, Y)$ *be a Walrasian production economy with* $\mathscr{E} : (A, \mathscr{A}, \nu) \to \mathcal{P}_{Ins} \times \mathbb{G}$ *and a Walrasian production* $Y \subset \mathbb{R}^\ell$*. Then, for any core allocation* $f : A \to \mathbb{G}$ *for the economy* $(\mathscr{E}, Y)$*, there exists a price vector* $p(\neq 0) \in Y^\circ$ *and a production vector* $y \in Y$ *such that* (p, f, y) *is a production subequilibrum (with survival) for* $(\mathscr{E}, Y)$ *satisfying*

(a) $f(a) \in X_a$ *for* a.e. $a \in A$*;*
(b) $f(a) \in D^*(X_a, \succ_a, e_a, p)$ *for* a.e. $a \in A$*;*
(c) $f(a) \in D(X_a, \succ_a, p \cdot e_a, p)$ *in (b) if* $p \cdot e_a \notin J_p(X_a)$*;*
(d) $\int f d\nu - \int e d\nu = y \in Y$ *and* $p \cdot y = 0$.

Proof Given a core allocation f for $(\mathscr{E}, Y)$, (a) is trivial, and there exists a production vector $y \in Y$ with $\int f d\nu - \int e d\nu = y$. For each $a \in A$, define $\succ_f(a) = \{x \in X_a \mid x \succ_a f(a)\}$ and $T(a) = \succ_f(a) - e_a$. By the local nonsatiation of the preference relation $(X_a, \succ_a)$ we have $\succ_f(a) \neq \emptyset$, and consequently $T(a) \neq \emptyset$ for every $a \in A$. The correspondence T as a subset of $A \times \mathbb{G}$ is $T = \{(a, \hat{x}) \in A \times \mathbb{G} \mid \hat{x} + e_a \in \succ_f(a)\}$. Let $G \subset \mathcal{P} \times \mathbb{G} \times \mathbb{G}$ be a subset $G = \{(X, \succ, x, y) \in \mathcal{P} \times \mathbb{G} \times \mathbb{G} \mid x \succ y\}$. We have $T = t^{-1}(G)$ for a map $t : A \times \mathbb{G} \to \mathcal{P} \times \mathbb{G} \times \mathbb{G}$ defined by $(a, \hat{x}) \mapsto (X_a, \succ_a, \hat{x} + e_a, f(a))$. Then, exactly as in the proof of Proposition 8.3, one can show that the correspondence T is measurable and $\int T d\nu \neq \emptyset$. Furthermore, defining a subset $\mathbb{T}$ of $\mathbb{R}^\ell$ by $\mathbb{T} = \cup_{S \in \mathscr{A}} \{ \int_S T d\nu \subset \mathbb{R}^\ell \mid \nu(S) > 0 \}$, we obtain $\mathbb{T} \cap Y = \emptyset$ as in the proof of Proposition 8.3, replacing rel.int Y by Y and a strictly improving set by an improving set to get a contradiction in the case of $\mathbb{T} \cap Y \neq \emptyset$.

It follows from Minkowski's Separating Hyperplane Theorem that the two convex sets $\mathbb{T}$ and $Y$are separated by a hyperplane in $\mathbb{R}^\ell$ such that

$$\mathbb{T} \subset H_{\geqq}(p, w) \text{ and } Y \subset H_{\leqq}(p, w) \text{ for some } p(\neq 0) \in \mathbb{R}^\ell \text{ and } w \in \mathbb{R}.$$

Since Y is a cone with its vertex at the origin, the second inclusion above implies that $w = 0$ and $p \in Y^\circ$. So we must have $p \cdot y \leqq 0$.

On the other hand, it follows from the first inclusion above that for every $S \in \mathscr{A}$ we have $p \cdot \overline{x} \geqq 0$ for each $\overline{x} \in \int_S T d\nu$, which implies from Mathematical Proposition 3.16, p. 118, that

$$\int_S (\inf p \cdot T) d\nu = \inf \left\{ p \cdot \overline{x} \,\middle|\, \overline{x} \in \int_S T d\nu \right\} \geqq 0 \text{ for each } S \in \mathscr{A}.$$

It follows that, a.e. $a \in A$, $p \cdot z \geqq 0$ for any $z \in T(a)$, which means that

$$p \cdot x \geqq p \cdot e_a \text{ for any } x \in X_a \text{ such that } x \succ_a f(a). \tag{8.7}$$

So if we check that

$$p \cdot f(a) \leqq p \cdot e_a, \text{a.e. } a \in A, \tag{8.8}$$

so that $f(a)$ is in the budget set $B(X_a, p \cdot e_a, p)$, a.e. $a \in A$, then we have $f(a) \in D^*(X_a, \succ_a, e_a, p)$ for a.e. $a \in A$, confirming (b).

Indeed, the local non-satiation of preference relations and (8.7) imply that

$$p \cdot f(a) \geqq p \cdot e_a, \text{a.e. } a \in A. \tag{8.9}$$

As a consequence, there is no set $S \in \mathscr{A}$ with $\nu(S) > 0$ for which we have $p \cdot f(a) > p \cdot e_a$, a.e. $a \in S$, otherwise there would be a contradiction that $0 < p \cdot \left(\int f d\nu - \int e d\nu \right) = p \cdot y \leqq 0$. This verifies (8.8), and by Proposition 4.10 (2), p. 152, property (c) is a consequence of (b).

To show that (p, f, y) is a subequilibrium for $(\mathscr{E}, Y)$ satisfying properties (a)–(d), it remains to verify $p \cdot y = 0$. Since (8.8) and (8.9) together imply $p \cdot f(a) = p \cdot e_a$, a.e. $a \in A$, we have $p \cdot y = p \cdot \left(\int f d\nu - \int e d\nu \right) = 0$. This completes the proof that (p, f, y) is a subequilibrium for $(\mathscr{E}, Y)$ that satisfies properties (a)–(d). □

Equivalenence Theorem for Non-Convex Commodity Space

We are now in a position to present the result of the equivalence between core and Walrasian allocations.

Theorem 8.2 (Equivalence Theorem)

Let $(\mathscr{E}, Y)$ be a Walrasian production economy with $\mathscr{E} : (A, \mathscr{A}, \nu) \to \mathcal{P}_{Ins} \times \mathbb{G}$ and a Walrasian production $Y \subset \mathbb{R}^\ell$. Then the set of all core allocations for the economy $(\mathscr{E}, Y)$ is identical to the set of all Walrasian allocations for $(\mathscr{E}, Y)$, i.e. $\mathscr{C}(\mathscr{E}, Y) = \mathscr{W}(\mathscr{E}, Y)$, if the wealth distribution of $\mathscr{E}$ is dispersed at each $p \in Y^\circ$.

Proof As we already have $\mathscr{W}(\mathscr{E}, Y) \subset \mathscr{C}(\mathscr{E}, Y)$ by Proposition 8.4, we show $\mathscr{C}(\mathscr{E}, Y) \subset \mathscr{W}(\mathscr{E}, Y)$. By Proposition 8.5 we have $\mathscr{C}(\mathscr{E}, Y) \subset \mathscr{W}_s(\mathscr{E}, Y)$. Moreover, if the wealth distribution of $\mathscr{E}$ is dispersed at each $p \in Y^\circ$, it follows from Proposition 4.10 (2), p. 152, and Proposition 5.1, p. 170, that we have $\mathscr{W}_s(\mathscr{E}, Y) = \mathscr{W}(\mathscr{E}, Y)$. Consequently, $\mathscr{C}(\mathscr{E}, Y) \subset \mathscr{W}(\mathscr{E}, Y)$. □

Remark 8.2 (Aumann's Equivalence Theorem) The pioneering and fundamental work in general equilibrium analysis with a non-atomic measure space of economic agents in which the equivalence theorem between core and Walrasian allocations was given is due to Aumann [6]. In terms of a production economy $(\mathscr{E}, Y)$ in the present book, *Aumann's economy with a continuum of traders* can be defined as follows:

Definition 8.5 (Aumann Economy with a Continuum of Traders) Let a production economy $(\mathscr{E}_{au}, Y)$ with $\mathscr{E}_{au} : (A, \mathscr{A}, \nu) \to \mathcal{P} \times \mathbb{G}$ and $Y \subset \mathbb{R}^\ell$ satisfy the following properties:

(i) $A = [0, 1]$, ν = the Lebesgue measure, $\mathbb{G} = \mathbb{R}^\ell$, and all consumption sets X are identical and $X = \mathbb{R}^\ell_+$.

(ii) All preference relations $(X, \succ)$ satisfy *monotonicity* in the sense that we have $x \succ y$ for any $x \neq y$ such that $x \geqq y \geqq 0$.

(iii) $Y = \mathbb{R}^\ell_-$.
(iv) $\int e d\nu \in \mathbb{R}^\ell_{++}$.

Verifying that the set of all preference relations satisfying monotonicity, denoted by $\mathcal{P}_{mo}$, is a measurable set and a subset of $\mathcal{P}_{lns}$ is a relatively straightforward mathematical exercise. By Proposition 8.4 we have $\mathcal{W}(\mathcal{E}_{au}, Y) \subset \mathcal{C}(\mathcal{E}_{au}, Y)$, and it follows from Proposition 8.5 that we also have $\mathcal{C}(\mathcal{E}_{au}, Y) \subset \mathcal{W}_s(\mathcal{E}_{au}, Y)$. Thus, for any given core allocation f for $(\mathcal{E}_{au}, Y)$, there exists a subequilibrium (p, f, y) for $(\mathcal{E}_{au}, Y)$ for some $p(\neq 0) \in Y^\circ = \mathbb{R}^\ell_+$ and $y \in Y$. Accordingly, for a.e. $a \in A$, $f(a) \in D^*(X_a, \succ_a, e_a, p)$ and $f(a) \in D(X_a, \succ_a, p \cdot e_a, p)$ if $e_a \notin C_p(X_a) = C_p(\mathbb{R}^\ell_+)$. Furthermore, we know that $p \cdot f(a) = p \cdot e_a$, a.e. $a \in A$, due to the monotonicity of the preference relations.

Given that $p \cdot \int e d\nu > 0$ by property (iv) and that $p(\neq 0) \in \mathbb{R}^\ell_+$, it follows that there exists a positive measure of points in A for which $p \cdot e_a > 0$, i.e. $\nu(\{a \in A \mid p \cdot e_a > 0\}) > 0$. The critical set of the consumption set $\mathbb{R}^\ell_+$ is given by the following: At $p \in \mathbb{R}^\ell_{++}$, the critical set is $C_p(\mathbb{R}^\ell_+) = \{0\}$. At $p \in \mathbb{R}^\ell_+ - \mathbb{R}^\ell_{++}$, the critical set is $C_p(\mathbb{R}^\ell_+) = \left\{x = (x_i)_{i=1,\ldots,\ell} \middle| x_i \geqq 0 \text{ if } p_i = 0, \text{ and } x_i = 0 \text{ if } p_i > 0\right\}$. The monotonicity of the preference relations implies that we must have $p \in \mathbb{R}^\ell_{++}$. Indeed, otherwise $f(a)$ cannot be a maximal element with respect to $\succ_a$ in the budget set $B(X_a, p \cdot e_a, p)$, which implies that $f(a) \notin D(X_a, \succ_a, p \cdot e_a, p)$ for $a \in A$ with $p \cdot e_a > 0$ whose wealth level $p \cdot e_a$ does not belong to $J_p(X_a)$, contrary to property (c) of Proposition 8.4. Then the critical set $C_p(\mathbb{R}^\ell_+)$ of the consumption set for each $a \in A$ consists only of the origin 0 in $\mathbb{R}^\ell$. This means that $p \cdot e_a \in J_p(X_a)$ if and only if $e_a = 0$. However, since we have $B(X_a, p \cdot e_a, p) = \{0\}$ for $a \in A$ with $e_a = 0$, we obtain $D^*(X_a, \succ_a, e_a, p) = D(X_a, \succ_a, p \cdot e_a, p)$ for every such $a \in A$. So we have $D^*(X_a, \succ_a, e_a, p) = D(X_a, \succ_a, p \cdot e_a, p)$, a.e. $a \in A$. This proves that f is indeed a Walrasian allocation and that $\mathcal{C}(\mathcal{E}_{au}, Y) \subset \mathcal{W}(\mathcal{E}_{au}, Y)$. □

Notes 8

Notes 8.1 [Section 8.1, p. 207]

[Perfect competition in the sense of Edgeworth, p. 207] See [35, pp. 16–20].
[Core competition, p. 208] See [113], [102], and [28].
[Existence of core allocations, p. 209] [103, 104] gave a direct proof of the determinacy of the core in economies with a finite number of individuals. [58] gave a proof of the existence of core allocation in the setting with a measure space of individuals. It provided a necessary and sufficient condition for the existence of core allocations. However, the economic interpretation of this condition seems difficult.

Notes Appendix: Some Historical Background on the Determinacy of Equilibrium in the Finite Population General Equilibrium Model

A brief historical background on the determinacy of equilibrium is in order. The first mathematical formulation of the competitive economy, now called the general equilibrium model, was given by Leon Walras (1874–1877) [122]. The intent of this theoretical model was to explain theoretically what equilibrium conditions would be achieved in an economy with many individuals interacting through the market when the market influence of each individual is negligible. Walras [122, Lesson 7] realized that he had to prove the existence of equilibrium for his theory to be meaningful. However, his arguments were limited to pointing out that the number of economic variables to be determined in a general equilibrium model is equal to the number of equations that make up the model, and the question of the determinacy of equilibrium in the strict sense was not explicitly addressed as a subject of study for more than half a century.

K. Menger's seminar in Vienna in the 1930s was actively involved in studies of the existence problem of Walrasian equilibria. Cassel's formulation of general equilibrium in [17] set the stage for active studies, and it reflects the state of knowledge of general equilibrium analysis in 1930. The papers in Menger's seminar include [92], [117], [130], [106], etc., culminating in the first rigorous formulation and proof of the existence of equilibrium in a Cassellian variant of the Walrasian system given by Wald [119, 120, 121]. J. von Neumann [93], who also attended the same seminar, was the first to introduce a generalization of Brouwer's fixed point theorem into economics by studying the existence problem of the optimal equilibrium growth path, although it was not the existence problem of Walrasian equilibrium. J. von Neumann's use of a generalized fixed point theorem later motivated mathematical economists to solve the Walrasian problem of the determinacy of general equilibrium.

Almost 20 years later, in the 1950s, serious research was carried out mainly by economic theorists in the United States. It was Arrow–Debreu [3] and McKenzie [83, 84] who filled in the gaps as a Walrasian model in the formulation of the 1930s and proved the first existence theorem in a general Walrasian model. These papers were presented at the annual meeting of the Econometric Society in Chicago in December 1952. Independently of these papers, Nikaido [94] also proved the existence of Walras equilibrium at almost the same time, but his paper was published in a journal about two years after the above papers. In addition to these papers, Gale [40], Debreu [21], Kuhn [69], and others also published papers on the existence problem, and McKenzie [86] and Debreu ([23] generalized the existence proofs in the series of papers in the 1950s. For further accounts of historical aspects of the existence of general equilibrium, see [124].

Bibliography

1. Aliprantis, C.D., Brown, D.J.: Equilibria in markets with a Riez space of commodities. Journal of Mathematical Economics **11**, 189–207 (1983)
2. Anderson, R.: Core theory with strongly convex preferences. Econometrica **49**, 1457–1468 (1981)
3. Arrow, K.J., Debreu, G.: Existence of an equilibrium for a competitive economy. Econometrica **22**, 265–290 (1954)
4. Arrow, K.J., Hahn, F.H.: General Competitive Analysis. Holden-Day, San Francisco (1971)
5. Artstein, Z.: Set valued measures. Trans. Amer. Math. Soc. **165**, 103–121 (1972)
6. Aumann, R.J.: Markets with a continuum of traders. Econometrica **32**, 39–50 (1964)
7. Aumann, R.J.: Integrals of set-valued functions. Jour. Math. Annal. Appl. **12**, 1–12 (1965)
8. Aumann, R.J.: Existence of competitive equilibria in markets with a continuum of traders. Econometrica **34**, 1–17 (1966)
9. Aumann, R.J.: Measurable utility and the measurable choice theorem. In: La Décision, Colloque Internationaux du C.N.R.S., Paris, 15–26 (1969)
10. Barvinok, A.: A Course in Convexity, *Graduate Studies in Mathematics*, vol. 54. American Mathematical Society, Providence, Rhode Island (2002)
11. Bauer, H.: Probability Theory and Elements of Measure Theory. Academic Press, New York (1981). This is a translation based on the third German edition.
12. Berge, C.: Topological Spaces. Oliver & Boyd, London (1963). This is a translation of the original French *Espace Topologiques Fonctions Multivoques*, published by Dunod, Paris (1962). The English translation is reprinted by Dover Publications, Inc., New York.
13. Bewley, T.: Existence of equilibria in economies with infinitely many commodities. Journal of Economic Theory **4**, 514–540 (1972)
14. Billingsley, P.: Convergence of Provability Measures. John Wiley & Sons, New York (1968)
15. Blackwell, D.: Idempotent Markoff chains. Annals Math. **43**, 560–567 (1942)
16. Broome, J.: Existence of equilibrium in economies with indivisible commodities. Journal of Economic Theory **5**, 224–250 (1972)
17. Cassel, G.: Grundriss einer Elementaren Preislehre. G. ges Staatswis. (1899). Translated as: *The Theory of Social Economy*, Harcourt Brace, New York (1923)
18. Cassels, J.W.S.: Measures of the non-convexity of sets and the Shapley-Folkman-Starr theorem. Mathematical Proceedings of the Cambridge Philosophical Society **78**, 433–436 (1975)
19. Castaing, C.: Sur les multi-applications measurables. Revue d'Informatique et de Recherche Opérationelle **1**, 91–126 (1967)
20. Cournot, A.: Recherche sur les Principes Mathématique de la Théorie des Richesse. Hachette, Paris (1838). Translated as: *Researches into the Mathematical Principles of the Theory of Wealth*, published by The MacMillan Company, New York (1927)
21. Debreu, G.: Market equilibrium. Proc Nat. Acad Sci. **42**, 876–878 (1956)
22. Debreu, G.: Theory of Value. John Wiley and Sons, New York (1959)

A. Yamazaki, *Mathematical Economics of General Equilibrium*, Monographs in Mathematical Economics 8, https://doi.org/10.1007/978-981-96-9497-6

23. Debreu, G.: New concepts and techniques for equilibrium analysis. International Economic Review **3**, 257–273 (1962)
24. Debreu, G.: Integration of correspondences. In: L. Lecam, J. Neyman, E.L. Scott (eds.) Proc. Fifth Berkeley Symposium Math. Stat. and Probability, no. II Part 1, 351–372. University of California Press, Berkeley, California (1967)
25. Debreu, G.: Neighboring economic agents. In: La Décision, Colloque Internationaux du C.N.R.S., Paris, 85–90 (1969)
26. Debreu, G.: Smooth preferences. Econometrica **40**, 603–615 (1972)
27. Debreu, G.: Existence of Competitive Equilibria, *Handbook of Mathematical Economics*, vol. II, chap. 15. North-Holland Publishing Co., Amsterdam (1982)
28. Debreu, G., Scarf, H.: A limit theorem on the core of an economy. International Economic Review **4**, 235–246 (1963)
29. De Finetti, B.: Probability, Induction and Statistics. Wiley, New York (1972)
30. De Finetti, B.: Theory of Probability. Wiley, New York (1974)
31. Dellacherie, C., Meyer, P.A.: Probabilities and Potential. North-Holland Publishing Co., Amsterdam (1978)
32. Dieudonne, J.: Foundations of Modern Analysis. Academic Press, New York (1969)
33. Dugundji, J.: Topology. Allyn and Bacon, Inc., Boston, Massachusetts (1966)
34. Dunford, N., Schwartz, J.T.: Linear Operators, *Pure and Applied Mathematics*, vol. VII. Interscience Publishers, New York (1958)
35. Edgeworth, F.Y.: Mathematical Psychics. C. Kegan Paul & Co., London (1881)
36. Edgeworth, F.Y.: On the determinateness of economic equilibrium. Giornare degli Economisti **40**, 603–615 (1891). Translated in: *Papers Relating to Political Economy II*, pp. 313–319, published by MacMillan & Co., London (1925)
37. Effros, E.G.: Convergence of closed subsets in a topological space. Proc. Amer. Math. Soc. **16**, 929–931 (1965)
38. Fell, J.M.G.: A Hausdorff topology for the closed subsets of a locally compact non-Hausdorff spaces. Proc. Amer. Math. Soc. **13**, 472–476 (1962)
39. Florenzano, M.: On the existence of equilibria in economies with an infinite dimensiional commodity space. Journal of Mathematical Economics **12**, 207–219 (1983)
40. Gale, D.: Law of supply and demand. Mathematica Scandinavica **3**, 155–169 (1955)
41. Garnir, H.G.: Solvay's axiom and functional analysis, pp. 189–204. No. 399 in Lecture Notes in Mathematics. Springer-Verlag, Berlin (1974)
42. Grodal, B.: A note on the space of preference relations. Journal of Mathematical Economics **1**, 279–294 (1974)
43. Halmos, P.R.: The range of a vector measure. Bull. Math. Soc. **54**, 416–421 (1948)
44. Halmos, P.R.: Measure Theory. D. Van Nostrand Co., New York (1950)
45. Halmos, P.R.: Finite-Dimensional Vector Spaces. Van Nostrand, Princeton, New Jersey (1958). Reprinted by Springer-Verlag New York Inc., New York (1974)
46. Halmos, P.R.: Naive Set Theory. Van Nostrand, Princeton, New Jersey (1960)
47. Hausdorff, F.: Set Theory, second edn. Chelsea Publishing Co., New York (1962). This is a translation of the original German *Grundzüge der Mengenlehre*, published in Leipzig (1914)
48. Henry, C.: Indivisibilité dans une economie d'echange. Econometrica **38**, 542–558 (1970)
49. Hicks, J.R.: Value and Capital. Oxford University Press, London (1939)
50. Hildenbrand, W.: The core of an economy with a measure space of economic agents. Review of Economic Studies **35**, 443–452 (1968)
51. Hildenbrand, W.: Pareto optimality for a measure space of economic agents. International Economic Review **10**, 363–372 (1969)
52. Hildenbrand, W.: Metric measure spaces of economic agents. In: L. Lecam, J. Neyman, E.L. Scott (eds.) Proc. Sixth Berkeley Symposium Math. Stat. and Probability, 81–95. University of California Press, Berkeley, California (1970)
53. Hildenbrand, W.: Existence of equilibria for economies with production and a measure space of consumers. Econometrica **38**, 608–623 (1970)
54. Hildenbrand, W.: On economies with many agents. Journal of Economic Theory **2**, 161–188 (1970)

55. Hildenbrand, W.: Core and Equilibria of a Large Economy. Princeton University Press, Princeton, New Jersey (1974)
56. Hildenbrand, W., Kirman, A.P.: Introduction to Equilibrium Analysis. North-Holland Publishing Co., Amsterdam (1976)
57. Hildenbrand, W., Mertens, J.F.: On Fatou's lemma in several dimensions. Zeitschrift für Wahrscheinlichkeitstheorie und Verwandte Gebiete **17**, 151–155 (1971)
58. Ichiishi, T., Weber, S.: Some theorems on the core of a non-sidepayment game with a measure space of players. International Journal of Game Theory **7**, 95–112 (1978)
59. Ichiishi, T., Yamazaki, A.: Interim core concepts for a bayesian pure exchange economiy. Journal of Mathematical Economics **40**, 347–370 (2004)
60. Ichiishi, T., Yamazaki, A.: Cooperative Extensions of the Bayesian Game. World Scientific Publishing Co., Singapore, New Jersey, and London (2006)
61. Jacobs, K.: Measure and Integral. Academic Press, New York (1978)
62. John Paul II, Pope: Centesimus Annus. Encyclical Letter, Vatican, Rome: Vatican (1991)
63. Jones, L.E.: A competitive model of commodity differentiation. Econometrica **52**, 507–530 (1984)
64. Kakutani, S.: A generalization of Brower's fixed point theorem. Duke Math. J. **8**, 457–459 (1941)
65. Kannai, Y.: Continuity properties of the core of a market. Econometrica **38**, 791–815 (1970)
66. Kelley, J.L.: General Topology. D. Van Nostrand Co., New York (1955)
67. Khan, M.A., Yamazaki, A.: On the cores of economies with indivisible commodities and a continuum of traders. Journal of Economic Theory **24**, 218–225 (1981)
68. Koopmans, T.C.: Three Essays on: The State of Economic Science. McGraw-Hill Book Co., New York (1957)
69. Kuhn, H.W.: On a theorem of Wald. In: Kuhn, H.W., Tucker, A.W. (eds.) Linear Inequalities and Related Systems, *Annal. Math. Stud.* no. 38, pp. 265–273. Princeton University Press, Princeton, New Jersey (1956)
70. Kuratowski, C.: Topology, vol. I. Academic Press, New York (1966). This is a translation of the original French *Topologie, volume I*, published by Polish Scientific Publishers (1958)
71. Kuratowski, C., Ryll-Nardzewski: A general theorem on selectors. Bull. Polish Acad. Sci. **13**, 397–403 (1965)
72. Landers, D.: Existenz und konsistenz von maximumlikelihood schätzern. Thesis. University of Cologne, Colongne (1968)
73. Lang, S.: Linear Algebra. Addison-Wesley Publishing Co., Reading, Massachusetts (1970)
74. Liapounoff, A.: Sur les fonctions-vecteurs complétement additives,. Bull. Acad. Sci. URSS Math. **4**, 465–478 (1940)
75. Lindenstrauss, J.: A short proof of liapounoff's convexity theorem. J. Math. Mec. **15**, 971–972 (1966)
76. Maharam, D.: Finitely additive measures on the integers. Indian J. Stat. **38**, 44–59 (1976)
77. Malinvaud, E.: Lectures on Microeconomic Theory. North-Holland, Amsterdam (1972)
78. Marshall, A.: Principles of Economics, eighth edn. The Macmillan Press, London (1920)
79. Mas-Colell, A.: An equilibrium existence theorem without complete or transitive preferences. Journal of Mathematical Economics **1**, 237–246 (1974)
80. Mas-Colell, A.: A model of equilibrium with differentiated commodities. Journal of Mathematical Economics **2**, 236–296 (1975)
81. Mas-Colell, A.: Indivisible commodities and general equilibrium theory. Journal of Economic Theory **16**, 443–456 (1977)
82. Mas-Colell, A.: The price equilibrium existence problem in topological vector lattice. Econometrica **54**, 1039–1053 (1986)
83. McKenzie, L.W.: On equilibrium in Graham's model of world trade and other competitive systems. Econometrica **22**, 147–161 (1954)
84. McKenzie, L.W.: Competitive equilibrium with dependent consumer preferences. In: H.A. Antosiewicz (ed.) Proc. The Second Symposium on Linear Programming, pp. 277–294. National Bureau of Standards, Washington D.C. (1955)

85. McKenzie, L.W.: Demand theory without a utility index. Review of Economic Studies **23–24**, 185–189 (1956–57)
86. McKenzie, L.W.: On the existence of general equilibrium for a competitive market. Econometrica **27**, 54–71 (1959)
87. McKenzie, L.W.: The classical theorem on existence of competitive equilibrium. Econometrica **49**, 819–841 (1981)
88. McKenzie, L.W.: Classical General Equilibrium Theory. The MIT Press, Cambridge, Massachusetts (2002)
89. Meyer, P.A.: Probability and Potentials. Blaisdell Publishing Co., Waltham (1966)
90. Michael, E.: Continuous selections I. Annal. of Math. **63**, 361–382 (1956)
91. Nachbin, L.: Topology and Order. Van Nostrand, Princeton (1965)
92. Neisser, H.: Lohnhöhe und beschäftigungsgrad im marktgleichgewict. Weltwirtschaftliches Archiv **36**, 415–455 (1932)
93. Neumann, J. von: Über eine ökonomisches gleichungssystem und eine verallgemeinerungdes des Brouwerscnen fixedpunktsatzes. Ergebnisse eines Matimatischen Kolloquiums **8**, 73–83 (1937). Translated in: Review of Economic Studies **13**, 1–9 (1945)
94. Nikaido, H.: On the classical multilateral exchange problem. Metroeconomica **8**, 135–145 (1956)
95. Nikaido, H.: Convex Structures and Economic Theory. vol. 51 in *Mathematics in Science and Engineering: a series of monographs and textbooks*. Academic Press, New York (1968)
96. Pareto, V.: Manuale di Economia Politica. Società Editrice Libraria, Milano (1906). Translated as: *Manual of Political Economy*, published by A. M. Kelly, New York (1971)
97. Parthasarathy, K.R.: Probability Measures on Metric Spaces. Academic Press, New York (1967)
98. Richter, H.: Verallegemeinerung eines in der statistik benötigten satzes der masstheorie. Math. Annalen **150**, 85–90 (1963)
99. Rockafellar, R.T.: Convex Analysis. Princeton University Press, Princeton, New Jersey (1970)
100. Ross, D.: An elementary proof of Lyapunov's theorem,. The Amer. Math. Month. **112**, 651–653 (2005)
101. Samuelson, P.A.: Foundations of Economic Analysis. Harvard University Press, Cambridge, Massachusetts (1947)
102. Scarf, H.: An analysis of markets with with a large number of participants. In: Recent Advances in Game Theory, pp. 127–155. Princeton University Press, Princeton, New Jersey (1962)
103. Scarf, H.: The core of an N-person game. Econometrica **35**, 50–69 (1967)
104. Scarf, H.: On the existence of a cooperative solution for a general class of N-person games. Journal of Economic Theory **3**, 169–181 (1971)
105. Schaefer, H.H.: Topological Vector Spaces. Graduate Texts in Mathematics. Springer-Verlag, New York (1980)
106. Schlesinger, K. Über die productions gleichungen der ökonomischen wertlehre. Ergebnisse eines Matimatischen Kolloquiums **6**, 10–11 (1935)
107. Schmeidler, D.: Competitive equilibria in markets with a continuum of traders and incomplete preferences. Econometrica **37**, 578–585 (1969)
108. Schmeidler, D.: Fatou's lemma in several dimensions. Proc. Amer. Math. Soc. **24**, 300–306 (1970)
109. Schumpeter, J.A.: History of Economic Analysis. (ed. by E.B. Schumpeter), Oxford University Press, New York (1954)
110. Sen, A.: Starvation and exchange entitlements: a general approach and its application to the great Bengal famine. Cambridge Journal of Economies **1**, 33–59 (1977)
111. Sen, A.: Ingredients of famine analysis: Availability and entitlements. The Quarterly Journal of Economics **96**, 433–464 (1981)
112. Shapley, L., Scarf, H.: On cores and indivisibility. Journal of Mathematical Economics **1**, 23–38 (1974)

113. Shubik, M.: Edgeworth market games. In: Luce, R.D., Tucker, A.W. (eds.) Contributions to the Theory of Games, vol. IV, *Annal. Math. Studi.* no. 40, pp. 267–278. Princeton University Press, Princeton, New Jersey (1959)
114. Sion, M., Bressler, D.W.: The current theory of analytic sets. Canadian J. Math. **16**, 207–230 (1964)
115. Smith, A.: An Inquiry into the Nature and Causes of the Wealth of Nations, v. 1 & 2 (ed. by E. Cannan), Methuen & Co., London (1920 [1776])
116. Sonnenshein, H.: Demand theory without transitive preferences, with applications to the theory of competitive equilibrium. In: J. Chipman et al. (eds.) Preferences, Utility and Demand, 214–223. Harcourt Brace Jovanovich, New York (1971)
117. Stackelberg, H. von: Zwei kritsche bimerkungen zur preistheorie Gustav Cassels. Zeitschrift für Nationalökonomie **4**, 456–472 (1933)
118. Vind, K.: Edgeworth allocations in an economy with many traders. International Economic Review **5**, 165–177 (1964)
119. Wald, A.: Über die eindeutige positive lösbarkeit der neuen productionsgleichungen. Ergebnisse eines Matimatischen Kolloquiums **6**, 12–20 (1935)
120. Wald, A.: Über die productionsgleichungen der ökonomischen wertlehre. Ergebnisse eines Matimatischen Kolloquiums **7**, 1–6 (1936)
121. Wald, A.: Über die productionsgleichungen der ökonomischen wertlehre. Zeitschrift für Nationalökonomie **7**, 637–670 (1936). Translated as: *On some systems of equations of mathematical economics.* Econometrica **19**, 368–403 (1951)
122. Walras, L.: Eléments d'Economic Politique Pure. Corbaz, Lausanne (1874–1877). Translated as: *Elements of Pure Economics*, published by Irwin, Chicago (1954)
123. Watson, P.D.: On the limits of sequences of sets. Quart. J. Math. **4**, 1–3 (1953)
124. Weintraub, E.R.: On the existence of a competitive equilibrium: 1930-1954. Journal of Economic Literature **21**, 1–39 (1983)
125. Willard, S.: General Topology. Addison-Wesley Publishing Co., Reading, Massachusetts (1970)
126. Yamazaki, A.: An equilibrium existence theorem without convexity assumptions. Econometrica **46**, 541–555 (1978)
127. Yamazaki, A.: Diversified consumption characteristics and conditionally dispersed endowment distribution: Regularizing effect and existence of equilibria. Econometrica **49**, 639–654 (1981)
128. Yamazaki, A.: Continuous preference relations which are observable in markets. Hitotsubashi Journal of Economics **23**, 40–47 (1983)
129. Yosida, K., Hewitt, E.: Finitely additive measures. Trans. Amer. Math. Soc. **72**, 46–66 (1952)
130. Zeuthen, F.: Das prinzip der knappheit, technische kombination, und ökonomische qualitat. Zeitschrift für Nationalökonomie **4**, 1–24 (1933)

Index

A. Yamazaki, *Mathematical Economics of General Equilibrium*, Monographs in Mathematical Economics 8, https://doi.org/10.1007/978-981-96-9497-6

D

MIX
Papier aus verantwortungsvollen Quellen
Paper from responsible sources
FSC® C105338

If you have any concerns about our products,
you can contact us on
ProductSafety@springernature.com

In case Publisher is established outside the EU,
the EU authorized representative is:
Springer Nature Customer Service Center GmbH
Europaplatz 3, 69115 Heidelberg, Germany

Printed by Libri Plureos GmbH
in Hamburg, Germany